Proceedings of the International Plant Sulfur Workshop

Series editors

Luit J. De Kok, University of Groningen, Groningen, The Netherlands
Heinz Rennenberg, Universität Freiburg, Freiburg, Germany
Malcolm J. Hawkesford, Rothamsted Research, Harpenden, United Kingdom

More information about this series at http://www.springer.com/series/8885

Luit J. De Kok • Malcolm J. Hawkesford
Silvia H. Haneklaus • Ewald Schnug
Editors

Sulfur Metabolism in Higher Plants - Fundamental, Environmental and Agricultural Aspects

Springer

Editors
Luit J. De Kok
Laboratory of Plant Physiology
Groningen Institute for Evolutionary Life Sciences
University of Groningen
Groningen, The Netherlands

Silvia H. Haneklaus
Federal Research Centre for Cultivated Plants
Institute for Crop and Soil Science, Julius Kühn-Institut (JKI)
Braunschweig, Germany

Malcolm J. Hawkesford
Plant Biology and Crop Science Department
Rothamsted Research
Harpenden, Hertfordshire, UK

Ewald Schnug
Federal Research Centre for Cultivated Plants
Institute for Crop and Soil Science, Julius Kühn-Institut (JKI)
Braunschweig, Germany

ISSN 2451-9073 ISSN 2451-9081 (electronic)
Proceedings of the International Plant Sulfur Workshop
ISBN 978-3-319-56525-5 ISBN 978-3-319-56526-2 (eBook)
DOI 10.1007/978-3-319-56526-2

Library of Congress Control Number: 2017943333

Printed on acid-free paper

This Springer imprint is published by Springer Nature
The registered company is Springer International Publishing AG
The registered company address is: Gewerbestrasse 11, 6330 Cham, Switzerland

Heinz Rennenberg

Foreword: There Is (Almost) No Way to Escape from Sulfur in Plant Research

This Sulfur Workshop in Plants series was initiated during a sabbatical stay in Groningen in 1988 as a result of discussions between Ineke Stulen, Luit De Kok, and myself. Therefore, it was logical that the first Sulfur Workshop on plants was held in Groningen in 1989. At this time, plant research on sulfur was largely focused on consequences of atmospheric pollution with SO_2 for plant growth and development, and studies on other aspects of sulfur metabolism in plants were not well developed (e.g., Rennenberg 1982) compared to studies in mammals (Meister and Anderson 1983). It is the merit of the highly successful first workshop on "Sulfur Nutrition and Sulfur Assimilation in Higher Plants" (subsequently abbreviated "Sulfur Workshop") held in Groningen that the door became wide open for studies on sulfur metabolism of plants including fundamental and applied aspects. In particular, it was the European plant science community that took advantage of this situation and soon played a leading role in this area of research.

When the first Sulfur Workshop was held in Groningen, I already could look back to more than a decade of studies on sulfur in plants. In my diploma and PhD thesis, I had worked on glutathione production in tobacco suspension cultures, a system that subsequently became recognized as a useful tool for in-depth analyses of glutathione synthesis and degradation in plants (Bergmann and Rennenberg 1993). At this time, it was also established that plants are not only a sink for atmospheric sulfur compounds but are also able to emit volatile sulfur compounds into the atmosphere (Rennenberg 1991). This new view of a bidirectional flux of sulfur between plants and the atmosphere initiated numerous studies on sulfur metabolism in terrestrial and aquatic plants that included volatile products.

Despite the multitude of valuable results obtained by studies with tissue cultures, it became obvious that sulfur compounds undergo long-distance transport (Bonas et al. 1982) and that regulation from the cellular scale-up to the seasonal dynamics of sulfur in plants requires studies at the whole-plant level (Herschbach and Rennenberg 1997; Herschbach et al. 2012). The significance of such studies was fueled by the generation of transgenic poplar plants with modified glutathione synthesis and reduction capacity (Noctor et al. 1998). With these tools, sulfur in plants entered the era of molecular biology. Molecular research on sulfur became soon extended to the analyses of sulfate transporters (Smith et al. 1997; Hawkesford et al. 2003) and the cross talk of sulfur metabolism with nitrogen and carbon metabolism (Kopriva and Rennenberg 2004) that constitute important areas of plant research until today. The current view on molecular studies has changed dramatically from the initial approaches that were focused on analyses of transcription of a set of enzymes and transporters: today, it is generally accepted that the characterization of metabolic processes and metabolic cross talk requires more than quantification of the transcriptome and largely relies on an integrative view on mRNA, protein, and metabolite abundances, as well as metabolite fluxes (Rennenberg and Herschbach 2014; Kalloniati et al. 2015).

Transgenic poplars with modified glutathione synthesis and reduction capacity became a useful tool to analyze the role of sulfur metabolism in the compensation of abiotic and biotic stress. The multiple stress compensation reactions relying on sulfur metabolism include reactive oxygen species (ROS) scavenging, heavy metal detoxification, and hypersensitive responses to pathogen attack (Noctor et al. 1998; Foyer and Rennenberg 2000; Peuke and Rennenberg 2006; Rennenberg and Herschbach 2012; He et al. 2015). Among the stress factors studied, the consequences of SO_2 exposure of plants turned out to be of particular complexity, but surprisingly, this was only recognized in recent years (Hänsch et al. 2007). As a radical, SO_2 can mediate ROS formation that requires reduced sulfur in the form of glutathione for scavenging in the Foyer-Halliwell-Asada cycle (Foyer and Rennenberg 2000); if involved in ROS formation in the apoplastic space, SO_2 can interact with lignin formation and requires scavenging by peroxidase activity for ROS homeostasis (Hamisch et al. 2012); as an essential intermediate, sulfite derived from SO_2 interacts with assimilatory sulfate reduction and, as a product of this pathway, also with sulfur nutrition (Rennenberg 1984; Takahashi et al. 2011; Herschbach et al. 2012). Even subsequent to oxidative and reductive SO_2 detoxification (Hamisch et al. 2012), the detoxification products, i.e., sulfate and sulfide, will interact with signaling processes at the cellular and whole-plant level (Leitner et al. 2009; Lisjak et al. 2010; García-Mata and Lamattina 2013; Hancock and Whiteman 2014; Calderwood and Kopriva 2014). This was indicated already by early H_2S fumigation studies with different plant species (De Kok et al. 1991; Herschbach et al. 1995a, b, 2000) but only recently connected to the consequences of SO_2 exposure (Hamisch et al. 2012). In the research area of sulfur-mediated signaling, the proposed role of sulfate as a root-to-shoot signal controlling stomatal aperture upon drought (Malcheska et al. 2017) provides a new notion, why excess sulfur in the form of sulfate has to be sequestered in the vacuole but also needs to be mobilized from this pool under particular environmental conditions, processes that were already observed in early studies on sulfur in plants (Rennenberg 1984).

Over the years, I made several attempts to escape from sulfur, e.g., by focusing on N and P nutrition, on radiatively active biogenic trace gases in the atmospheres, and recently on plant carnivory. At the end, (almost) all of these attempts ended up with studies on sulfur in plants. Even for plant carnivory, sulfur metabolism turned out to be of pivotal importance. In the Venus flytrap, a plant that actively catches its animal prey with snap traps, the prey is digested in hermetically closed traps by the release of an acidic, sulfur-rich enzyme cocktail from the gland-based secretory system on the inner surface of the traps (see Fasbender et al. 2017 and literature cited therein). The production of the sulfur-rich hydrolytic enzyme cocktail requires stimulation of thiol synthesis from assimilatory sulfate reduction during prey digestion (Scherzer et al. 2017). Thus, whatever process in plants I found interesting to study (almost) always ended up to be connected to sulfur. Therefore, finally, sulfur in plants accompanied me from my diploma, PhD, and habilitation thesis to my position as research associate at the University of Cologne and the DOE Plant Research Laboratory at Michigan State University in East Lansing, and to my professor positions at the University of Cologne, the Technical University of Munich, the Fraunhofer Institute of Atmospheric Environmental Research, and the University of Freiburg up to my retirement in 2017.

The broad range of different aspects of sulfur metabolism could only be studied in more than 40 years of my research activities in collaboration with numerous colleagues and friends, including, among many others (in alphabetical order), Ludwig Bergmann, Christian Brunold, Luit J. De Kok, Manolis Flemetakis, Christine Foyer, Dieter Grill, Robert Hänsch, Rainer Hedrich, Rüdiger Hell, Stanislav Kopriva, Ralf R. Mendel, Andrea Polle, Winfried Rauser, Kazuki Saito, Andreas Weber, and Marcus Wirtz that in several cases stayed in my group for a period of time. In addition, work on sulfur in plants in my group would not have been possible without additional strong partners such as Cornelia Herschbach, Jürgen Kreuzwieser, and Monika Eiblmeier who have accompanied me in my research at the University of Freiburg until today. It is time to thank them all for creating such a fruitful and pleasant working atmosphere.

Institut für Forstwissenschaften
Universität Freiburg
Freiburg, Germany

Heinz Rennenberg

References

Bergmann L, Rennenberg H (1993) Glutathione metabolism in plants. In: De Kok LJ, Stulen I, Rennenberg H, Brunold C, Rauser W (eds) Sulfur nutrition and sulfur assimilation in higher plants. SPB Academic Publishing, The Hague, pp 109–123

Bonas U, Schmitz K, Rennenberg H, Bergmann L (1982) Phloem transport of sulfur in *Ricinus*. Planta 155:82–88

Calderwood A, Kopriva S (2014) Hydrogen sulfide in plants: from dissipation of excess to signaling molecule. Nitric Oxide 41:72–78

De Kok LJ, Rennenberg H, Kuiper PJC (1991) The internal resistance in spinach shoots to atmospheric H_2S deposition is determined by metabolism. Plant Physiol Biochem 29:463–470

Fasbender L, Maurer D, Kreuzwieser J, Kreuzer I, Schulze WX, Kruse J, Becker D, Alfarraj S, Hedrich R, Werner C, Rennenberg H (2017) The carnivorous Venus flytrap uses prey derived amino acid carbon to fuel respiration. New Phytol (in press)

Foyer CH, Rennenberg H (2000) Regulation of glutathione synthesis and its role in abiotic and biotic stress defense. In: Brunold C, Rennenberg H, De Kok LJ, Stulen I, Davidian J-C (eds) Sulfur nutrition and sulfur assimilation in higher plants: molecular, biochemical and physiological aspects. Paul Haupt, Bern, pp 127–153

García-Mata C, Lamattina L (2013) Gasotransmitters are emerging as new guard cell signaling molecules and regulators of leaf gas exchange. Plant Sci 201–201:66–73

Hamisch D, Randewig D, Schliesky S, Weber APM, Geffers R, Herschbach C, Rennenberg H, Mendel RR, Hänsch R (2012) Impact of SO_2 on *Arabidopsis thaliana* transcriptome in wildtype and sulfite oxidase knock-out plants analyzed by RNA deep sequencing. New Phytol 196:1074–1085

Hancock JT, Whiteman M (2014) Hydrogen sulfide and cell signaling: team player or referee? Plant Physiol Biochem 78:37–42

Hänsch R, Lang C, Rennenberg H, Mendel RR (2007) Significance of plant sulfite oxidase. Plant Biol 9:589–595

Hawkesford MJ, Buchner P, Hopkins L, Howarth JR (2003) The plant sulfate transporter family: specialized functions and integration with whole plant nutrition. In: Davidian J-C, De Kok LJ, Stulen I, Hawkesford MJ, Schnug E, Rennenberg H (eds) Sulfur transport and assimilation in plants: regulation, Interaction, Signaling. Backhuys Publishers, Leiden, pp 1–10

He J, Li H, Ma C, Zhang Y, Polle A, Rennenberg H, Cheng X, Luo Z-B (2015) Over-expression of γ-glutamylcysteine synthetase mediates changes in cadmium influx, allocation, and detoxification in poplar. New Phytol 205:240–254

Herschbach C, Rennenberg H (1997) Sulfur nutrition of conifers and deciduous trees. In: Rennenberg H, Eschrich W, Ziegler H (eds) Trees – Contributions to modern tree physiology. Backhuys Publishers, Leiden, pp 293–311

Herschbach C, De Kok LJ, Rennenberg H (1995a) Net uptake of sulphate and its transport to the shoot in tobacco plants fumigated with H_2S or SO_2. Plant Soil 175:75–84

Herschbach C, De Kok LJ, Rennenberg H (1995b) Net uptake of sulfate and its transport to the shoot in spinach plants fumigated with H_2S or SO_2: does atmospheric sulfur affect the 'inter-organ' regulation of sulfur nutrition? Bot Acta 108:41–46

Herschbach C, van der Zalm E, Schneider A, Jouanin L, De Kok LJ, Rennenberg H (2000) Regulation of sulphur nutrition in wildtype and transgenic poplar overexpressing γ-glutamylcysteine synthetase in the cytosol as affected by atmospheric H_2S. Plant Physiol 124:461–473

Herschbach C, Gessler A, Rennenberg H (2012) Long-distance transport and plant internal cycling of N- and S-compounds. Progr Bot 73:161–188

Kalloniati C, Krompas P, Karalias G, Udvardi MK, Rennenberg H, Herschbach C, Flemetakis M (2015) Nitrogen-fixing nodules as a new strong source of reduced sulfur trigger global changes in sulfur metabolism in *Lotus japonicus*. Plant Cell 27:2384–2400

Leitner M, Vandelle E, Gaupels F, Bellin D, Belledonne M (2009) NO signals in the haze – nitric oxide signalling in plant defense. Curr Opin Plant Biol 12:451–458

Lisjak M, Srivastava N, Teklic T, Civale L, Lewandowski K, Wilson I, Wood ME, Whiteman M, Hancock JT (2010) A novel hydrogen sulphide donor causes stomatal opening and reduces nitric oxide accumulation. Plant Physiol Biochem 48:931–935

Malcheska F, Ahmad A, Batoo S, Ludwig-Müller J, Kreuzwieser J, Randewig D, Hedrich R, Hell R, Wirtz M, Herschbach C, Rennenberg H (2017) Drought enhanced xylem sap sulfate closes stomata by affecting ALMT12 and guard cell ABA synthesis. Plant Physiol, revised submitted

Meister A, Anderson ME (1983) Glutathione. Annu Rev Biochem 52:711–760

Noctor G, Arisi A-CM, Jouanin L, Kunert KJ, Rennenberg H, Foyer CH (1998) Glutathione: biosynthesis, metabolism and relationship to stress tolerance explored in transformed plants. J Exp Bot 49:623–647

Peuke A, Rennenberg H (2006) Heavy metal resistance and phytoremediation with transgenic trees. In: Fladung M, Ewald D (eds) Tree transgenesis-recent developments. Springer Publishers, Heidelberg, pp 137–155

Rennenberg H (1982) Glutathione metabolism and possible biological roles in higher plants. Phytochemistry 21:2771–2781

Rennenberg H (1984) The fate of excess sulfur in higher plants. Annu Rev Plant Physiol 35:121–153

Rennenberg H (1991) The significance of higher plants in the emission of sulfur compounds from terrestrial ecosystems. In: Sharkey T et al (eds), Trace gas emissions by plants. Academic, New York, pp 217–265

Rennenberg H, Herschbach C (2012) Sulfur compounds in multiple compensation reactions of abiotic stress responses. In: De Kok LJ, Tausz M, Hawkesford MJ, Höfgen R, McManus MT, Norton RM, Rennenberg H, Saito K, Schnug E, Tabe L (eds) Sulfur metabolism in plants: mechanisms and application to food security and responses to climate change. Springer, Berlin, pp 203–215

Rennenberg H, Herschbach C (2014) A detailed view on sulphur metabolism at the cellular and whole plant level illustrates challenges in metabolite flux analyses. J Exp Bot 65:5711–5724

Scherzer S, Shabala L, Hedrich B, Fromm J, Bauer H, Munz E, Jakob P, Al-Rascheid K, Kreuzer I, Becker D, Eiblmeier M, Rennenberg H, Shabala S, Bennett M, Hedrich R, Neher E (2017) Insect hapto-electrical stimulation of Venus flytrap triggers exocytosis in gland cells. Proc Natl Acad Sci U S A, submitted

Smith FW, Hawkesford MJ, Ealing PM, Clarkson DT, Vanden Berg PJ, Belcher AR, Warrilow AGS (1997) Regulation of expression of a cDNA from barley roots encoding a high affinity sulphate transporter. Plant J 12:875–884

Takahashi H, Kopriva S, Giordano M, Saito K, Hell R (2011) Sulfur assimilation in photosynthetic organisms; molecular functions and regulations of transporters and assimilatory enzymes. Annu Rev Plant Biol 62:157–184

Preface

The International Plant Sulfur Workshop series was initiated in order to bring together scientists from various research disciplines and to discuss all aspects of sulfur metabolism, from molecular biology, biochemistry, and physiology to ecology and agriculture. The first workshop in the series entitled "Sulfur Nutrition and Sulfur Assimilation in Higher Plants: Fundamental Environmental and Agricultural Aspects" was held in Haren, the Netherlands, 1989. The following workshops were held in Garmisch-Partenkirchen, Germany, 1992; Newcastle upon Tyne, United Kingdom, 1996; Wengen, Switzerland, 1999; Montpellier, France, 2002; Kisarazu, Chiba, Japan, 2005; Warsaw, Poland, 2008; Creswick, Victoria, Australia, 2010; and Freiburg-Munzingen, Germany, 2014. Contents of the respective proceedings are included in this volume.

This proceedings volume contains a selection of invited and contributed papers of the 10th Jubilee Plant Sulfur Workshop, which was held in Goslar, Germany, from September 1 to 4, 2015. During this workshop, the outcome of the previous workshops was summarized, and the still existing gaps and prospects for future research were highlighted by a selection of speakers who have significantly contributed to plant sulfur research during the last 25 years.

We are delighted to dedicate this volume to our dear colleague Heinz Rennenberg from the University of Freiburg, Germany, who together with Ineke Stulen, Christian Brunold, and Luit J. De Kok initiated the workshop series and furthermore was involved in the organization and issuing of proceedings volumes of all previous plant sulfur workshops. In addition, he has significantly contributed to the understanding of the regulation of uptake and assimilation of sulfur and the significance of sulfur metabolites in stress tolerance of higher plants over more than four decades thus leaving a durable "sulfur footprint".

Groningen, The Netherlands — Luit J. De Kok
Braunschweig, Germany — Silvia H. Haneklaus
Harpenden, UK — Malcolm J. Hawkesford
Braunschweig, Germany — Ewald Schnug

Acknowledgment

The editors thank Sue Steele for grammatical editing of the chapters.

Contents

Part I
Overview Papers

Sulfate Transport in Plants: A Personal Perspective

Malcolm J. Hawkesford

Abstract Early key research milestones for sulfate transport in plants include the first description of kinetics of sulfate uptake into plant roots (Leggett and Epstein, Plant Physiol 31:222–226, 1956), nutritionally regulated sulfate uptake into plants (Clarkson et al., J Exp Bot 34:1463–1483, 1983), and the first gene for a plant sulfate transporter (Smith et al., Proc Natl Acad Sci U S A 92:9373–9377, 1995a). Since then a well-described gene family encoding putative sulfate transporters has been characterized in multiple species, initially most notably in Arabidopsis but subsequently for a number of other models or important crops (examples: Brassica, wheat, rice, poplar and Medicago, see Buchner et al., Genome 47:526–534, 2004a; Buchner et al., Plant Physiol 136:3396–3408, 2004b; Buchner et al., Mol Plant 3:374–389, 2010; Kumar et al., Plant Signal Behav 10:e990843, 2015; Dürr et al., Plant Mol Biol 72:499–517, 2010; Gao et al., Planta 239:79–96, 2014). Regulation of expression has been well documented and this regulation is both a useful marker of sulfur-nutritional status and a model for the elucidation of control pathways. The complexity of the gene family in relation to functional, regulatory and spatial distribution indicates an apparent whole plant management system for sulfur, coordinated with growth and demand and interacting with nutrient availability. In addition to sulfate, there is direct involvement of this transporter family in the uptake and accumulation of both selenate and molybdate, with clear consequences for nutritional quality. Is the story now complete almost 60 years since the first transport description and 20 years since the first sulfate transporter gene isolation, and a plethora of research projects and publications? Do we know how sulfur is acquired and appropriately distributed within the plant? Do we know the critical signals that control these processes? Are we even sure that these processes are coordinated? This review documents research progress and assesses to what extent the key questions have been addressed.

M.J. Hawkesford (✉)
Plant Biology and Crop Science Department, Rothamsted Research, Harpenden, Hertfordshire AL5 2JQ, UK
e-mail: malcolm.hawkesford@rothamsted.ac.uk

L.J. De Kok et al. (eds.), *Sulfur Metabolism in Higher Plants - Fundamental, Environmental and Agricultural Aspects*, Proceedings of the International Plant Sulfur Workshop, DOI 10.1007/978-3-319-56526-2_1

Introduction

All plants require sulfur for growth and for land plants is most is acquired from the external environment as sulfate. For land plants this from the soil via the roots. Typically concentrations are low and often extremely variable. Thus, transport needs to be active to facilitate uptake against a concentration gradient, specific for sulfate and regulated to optimize uptake to growth and ensure optimal energy utilization in this process. For vascular plants transport is not only across a single membrane at the soil-root interface but also across many other plasma membranes to facilitate distribution, across the chloroplast membrane to the site of reduction and also across the tonoplast to allow transport in and out of the vacuole for the transient storage of excess sulfate taken up.

Progress on the understanding of plant sulfate transporters has been substantial and reported in successive volumes of the Sulfur Workshop series, with key landmark papers from a number of groups being published throughout this period. Some early key milestones in the development of the plant sulfate transporter research field are illustrated as a timeline in Fig. 1.

The first suggestion for active absorption was in an analysis of whole plant uptake of sulfate into barley roots. An enzyme based description of affinities and competition by selenate but not nitrate or phosphate unequivocally demonstrated the activity of a transmembrane ion transporter (Leggett and Epstein 1956). It would be 40 years before the molecular components would be identified in plants (Smith et al. 1995b). Prior to this key elements of regulation by de-repression (that is induction upon starvation) were described in a topical legume (Clarkson et al. 1983) and suggestion for involvement of a metabolite linking N and S metabolism, namely *O*-acetylserine (OAS), was described in maize (Clarkson et al. 1999). The importance of OAS as a regulator of gene expression for a cluster of genes has been described, separating S-related and other regulation (Hubberten et al. 2012, 2015). Mechanistic evidence for transport being driven by proton gradients was obtained in a duckweed (Lass and Ullrich-Eberius 1984).

The first substantial progress on the identification of sulfate transporters genes was inevitably for bacteria (Ohta et al. 1971; Sirko et al. 1990), fungi and yeast (Ketter et al. 1991; Smith et al. 1995b; Cherest et al. 1997), mammalian systems (Schweinfest et al. 1993; Hästbacka et al. 1994; Silberg et al. 1995) and finally in plants (Kouchi and Hata 1993; Smith et al. 1995a, 1997 ; Takahashi et al. 1996). Similarities in the sequence of many of the genes, some not identified as sulfate transporters was first noted by Sandal and Marcker (1994). Much of this work has been reviewed in previous volumes in this series: Kredich 1993 (bacteria); Thomas et al. 1997 (yeast); Davidian et al. 2000, Hawkesford et al. 2003, Buchner et al. 2010, Hawkesford 2012 (plants) and elsewhere (Markovich 2001) for mammalian transporters. These transporters are now recognised to be part of a large family of transmembrane ion transporters known as SulP (see also Price et al. 2004).

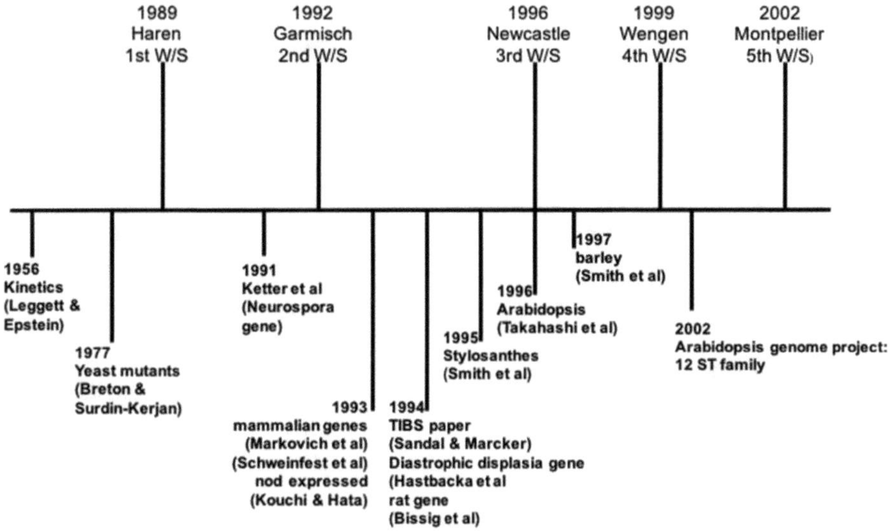

Fig. 1 Key early milestones in the development of the understanding of sulfate transport in plants placed in relation to the first 5 Sulfur Workshops

A Family of Sulfate Transporters

In a series of papers predominantly from the Takahashi group but with notable contributions from a number of others including the Davidian group it became apparent that a gene family of up to 14 genes encoded a group of related proteins in Arabidopsis (Takahashi et al. 1996, 1997; Vidmar et al. 2000). Similar gene families were subsequently identified in Brassica (Buchner et al. 2004b), in rice (Kumar et al. 2015), poplar (Dürr et al. 2010), Medicago (Gao et al. 2014) and in wheat (Buchner et al. 2004a).

Phylogenetic analysis of plant sulfate transporter sequences indicates discrete clades within the family (Fig. 2) and it has been proposed that these align with discrete functions and that within clades there may be some functional redundancy (Hawkesford 2003). In summary, Group 1 represents high affinity types responsible for up take into the cell, particularly in the roots, and are subject to nutritional regulation. Group 2 are lower affinity, less regulated and distributed throughout the plant. Group 3 are somewhat more enigmatic (see below), Group 4 are uniquely tonoplast located and responsible to vacuolar efflux. Group 5 are the most distantly related to the rest of the family and the 2 members are quite distinct from each other, lack a STAS domain and remain something of a puzzle. They seem to be involved in Mo accumulation, perhaps transport and as such have been name mot1 and mot2 (Tomatsu et al. 2007; Baxter et al. 2008; Gasber et al. 2011).

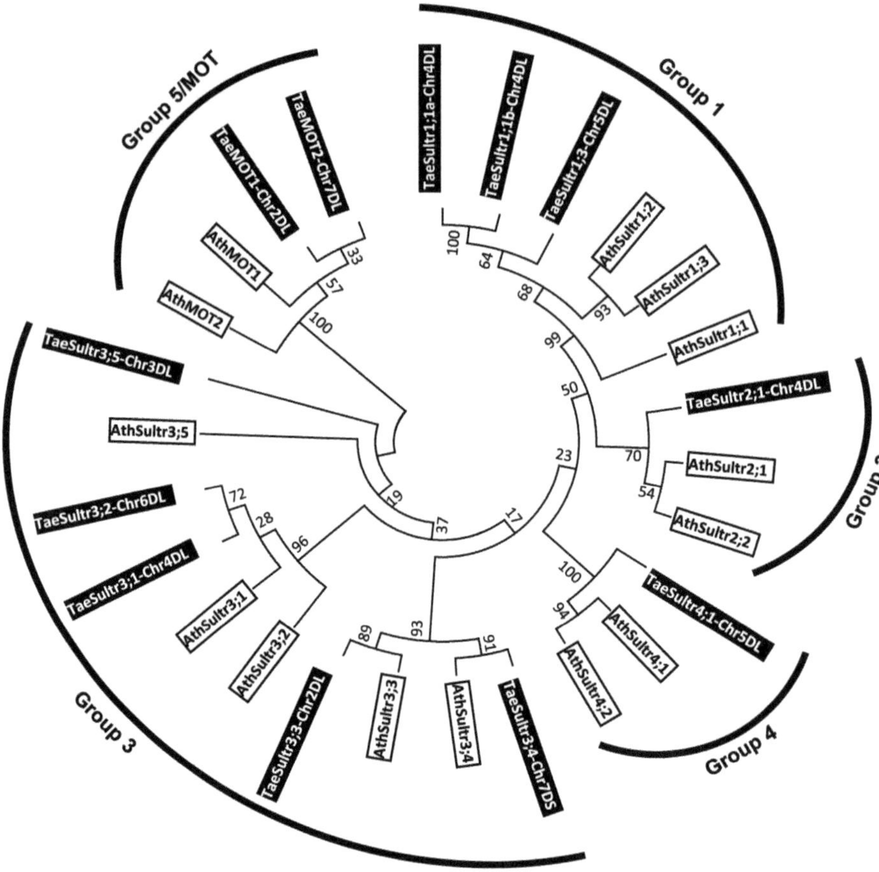

Fig. 2 Phylogenetic relationship of the wheat and Arabidopsis sulfate transporter gene families: Neighbour-Joining Tree (Mega 6, Tamura et al. 2013) from Multiple Alignment (ClustalX V.2.1, Larkin et al. 2007) of coding cDNAs of the *Triticum aestivum* cv. Chinese spring D-genome (*white bold – black highlighted*) and *Arabidopsis thaliana* (*square framed white highlighted*) sulfate transporter gene family. The bootstrap values, expressed as percentage, were obtained from 1000 replicate trees (Courtesy of Peter Buchner)

The Transporter Itself: Recent Structural Insights

Early analysis of the amino acid sequences of the transporter was suggestive of 12 transmembrane domains, based on hydrophobicity plots and occurrence of charged amino acids (Clarkson et al. 1993; Smith et al. 1995a; Takahashi et al. 1996). More recent analysis of transporters in the same superfamily (SulP/SLC26 family) combining both topology mapping of for example the BicA transporter (see Price and Howitt 2014) and for prestin, homology modelling, molecular dynamics simulations and cysteine accessibility scanning are strongly supportive of a complex 14 transmembrane model (Gorbunov et al. 2014). In this analysis a

3-dimensional model has been derived which also proposes a central cavity as the substrate-binding site, midway in an anion permeation channel. Features of this cavity are almost certainly involved in substrate specificity and could potentially be modified to further increase selectivity, for example between sulfate and selenate, opening up the potential for designer crops.

An additional feature of members of this family is the STAS domain (Aravind and Koonin 2000; Rouached et al. 2005). Activity is totally dependent upon its presence and it is strongly suggested that is it involved in protein:protein interaction regulating activity, probably involving phosphorylation of a threonine residue.

The question of whether the transport acts as a monomer or oligomer is of interest and it has been suggested that heterodimers are required for activity or may have an import regulatory role. Maximal sulfate uptake and growth were obtained when a Group 3 transporter was co-expressed with a Group 2 transporter from Arabidopsis in yeast complementation approach, suggestive of the activity of a heterodimer (Kataoka et al. 2004a). No activity of the Group 3 when expressed alone was seen in this study. In contrast Group 3 transporter isolated from Lotus root nodules was able to complement a yeast mutant when expressed by itself (Krussell et al. 2005) indicating some variability for this oligomer requirement.

Specificity for Sulfate, Selenate and Molybdate

The non-specificity of the transporter was exploited in early studies, particularly with yeast, to obtain sulfate transporter-less mutants by harassing the toxic nature of oxyanion analogues of sulfate, particularly selenate but also chromate (Breton and Surdin-Kerjan 1977; Smith et al. 1995b). Selenate has also been applied as a selection agent for the isolation of Arabidopsis mutants by several groups (see for example, Shibagaki et al. 2002).

As the anions sulfate, selenate and molybdate are all transported by the same transporters, it is not surprising that their respectively accumulations in plant tissues are connected. Analysis of what grain from mildly sulfate deficient plots at Rothamsted showed a remarkable accumulation of Se and Mo (Shinmachi et al. 2010; Stroud et al. 2010). The S-deficient plots had a reduced grain yield and reduced grain S-content, both decreasing by about 10%, but several-fold increases in Se and Mo content. This could be partially explained by the observed induction of sulfate transporters in the roots of these field-grown plants, increasing uptake, and a more favourable ration of selenate and molybdate relative to sulfate in the soil solution. Whilst Se generally flows the distribution of sulfate in term of redistribution during grain filling and in relation to storage protein distribution in the grain, some enriched sub-cellular regions were indicative of specific accumulation on non-protein Se, possibly in vacuoles (Moore et al. 2010). Mo was less efficiently remobilized to the grain than Se during grain filling indicating either a fixation of the mineral in the vegetative tissue or a limitation to its later transportation (Shinmachi et al. 2010; Stroud et al. 2010).

Where Now?

Much has been determined about the nature of sulfate transporters in plants, not only in model species but also in crops. A knowledge of the regulation and properties of the transporters helps explain many physiological phenomena and some agronomic responses of crops. The question remains of how may this aid in breeding better genotypes or in informing agronomic treatments.

A previously stated ideotype for optimum S use involves uptake and storage during fluctuating supply, effective remobilization upon demand and appropriate partitioning to ensure healthy and nutritious crops (Hawkesford 2012). Breeding or biotechnology may help deliver such germplasm and the acquired knowledge is an essential prerequisite for such developments. Sulfur will always be required for crop growth so effective capture and utilization are worthy targets.

Acquisition is an important issue. Certainly the adaptation of de-repression will aid scavenging, but only in conjunction with root proliferation. Prospects for improving efficiency of uptake are limited, although constitutive uptake and over-accumulation, followed by storage and effective remobilization remains one key strategy. In relation to this strategy, challenges still exist in the understanding of movement of sulfate within the plant from organ to organ, distribution within specific tissues and finally within individual cells between organelles. It is still unclear as to how S moves into and out of the chloroplast, the key point of entry into the biosynthetic pathway. One reports indicated a chloroplast localizing isoform of the family (Takahashi et al. 1999) but this remains to be corroborated. Clearer is the involvement of Group 3 transporters in release of sulfate from vacuoles, a key storage site (Kataoka et al. 2004b).

David Clarkson proposed the idea of a black box (Fig. 3) in a foreword to the proceedings of the 3rd Workshop (Clarkson 1997). Substantial progress has been made in determining detail within this box since then, but the question of how to improve sulfur nutrient use efficiency remains. Sulfur is required for growth and health, for resistance to biotic and abiotic stresses, and contributes to nutritional properties of food and feed. Decreasing requirements is unlikely to be an option, optimizing agronomic inputs remains the key practical approach, although in the future this may be complemented with plants optimized genetically for specific qualities. Some investigations into natural variation in Arabidopsis have been made

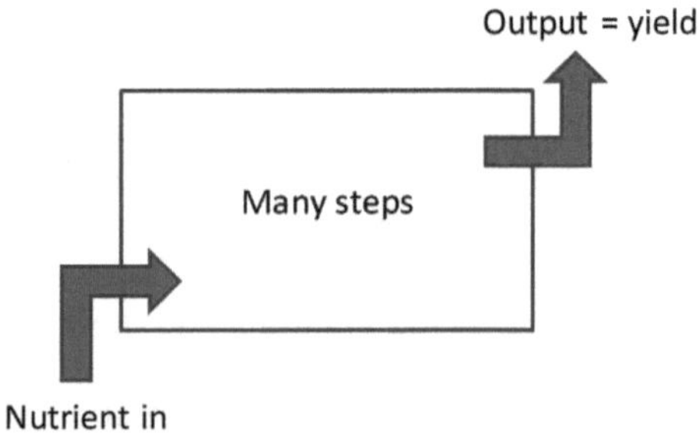

Fig. 3 Clarkson's Black Box (Clarkson 1997). A modified representation of the simple representation of nutrient use efficiency, but inevitably complex in the detail

(Loudet et al. 2007) but there has been little investigation in crop plants and this is a key area for development.

Acknowledgements Rothamsted Research receives support from the Biotechnology and Biological Sciences Research Council (BBSRC) of the UK as part of the 20:20 Wheat project.

References

Aravind L, Koonin EV (2000) The STAS domain – a link between anion transporters and antisigma-factor antagonists. Curr Biol 10:R53–R55

Baxter I, Muthukumar B, Park HC, Buchner P, Lahner B, Danku J, Zhao K, Lee J, Hawkesford MJ, Guerinot ML, Salt DE (2008) Variation in molybdenum content across broadly distributed populations of *Arabidopsis thaliana* is controlled by a mitochondrial molybdenum transporter (MOT1). PLoS Genet 4(2):e1000004. doi:10.1371/journal.pgen.1000004

Bissig M, Hagenbuch B, Stieger B, Koller T, Meier P (1994) Functional expression cloning of the canalicular sulfate transport system of rat hepatocytes. J Biol Chem 269:3017–3021

Breton A, Surdin-Kerjan Y (1977) Sulfate uptake in *Saccharomyces cerevisiae*: biochemical and genetic study. J Bacteriol 132:224–232

Buchner P, Prosser IM, Hawkesford MJ (2004a) Phylogeny and expression of paralogous and orthologous sulphate transporter genes in diploid and hexaploid wheats. Genome 47:526–534

Buchner P, Stuiver CEE, Westerman S, Wirtz M, Hell R, Hawkesford MJ, De Kok LJ (2004b) Regulation of sulfate uptake and expression of sulfate transporter genes in *Brassica oleracea* as affected by atmospheric H_2S and pedospheric sulfate nutrition. Plant Physiol 136:3396–3408

Buchner P, Parmar S, Kriegel A, Carpentier M, Hawkesford MJ (2010) The sulfate transporter family in wheat: tissue-specific gene expression in relation to nutrition. Mol Plant 3:374–389

Cherest H, Davidian J-C, Thomas D, Benes V, Ansorge W, Surdin-Kerjan Y (1997) Molecular characterization of two high affinity sulfate transporters in *Saccharomyces cerevisiae*. Genetics 145:627–635

Clarkson D (1997) Foreword: Sulphate transport and its regulation: a personal view. In: Cram WJ, De Kok LJ, Stulen I, Brenhold C, Rennenberg H (eds) Proceedings of the 3rd workshop on sulphur metabolism in higher plants, Newcastle., Backhuys, Leiden, pp xi–xiii

Clarkson DT, Smith FW, Vanden Berg PJ (1983) Regulation of sulphate transport in a tropical legume, *Macroptilium atropurpureum*, cv Siratro. J Exp Bot 34:1463–1483

Clarkson DT, Hawkesford MJ, Davidian J-C (1993) Membrane and long distance transport of sulfate. In: De Kok LJ, Stulen I, Rennenberg H, Brunold C, Rauser WE (eds) 2nd workshop on sulfur metabolism in higher plants, Garmisch. SPB Publishing, The Hague, pp 3–20

Clarkson DT, Diogo E, Amancio S (1999) Uptake and assimilation of sulphate by sulphur deficient *Zea mays* cells: the role of O-acetyl-L-serine in the interaction between nitrogen and sulphur assimilatory pathways. Plant Physiol Biochem 37:283–290

Davidian JC, Hatzfeld Y, Cathal N, Tagmount A, Vidmar JJ (2000) Sulfate uptake and transport in plants. In Brunhold C, Rennenberg H, De Kok LJ, Stulen I, Davidian JC (eds) 4th workshop on sulfur nutrition and sulfur assimilation in higher plants, Wengen. Paul Haupt, Bern, pp 19–40

Dürr J, Bücking H, Mult S, Wildhagen H, Palme K, Rennenberg H, Dittengou F, Herschbach C (2010) Seasonal and cell type specific expression of sulfate transporters in the phloem of *Populus* reveals tree specific characteristics for SO_4^{2-} storage and mobilization. Plant Mol Biol 72:499–517

Gao Y, Tian Q, Zhang WH (2014) Systemic regulation of sulfur homeostasis in *Medicago truncatula*. Planta 239:79–96

Gasber A, Klaumann S, Trentmann O, Trampczynska A, Clemens S, Schneider S, Sauer N, Feifer I, Bittner F, Mendel RR, Neuhaus HE (2011) Identification of an Arabidopsis solute carrier critical for intracellular transport and inter-organ allocation of molybdate. Plant Biol 13:710–718

Gorbunov D, Sturlese M, Nies F, Kluge M, Bellanda M, Battistutta R, Oliver D (2014) Molecular architecture and the structural basis for anion interaction in prestin and SLC26 transporters. Nat Commun 5:3622

Hästbacka J, de la Chapelle A, Mahtani MM, Clines G, Reeve-Daly MP, Daly M, Hamilton BA, Kusumi K, Trivedi B, Weaver A, Coloma A, Lovett M, Buckler A, Kaitila I, Lander ES (1994) The diastrophic dysplasia gene encodes a novel sulfate transporter: positional cloning by fine-structure linkage disequilibrium mapping. Cell 78:1073–1087

Hawkesford MJ (2003) Transporter gene families in plants: the sulphate transporter gene family – redundancy or specialization? Physiol Plant 117:155–163

Hawkesford MJ (2012) Sulfate uptake and assimilation: whole plant regulation. In: De Kok LJ, Tausz M, Hawkesford MJ, Hoefgen R, McManus MT, Norton RM, Rennenberg H, Saito K, Schnug E, Tabe L (eds) Sulfur metabolism in plants, proceedings of the international plant sulfur workshop, Creswick. Springer, Dordrecht, pp 11–24

Hawkesford MJ, Buchner P, Hopkins L, Howarth JR (2003) The plant transporter family: specialised functions and integration with whole plant nutrition. In: Davidian J-C, Grill D, De Kok LJ, Stulen I, Hawkesford MJ, Schnug E, Rennenberg H (eds) Sulfur transport and assimilation in plants: regulation, interaction and signaling workshop, Montpellier. Backhuys, Leiden, pp 1–10

Hubberten HM, Klie S, Caldana C, Degenkolbe T, Willmitzer L, Hoefgen R (2012) Additional role of O-acetylserine as a sulfur status-independent regulator during plant growth. Plant J 70:666–677

Hubberten H-M, Watanabe M, Bielecka M, Heyneke E, Aarabi F, Hoefgen R (2015) More than a substrate: the *O*-acetylserine responsive transcriptome. In: De Kok LJ, Hawkesford MJ, Rennenberg H, Saito K, Schnug E (eds) Molecular physiology and ecophysiology of sulfur, proceedings of the international plant sulfur workshop. Springer, Cham, pp 133–143

Kataoka T, Hayashi N, Yamaya T, Takahashi H (2004a) Root-to-shoot transport of sulfate in Arabidopsis. Evidence for the role of SULTR3;5 as a component of low-affinity sulfate transport system in the root vasculature. Plant Physiol 136:4198–4204

Kataoka T, Watanabe-Takahashi A, Hayashi N, Ohnishi M, Mimura T, Buchner P, Hawkesford MJ, Yamaya T, Takahashi H (2004b) Vacuolar sulfate transporters are essential determinants controlling internal distribution of sulfate in Arabidopsis. Plant Cell 16:2693–2704

Ketter JS, Jarai G, Fu YH, Marzluf GAB (1991) Nucleotide-sequence, messenger-RNA stability, and DNA recognition elements of Cys-14, the structural gene for sulfate permease-II in Neurospora-crassa. Biochemistry 30:1780–1787

Kouchi H, Hata S (1993) Isolation and characterization of novel nodulin cDNAs representing genes expressed at early stages of soybean nodule development. Mol Gen Genet 238:106–119

Kredich NM (1993) Gene regulation of sulfur assimilation. In: De Kok LJ, Stulen I, Rennenberg H, Brunold C, Rauser WE (eds) 2nd workshop on sulfur metabolism in higher plants, Garmisch. SPB Publishing, The Hague, pp 37–47

Krusell L, Krause K, Ott T, Desbrosses G, Kramer U, Sato S, Nakamura Y, Tabata S, James EK, Sandal N, Stougaard J, Kawaguchi M, Miyamoto A, Suganuma N, Udvardi MK (2005) The sulfate transporter SST1 is crucial for symbiotic nitrogen fixation in *Lotus japonicus* root nodules. Plant Cell 17:1625–1636

Kumar S, Asif MH, Chakrabarty D, Tripathi RD, Dubey RS, Trivedi PK (2015) Comprehensive analysis of regulatory elements of the promotors of rice sulfate transporter gene family and functional characterization of *OsSul1;1* promoter under different metal stress. Plant Signal Behav 10:e990843

Larkin MA, Blackshields G, Brown NP, Chenna R, McGettigan PA, McWilliam H, Valentin F, Wallace IM, Wilm A, Lopez R, Thompson JD, Gibson TJ, Higgins DG (2007) Clustal W and Clustal X version 2.0. Bioinformatics 23:2947–2948

Lass B, Ullrich-Eberius CI (1984) Evidence for proton/sulfate cotransport and its kinetics in *Lemna gibba* G1. Planta 161:53–60

Leggett JE, Epstein E (1956) Kinetics of sulphate absorption by barley roots. Plant Physiol 31:222–226

Loudet O, Saliba-Colombani V, Camilleri C, Calenge F, Gaudon V, Koprivova A, North KA, Kopriva S, Daniel-Vedele F (2007) Natural variation for sulfate content in *Arabidopsis thaliana* is highly controlled by APR2. Nat Genet 39:896–900

Markovich D (2001) Physiological roles and regulation of mammalian sulfate transporters. Physiol Rev 81:1499–1533

Moore KL, Schroder M, Lombi E, Zhao FJ, McGrath SP, Hawkesford MJ, Shewry PR, Grovenor CRM (2010) NanoSIMS analysis of arsenic and selenium in cereal grain. New Phytol 185:434–445

Ohta N, Galsworthy PR, Pardee AB (1971) Genetics of sulfate transport by *Salmonella-typhimurium*. J Bacteriol 105:1053–1062

Price GD, Howitt SM (2014) Topology mapping to characterize cyanobacterial bicarbonate transporters: BicA (SulP/SLC26 family) and SbtA. Mol Membr Biol 31:177–182

Price GD, Woodger FJ, Badger MR, Howitt SM, Tucker L (2004) Identification of a SulP-type bicarbonate transporter in marine cyanobacteria. Proc Natl Acad Sci U S A 101:18228–18233

Rouached H, Berthomieu P, El Kassis E, Cathala N, Catherinot V, Labesse G, Davidian J-C, Fourcroy P (2005) Structural and functional analysis of the C-terminal STAS (sulfate transporter and anti-sigma antagonist) domain of the *Arabidopsis thaliana* sulfate transporter SULTR1.2. J Biol Chem 280:15976–15983

Sandal NN, Marcker KA (1994) Similarities between a soybean nodulin, *Neurospora crassa* sulphate permease II and a putative human tumour suppressor. Trends Biochem Sci 19:19

Schweinfest CW, Henderson KW, Suster S, Kondoh N, Papas TS (1993) Identification of a colon mucosa gene that is down-regulated in colon adenomas and adenocarcinomas. Proc Natl Acad Sci U S A 90:4166–4170

Shibagaki N, Rose A, McDermott JP, Fujiwara T, Hayashi H, Yoneyama T, Davies JP (2002) Selenate-resistant mutants of *Arabidopsis thaliana* identify Sultr1;2, a sulfate transporter required for efficient transport of sulfate into roots. Plant J 29:475–486

Shinmachi F, Buchner P, Stroud JL, Parmar S, Zhao FJ, McGrath SP, Hawkesford MJ (2010) Influence of sulfur deficiency on the expression of specific sulfate transporters and the distribution of sulfur, selenium and molybdenum in wheat. Plant Physiol 153:327–336

Silberg DG, Wang W, Moseley R, Traber PG (1995) The Down-regulated in adenoma gene (Dra) encodes an intestine-specific membrane sulfate transport protein. Gastroenterology 108:A325–A325

Sirko A, Hryniewicz M, Hulanicka D, Bock A (1990) Sulfate and thiosulfate transport in *Escherichia coli* K-12 – nucleotide-sequence and expression of the cysTWAM gene-cluster. J Bacteriol 172:3351–3357

Smith FW, Ealing PM, Hawkesford MJ, Clarkson DT (1995a) Plant members of a family of sulfate transporters reveal functional subtypes. Proc Natl Acad Sci U S A 92:9373–9377

Smith FW, Hawkesford MJ, Prosser IM, Clarkson DT (1995b) Isolation of a cDNA from *Saccharomyces cerevisiae* that encodes a high-affinity sulfate transporter at the plasma-membrane. Mol Gen Genet 247:709–715

Smith FW, Hawkesford MJ, Ealing PM, Clarkson DT, Vanden Berg PJ, Belcher AR, Warrilow AGS (1997) Regulation of expression of a cDNA from barley roots encoding a high affinity sulphate transporter. Plant J 1:875–884

Stroud JL, Zhao FJ, Buchner P, Shinmachi F, McGrath SP, Abecassis J, Hawkesford MJ, Shewry PR (2010) Impacts of sulphur nutrition on selenium and molybdenum concentrations in wheat grain. J Cereal Sci 52:111–113

Takahashi H, Sasakura N, Noji M, Saito K (1996) Isolation and characterization of a cDNA encoding a sulfate transporter from *Arabidopsis thaliana*. FEBS Lett 392:95–99

Takahashi H, Yamazaki M, Sasakura N, Watanabe A, Leustek T, De Almeida Engler J, Engler G, Van Montagu M, Saito K (1997) Regulation of sulfur assimilation in higher plants: a sulfate transporter induced in sulfate-starved roots plays a central role in *Arabidopsis thaliana*. Proc Natl Acad Sci U S A 94:11102–11107

Takahashi H, Asanuma W, Saito K (1999) Cloning of an Arabidopsis cDNA encoding a chloroplast localizing sulphate transporter isoform. J Exp Bot 50:1713–1714

Tamura K, Stecher G, Peterson D, Filipski A, Kumar S (2013) MEGA6: molecular evolutionary genetics analysis version 6.0. Mol Biol Evol 30:2725–2729

Thomas D, Kuras L, Cherest H, Blaiseau PL, Surdin-Kerjan Y (1997) Regulation of gene expression in yeast sulphur metabolism. In: Cram WJ, De Kok LJ, Stulen I, Brenhold C, Rennenberg H (eds), Proceedings of the 3rd workshop on sulphur metabolism in higher plants, Newcastle. Backhuys, Leiden, pp 27–38

Tomatsu H, Takano J, Takahashi H, Watanabe-Takahashi A, Shibagaki N, Fujiwara T (2007) An *Arabidopsis thaliana* high-affinity molybdate transporter required for efficient uptake of molybdate from soil. Proc Natl Acad Sci U S A 104:18807–18812

Vidmar JJ, Tagmount A, Cathala N, Touraine B, Davidian JC (2000) Cloning and characterization of a root specific high-affinity sulfate transporter from *Arabidopsis thaliana*. FEBS Lett 475:65–69

Quo Vadis Sulfur Investigation?: 25 Years of Research into Plant Sulfate Reduction

Stanislav Kopriva

Abstract Sulfate assimilation is an essential pathway of plant primary metabolism providing cysteine for methionine and protein synthesis and as a source of reduced sulfur for synthesis of numerous other essential metabolites. The control of the pathway has long been a focus of plant sulfur research. Significant progress in understanding the pathway and its regulation has been made since the first Sulfur Workshop in 1989. All enzymatic steps have been identified, and the corresponding genes cloned. The investigations of sulfate assimilation were always quickly adopting newly developed approaches and the introduction of molecular biology led to a rapid switch to Arabidopsis as the main model for sulfur research. We learned a lot about the regulation of the pathway and identified genes affected by various environmental conditions or chemical signals. Sulfur research was one of the first areas of plant science to make use of systems biology with several seminal studies of major importance. The increased use of genetic approaches resulted in identification of new regulatory factors and most recently in finding genes responsible for genetic variation of sulfur-related traits in natural populations. Nevertheless, many questions remain open. This overview of the different approaches to study sulfur metabolism will highlight the success stories. The major progress since the first workshop will be summarised, and a new set of open questions will focus on how plant sulfur research will develop over the next 25 years.

Introduction

Plants take up the essential nutrient sulfur in the oxidized form of sulfate, reduce and incorporate it into various metabolites and cellular components (Kopriva 2006; Takahashi et al. 2011). The assimilation of sulfate into cysteine is therefore an essential pathway of plant primary metabolism, with the reductive part being key not just for the pathway but for life. Therefore sulfate reduction has been in the focus of plant sulfur research for many years, including being a major topic from

S. Kopriva (✉)
Botanical Institute, Cluster of Excellence on Plant Sciences (CEPLAS), University of Cologne, Zülpicher Str. 47b, 50674 Cologne, Germany
e-mail: skopriva@uni-koeln.de

L.J. De Kok et al. (eds.), *Sulfur Metabolism in Higher Plants - Fundamental, Environmental and Agricultural Aspects*, Proceedings of the International Plant Sulfur Workshop, DOI 10.1007/978-3-319-56526-2_2

the beginning of the Plant Sulfur Workshops (Brunold 1990). The final pathway (Fig. 1), however, has been formulated quite recently, after years of controversy (Kopriva and Koprivova 2004). Sulfate is firstly activated by adenylation catalyzed by ATP sulfurylase to adenosine 5′-phosphosulfate (APS). The activated sulfate is reduced in two steps, first through a two electron reduction to sulfite and subsequently by six more electrons to sulfide. Sulfide is then incorporated into O-acetylserine to form cysteine (Takahashi et al. 2011). The pathway in plants is thus very similar to sulfate assimilation in bacteria and fungi, the original models for dissection of the pathway (Kopriva 2015), which was one of the main reasons for the controversy and discussions regarding the plant pathway. As the model microorganisms *Escherichia coli* and *Saccharomyces cerevisiae* require a second activation step for the sulfate to be reduced (phosphorylation of APS to 3′-phosphoadenosine 5′-phosphosulfate (PAPS)), it was believed that plant sulfate assimilation proceeds in the same way with PAPS as an intermediate (Thompson 1967). However, in the beginning of the 1970s it was shown that APS is used directly in plants and algae and the corresponding enzyme was named APS sulfotransferase (Schmidt 1972). The reduction of APS was shown to result in a carrier-bound intermediate and a thiosulfonate reductase was described as a component of plant sulfate assimilation (Schmidt 1973). In the following years, APS sulfotransferase was shown to be highly regulated and a key enzyme for the control of the pathway, however, PAPS sulfotransferases have also been part of the literature (Schmidt and Jäger 1992; Brunold and Suter 1984). Thus, at the time of the first Plant Sulfur Workshop in Groningen, in 1989, many fundamental questions about sulfate reduction remained open (Brunold 1990) which are summarized in the following:

1. the physiological significance of the enzyme reactions detected *in vitro*, especially of the APS sulfotransferase versus the PAPS sulfotransferase pathway and of the organic thiosulfate reductase versus the sulfite reductase mechanism,
2. the detailed characterization of APS sulfotransferase and PAPS sulfotransferase,
3. the molecular basis of the regulatory phenomena observed,
4. the contribution of the root system to assimilatory sulfate reduction.

In this review the progress in answering these questions will be summarized, an overview of the different approaches to study sulfur metabolism presented, and new set of open questions formulated on how plant sulfur research will develop in the next 25 years.

Current Understanding of the Sulfate Assimilation Pathway

As evident from the first question, at the time of the first Sulfur workshop, two alternative enzymes for the first step of sulfate reduction were discussed, APS sulfotransferase and PAPS sulfotransferase (Brunold 1990). The situation however

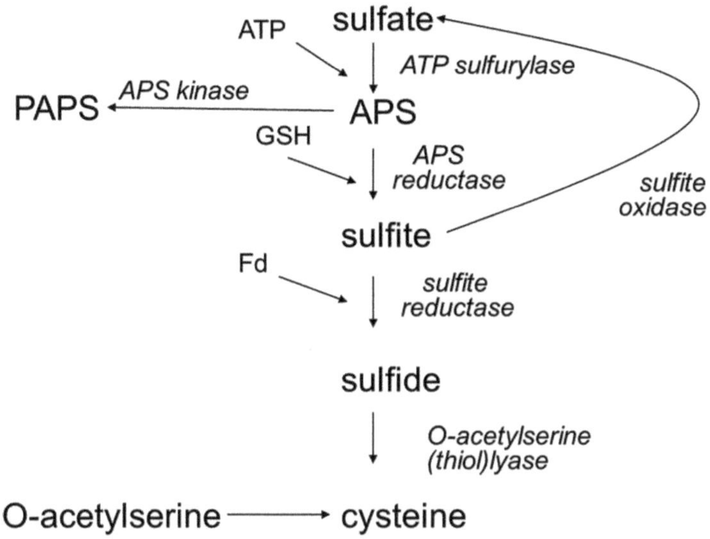

Fig. 1 Current understanding of plant sulfate assimilation

became even more complicated after the cloning of an Arabidopsis enzyme, which produced sulfite from APS and was named APS reductase (Gutierrez-Marcos et al. 1996; Setya et al. 1996). Three possible routes of sulfate to sulfide have been proposed, APS dependent bound-sulfite way (through APS sulfotransferase and thiosulfonate reductase), APS dependent reduction (APS reductase, sulfite reductase) or PAPS-dependent pathway (Fig. 2). The controversy has been clarified when the enzyme characterized as APS sulfotransferase was purified from *Lemna minor*, its N-terminal sequence was determined and the corresponding cDNA cloned (Suter et al. 2000). The sequence of the APS sulfotransferase was highly similar to the sequence of APS reductase, which together with identification of free sulfite as reaction product resulted in postulating APS reductase as the component of sulfate assimilation. In addition, experiments in oxygen free atmosphere showed that the bound sulfite is the result of reaction of sulfite with oxidized thiols, and therefore probably irrelevant for the sulfate reduction pathway, which proceeds through APS reductase and sulfite reductase (Suter et al. 2000). However, the question of PAPS reductase/sulfotransferase in plants has not been completely resolved. Even though everything suggests that APS reductase is the sole reducing enzyme, it is impossible to completely rule out the existence of PAPS-dependent enzyme. The crucial evidence, which would show that disruption of all APS reductase isoforms is lethal, is not available and thus the plant PAPS reductase is still possibly out there.

Although the question of the first reductive step in sulfate assimilation was resolved, the pathway has changed further. A new enzyme was added to the pathway, which had not been known in plants, sulfite oxidase (Eilers et al. 2001). Sulfite oxidase acts against the flow of sulfur in sulfate assimilation, therefore, to

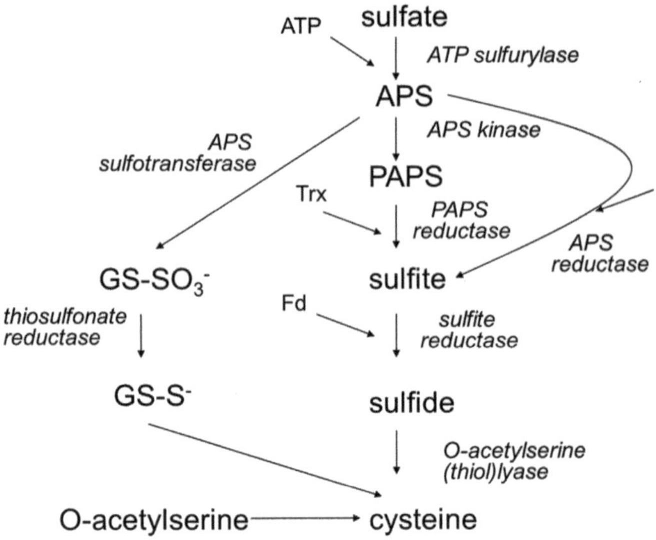

Fig. 2 Three possible routes of sulfate reduction in plants as discussed 1996–2000

prevent a futile cycle it is spatially separated from the reducing enzymes in peroxisomes. The enzyme protects plants against elevated sulfur dioxide and appears to be important for sulfite homeostasis, but whether this is its main physiological function is still under discussion (Lang et al. 2007; Brychkova et al. 2007, 2013).

In addition, in the last decade it became apparent that a second non-reductive path of sulfate assimilation, into PAPS and further into sulfated compounds, is essential also in plants (Mugford et al. 2010). APS kinase, producing PAPS from APS, has been shown to be important for synthesis of glucosinolates, a group of sulfated secondary compounds with plethora of functions in plant defence (Mugford et al. 2009). However, while the loss of glucosinolates does not compromise Arabidopsis in the absence of pathogens, a mutant disrupted in *APK1* and *APK2* isoforms of APS kinase showed a clear semi-dwarf phenotype (Mugford et al. 2009). These experiments showed the importance of PAPS for plant performance, however, the exact nature of the essential metabolites is not known.

In conclusion, the first question concerning the pathway of sulfate assimilation now seems to be largely answered (Fig. 1).

Biochemistry of APS Reductase

The second question, on the detailed characterization of APS and PAPS sulfotransferases, has naturally been limited to the APS dependent enzyme. The analysis of purified APS sulfotransferase revealed that it is identical to recombinant

APS reductase, identified by molecular cloning (Suter et al. 2000). The enzyme uses APS as a substrate with affinity of 10–20 μM, and glutathione or other thiols as reductant. Sequence analysis has shown that the enzyme is composed from three domains, a plastid targeting peptide, a reductase domain similar to bacterial PAPS reductases, and a C-terminal thioredoxin-like domain (Setya et al. 1996; Gutierrez-Marcos et al. 1996). The enzyme can be split into these two domains and the activity can be reconstituted (Bick et al. 1998). This makes sense because bacterial PAPS reduction is also dependent on thioredoxin or glutaredoxin. The two domains have distinct functions: the reductase domain is responsible for the interaction with APS and the formation of sulfite, whereas the C-terminal domain interacts with the reductant and resumes the function of glutaredoxin (Bick et al. 1998). APS reductase contains an iron-sulfur cluster, which was shown to be diamagnetic and asymmetric, as in the enzyme isoform studied (APR2 from *Arabidopsis thaliana*) it is most probably bound only by three cysteine residues in the reductase domain (Kopriva et al. 2001). When the enzyme is incubated with APS in the absence of reductants, a stable reaction intermediate can be detected with sulfite covalently bound to a cysteine residue in the reductase domain (Weber et al. 2000). The sulfite can be released by treatment with free thiols or sulfite, or by addition of the recombinant C-domain, confirming that the two domains have different functions in the catalysis, but raising a question on the function of the iron sulfur cluster.

Genes for the $[Fe_4S_4]$ cluster containing APS reductase have been found in all seed plants, basal plants, and green algae. However, this is not the only form of this enzyme (Fig. 3). Comparison of plant APS reductase with its bacterial counterparts revealed that the bacteria possess two classes of enzymes, similar to the reductase domain. The "classical" PAPS reductases from *E. coli* and *Salmonella* (and also yeast and fungi) differ from the plant enzyme mainly due to the absence of two conserved cysteine pairs, which are responsible for binding of the FeS cluster (Kopriva et al. 2002). Correspondingly, the cofactor has never been reported for PAPS reductases. However, another form of bacterial enzyme has been identified, more similar in sequence to the plant APS reductase, including the two cysteine pairs (Bick et al. 2000; Williams et al. 2002). These bacterial enzymes use APS rather than PAPS and contain the FeS cluster (Kopriva et al. 2002). Due to their similarities, the APS and PAPS reductases have been both referred to as sulfonucleotide reductases and shown to have a very similar reaction mechanism (Carroll et al. 2005). The specificity for APS or PAPS seems to be linked to the presence or absence of the cluster, respectively (Bhave et al. 2012). Incidentally this is also true for the dissimilatory APS reductase in sulfur oxidizing bacteria, which, although having a completely different primary structure, possess FeS centre (Frigaard and Dahl 2009).

The variation in APS reductase forms is even greater. Another isoform of APS reductase has been found in the moss *Physcomitrella patens*, which in contrast to the "classical" plant APS reductase does not bind the FeS cluster, and does not possess the thioredoxin-like domain but still reduces APS and not PAPS (Kopriva et al. 2007). This challenges the simple link between FeS cluster and APS reduction, which seems to be valid only in prokaryotes, and reinforces the question

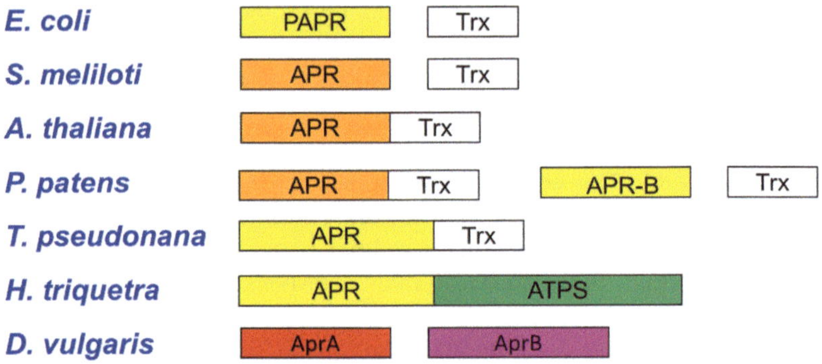

Fig. 3 Diversity of APS reductase forms in different organisms. *Yellow marks* PAPS reductases or APS reductases type B without FeS cluster, the presence of the cluster is marked by *orange colour*. The AprA and AprB isoforms of dissimilatory reductase from *Desulfovibrio vulgaris* are shown for comparison

regarding the function of the FeS cluster. This isoform of APS reductase, named APR-B, is present in several other basal plants, including *Selaginella* and *Marchantia* (Kopriva et al. 2007). A highly similar enzyme is dominating in the Eukaryotic microalgae, such as diatoms and haptophytes. These organisms, e.g. *Thalassiosira pseudonana* or *Emiliania huxleyi* possess genes for APS reductase of the APR-B type, and with the thioredoxin-like domain (Fig. 3) (Patron et al. 2008). Interestingly, the APS reductase activity in the microalgae is approximately two orders of magnitude higher than in higher plants (Bochenek et al. 2013). Thus, although a lot of progress in understanding the biochemistry of APS reductase has been made, the reaction mechanism is not yet completely resolved, since structure is available from the bacterial enzymes, or the APR-B, but not the full plant enzyme with both domains. It is however remarkable, that the same reaction can be catalyzed by enzymes with a similar primary structure but with, and without, the FeS cofactor.

Molecular Mechanisms of Regulation

At the time of the first Sulfur Workshop, the regulation of sulfate assimilation, particularly the reductive part, was well described at the physiological level. It was shown that sulfate reduction is induced by limiting sulfur availability and cadmium, and decreased by reduced sulfur, both by thiols and atmospheric SO_2 (Brunold 1990). Regulatory interactions with nitrogen nutrition had been described: a decrease of sulfate reduction capacity at low N and induction by ammonium (Brunold and Suter 1984). The developmental regulation of the pathway was described, showing an increase in activity from etiolated seedlings to green tissues and in the development of leaves until maturity was reached (Brunold 1990). In addition, feedback inhibition of individual enzymes of the pathway was described, such as APS inhibiting ATP sulfurylase and AMP acting negatively on APS sulfotransferase. Generally, it was believed that the regulation was mainly at the level of ATP sulfurylase, with many conditions affecting additionally APS reductase.

The progress in understanding the regulation of sulfate assimilation has been significant over the past 25 years. It is impossible to describe all new findings on compounds and conditions affecting the pathway, the mechanisms, signalling pathways, transcription factors, etc.; these have been described in many reviews (Davidian and Kopriva 2010; Kopriva 2006; Koprivova and Kopriva 2014; Takahashi et al. 2011; Leustek et al. 2000; Hell 1997; Rausch and Wachter 2005; Chan et al. 2013). Instead, a personal selection of the major findings will be given, not only showing the major concepts but also the variety of links between sulfate assimilation and plant metabolism. The progress in uncovering the mechanisms of regulation is highly interconnected with the variety of methods and approaches used and with a shift in focus of the studies towards APS reductase.

The classical physiological studies led to formulation of the major concept in regulation of sulfate assimilation, the demand driven control, which explained most of the known regulatory events (Lappartient et al. 1999; Lappartient and Touraine 1996). The demand driven regulation is an overarching concept that still helps to explain observed regulation of sulfate assimilation by many environmental conditions. Another major impact on studies of the control of the pathway was the identification of signalling function of O-acetylserine (Neuenschwander et al. 1991), leading to some controversies but also to formulation of function on different levels (Hubberten et al. 2012b; Wirtz and Hell 2006). Other signals have been found with a more or less clear contribution to the regulation, such as sugars and phytohormones, particularly jasmonate and cytokinins (Jost et al. 2005; Maruyama-Nakashita et al. 2004). Most important was the finding of sulfur starvation inducible microRNA miR395, which targets three isoforms of ATP sulfurylase and the sulfate transporter SULTR2;1 and contributes strongly not only to response to sulfate starvation but also to regulation of sulfate partitioning (Kawashima et al. 2009, 2011; Liang et al. 2010).

The first major step in understanding the mechanisms of regulation was the introduction of molecular biology methods to complement the physiological and biochemical studies. These showed in the beginning that regulation of the pathway, e.g. by sulfate starvation, light, nitrogen sources, or reduced sulfur compounds, is mostly on the transcriptional level (Kopriva et al. 1999; Koprivova et al. 2000; Vauclare et al. 2002; Takahashi et al. 1997). APS reductase was more highly regulated than other components of the pathway, suggesting its central role in the control of the pathway. The key role of this enzyme was confirmed using flux control analysis, at least for the feedback regulation by thiols (Vauclare et al. 2002). These studies were followed by numerous reports showing transcriptional regulation of one or multiple genes of sulfate assimilation by a large variety of metabolites or environmental conditions (reviewed e.g. in Davidian and Kopriva 2010; Kopriva and Rennenberg 2004).

Numerous expression analyses showed that multiple genes are regulated by a single treatment. The availability of Arabidopsis genome then enabled the design of the first global transcriptomics studies and with three seminal papers on response to sulfate starvation sulfur research entered the systems biology era (Maruyama-Nakashita et al. 2003; Nikiforova et al. 2003; Hirai et al. 2003). The sulfate starvation response is one of the best studied environmental conditions by systems biology approaches, pioneering the combination of metabolite-gene networks and producing some of the best examples of mechanistic understanding derived from –omics data (Nikiforova et al. 2005; Hirai et al. 2005; Hubberten et al. 2012b). These studies also delivered first hints to the connection between sulfate assimilation and synthesis of glucosinolates (Hirai et al. 2005; Malitsky et al. 2008). Altogether these studies revealed a core cluster of genes highly regulated by sulfate starvation, some belonging to the pathway, e.g. sulfate transporters *SULTR1;1*, *1;2*, *4;1*, *4;2* or the APS reductase and some with unknown function, which is now only starting to be understood. They showed that sulfate starvation response includes modulation of phytohormone synthesis, particularly jasmonate and auxin but, while they identified potential regulatory hubs and factors, the progress in deciphering mechanisms of regulation has been rather limited.

Comparing results of expression analyses with activity measurements or metabolite concentrations it became soon apparent that not all regulation can be explained by transcriptional control and that various post-transcriptional mechanisms contribute to control of sulfate assimilation (Takahashi et al. 2011). Obviously, the miR395 is one of these mechanisms, restricting expression of the SULTR2;1 transporter to the xylem during sulfate deficiency and so increasing efficiency of sulfate root to shoot translocation (Kawashima et al. 2009). Another regulatory mechanism is the fluid dissociation and association of the cysteine synthase complex by changes in concentration of sulfide and O-acetylserine, which is probably an important player in sensing sulfur availability (Wirtz and Hell 2006). However, the most common and fundamental post-transcriptional mechanism of regulation of sulfur metabolism is redox. The first enzyme of the pathway shown to be redox regulated was, not surprisingly, APS reductase. This regulatory mechanism was initially based on *in vitro* observations of activation of recombinant APS reductase by oxidative agents and inactivation by reduction and corroborated *in vivo* by observed discrepancy between response of APS reductase protein and enzyme activity to oxidative stress (Bick et al. 2001; Kopriva and Koprivova 2004). The next target of redox regulation is the γ-glutamylcysteine synthetase (γECS), the first of two enzymes catalyzing glutathione synthesis. The inhibition of this enzyme by glutathione has been shown to be caused by redox regulation rather than product inhibition (Hothorn et al. 2006). The latest acquisition in the list of redox regulating enzymes of sulfur metabolism is APS kinase (Ravilious et al. 2012). Interestingly, the redox regulation has been recognized only after the structure of the enzyme has been solved, whereas previous biochemical characterizations had not found any hints. In contrast to APS reductase, APS kinase is activated in its reduced form and inactivated by oxidation (Ravilious et al. 2012). Since APS kinase is important for control of sulfur partitioning between primary reductive assimilation and PAPS and secondary products (Mugford et al. 2011), the redox regulation may be important for the mechanism, but its biological relevance needs to be shown *in planta*.

The increasing availability of molecular and genetic resources in Arabidopsis allowed the use of genetic tools to further dissect the regulation of the sulfate assimilation pathway. Reverse genetic approaches helped, e.g., to understand the differences in functions of individual isoforms of the pathway components. The main contributions if this approach were linking the two branches of sulfate assimilation with glucosinolates synthesis through analysis of mutants in APS kinase (Mugford et al. 2009) and evidence that also other components of sulfate assimilation pathway than APS reductase may represent a bottleneck, namely sulfite reductase and serine acetyltransferase (Haas et al. 2008; Khan et al. 2010). To find new regulatory factors, however, other genetic methods have been applied. The biggest success in unravelling the mechanisms of regulation has been the identification of SULFATE LIMITATION1 (SLIM1), a transcription factor of the Ethylene-Insensitive-3-Like family (EIL3), a central regulator of the response to sulfate starvation (Maruyama-Nakashita et al. 2006). SLIM1 is responsible for the induction of transcripts for sulfate transporters, miR395, and many other genes by sulfate deficiency (Maruyama-Nakashita et al. 2006; Wawrzynska and Sirko 2014). The mechanism of its action is unknown yet, as *SLIM1* mRNA is not affected by sulfate starvation, and it is also not the only factor responsible for the response, since upregulation of transcripts for APS reductase is SLIM1 independent (Wawrzynska and Sirko 2014). Other transcription factors have been shown to regulate sulfate assimilation, e.g. the LONG HYPOCOTYL 5 (HY5) (Lee et al. 2011), but the understanding of the transcriptional machinery controlling sulfate assimilation is limited. This is in contrast to the regulation of glucosinolates synthesis, where a complex interplay of six MYB factors and three bHLH (MYC) factors has been described (Schweizer et al. 2013; Sonderby et al. 2010; Frerigmann and Gigolashvili 2014). Interestingly, these MYB factors also regulate genes of primary sulfate assimilation, showing that the glucosinolates biosynthesis should be considered within the core sulfur metabolism, at least in Arabidopsis and other Brassicaceae (Yatusevich et al. 2010).

All the approaches mentioned above gave some insights into the regulation of sulfur metabolism. The next set of experiments, exploiting quantitative genetics, identified further genes that are responsible for the natural variation of sulfur related traits and that can be potentially used for their modification in crops. These approaches profited from the availability of well defined Arabidopsis ecotypes and the progress in their genotyping. Two areas of sulfur research have been focus of the natural variation studies, glucosinolates and sulfate homeostasis. The glucosinolates diversity in Arabidopsis covers both quantitative and qualitative differences, particularly among the aliphatic, methionine derived glucosinolates. Five loci were identified that explain most of the qualitative variation among multiple accessions, primarily in the chain length, and one of them controls up to 75% of the quantitative differences (Kliebenstein et al. 2001b). In a more refined QTL study 20 loci were found to control the variation in glucosinolates between Ler and Cvi accessions (Kliebenstein et al. 2001a). Among the most important loci is the GS-ELONG locus, containing multiple methylthioalkylmalate synthase genes, which plays a key role in the methionine elongation step of glucosinolate synthesis (Kroymann et al. 2001). The second major QTL is the GS-AOP, which is formed by two genes for 2-oxoglutarate–dependent dioxygenases, responsible for chain modification of the aliphatic glucosinolates and contributing greatly to the diversity (Kliebenstein et al. 2001c). The same loci were identified in a genome wide association study (GWAS), along with other, so far unknown ones (Chan et al. 2011). Therefore, it can be expected that the current view of control of glucosinolates synthesis remains incomplete.

The second theme approached by quantitative genetics is the accumulation of sulfate and sulfur. A QTL analysis of sulfate content in leaves of recombinant inbred lines from Bay-0 and Shahdara ecotypes revealed a non-synonymous single nucleotide polymorphism (SNP) in gene for APR2 isoform of APS reductase, resulting in almost complete inactivation of the corresponding enzyme (Loudet et al. 2007). Loss of APR2 leads to diminishing of total enzyme activity by 75%, consequently the flux through sulfate assimilation is reduced and sulfate accumulates. Interestingly, a second QTL from the same screen has been identified as ATPS1 isoform of ATP sulfurylase, additive to the effect of APR2 (Koprivova et al. 2013). The phenotype is caused by variation in mRNA accumulation, which correlates with a deletion in the first intron of *ATPS1* genes, leading to lower enzyme activity, reduced flux, and accumulation of sulfate. The QTL analysis, however, can only reveal variation in the two parent ecotypes, whereas much larger variation exists among the different accessions. When a wider variation in sulfur homeostasis has been assessed in 350 accessions, particularly high total sulfur content was detected in the Hod ecotype. The high sulfur content was caused mainly by high accumulation of sulfate, and exactly as in the Bay-0 × Shahdara population the causal gene was shown to be *APR2* (Chao et al. 2014). A further analysis of links between high sulfate/sulfur accumulation and APR revealed another small group of ecotypes related to Lov-1, with naturally inactive APR2. All three accessions (Shahdara, Hod, and Lov-1) possess different amino acid substitutions that lead to at least 1000-fold reduction of enzyme activity (Chao et al. 2014). Thus, in the Arabidopsis population, APR2 was at least three times independently inactivated, with slightly different consequences for accumulation of sulfate and sulfur.

Three more findings are considered potentially important for understanding the regulation of sulfate assimilation and its integration within general metabolic networks. Firstly, the unexpected finding of coordinated enrichment of expression of sulfate assimilation and glucosinolates synthesis in bundle sheath cells in Arabidopsis (Aubry et al. 2014). This resembles the localization of sulfate assimilation in plants with C4 photosynthesis, which is one of the unsolved questions of sulfur research. Secondly, several reports indicated an important role of the interplay between sulfite reductase and sulfite oxidase in "sulfite network" for sulfite homeostasis and regulation of sulfate assimilation (Brychkova et al. 2013). Finally, linking glucosinolates into core sulfur metabolism brought a connection to an enigmatic enzyme, 3′(2′),5′-bisphosphate nucleotidase, known as SAL1 or Fiery1, which has been found in many genetic screens targeting mutants in stress response or leaf morphology (Robles et al. 2010). The enzyme metabolizes phosphoadenosine phosphate (PAP), which is a by-product of sulfation reactions, e.g., in synthesis of glucosinolates. PAP accumulates during drought stress and was proposed as a retrograde signal of stress from plastids to the nucleus (Estavillo et al. 2011). The loss of FRY1 also has a direct effect on sulfur assimilation, the glucosinolate content is lower in the mutants and the plants accumulate desulfoglucosinolate precursors (Lee et al. 2012). In addition, *fry1* mutants show lower accumulation of sulfate and total sulfur, making the enzyme and PAP interesting targets for further investigations.

Overall, many mechanisms of regulation have been discovered, but since every step in the pathway can become limiting, the control of sulfur metabolism is far from being fully understood.

Sulfate Assimilation in Roots

The fourth question formulated at the first Sulfur Workshop is one with the least definite answer. Root can reduce sufficient sulfate for their own demand, as shown clearly in root cultures growing in sulfate as the sole sulfur source (Vauclare et al. 2002). However, sulfate reduction in roots cannot complement the reduced APS reductase activity in the shoots, as demonstrated by reciprocal grafting of Col-0 and Hod accessions (Chao et al. 2014). In a split root system, the regulation of sulfate transport seems to be dependent on local sulfate availability and sulfate is translocated to shoots but not to a sulfate deficient part of the root system (Hubberten et al. 2012a). Roots contribute to sulfate deficiency response by increased uptake and translocation of sulfate, and there seems to be a communication between shoots and roots regarding the sulfur status (Hubberten et al. 2012a). The contribution of roots to overall sulfur homeostasis, sensing, and signalling, however, needs to be systematically investigated, e.g., using more grafting experiments with various mutants in sulfate assimilation and signalling pathways.

Sulfur Research in the Next 25 Years

The progress in understanding the pathway of sulfur assimilation and the regulation has been overwhelming. However, there are still many open questions, some reformulating the "old" concepts, some derived from the emerged new insights. Thus, in analogy with the first Sulfur Workshop and the most intriguing open questions formulated there, the following questions can be considered key for the next 25 years of plant sulfur research:

How is the sulfur flux controlled? This question has not changed much in the past 25 years, but the concept has changed dramatically. Whilst searching for the rate limiting enzyme, it became obvious that control is shared and may lie on every step of the pathway, dependent on the actual metabolic and developmental status of the plant. Understanding the flux control will require consideration of sulfur fluxes in the context of the whole plant metabolism and employment of advanced mathematical tools and models (Calderwood et al. 2014). Solving this question will obtain a fundamental understanding about the control of the pathway and its integration with other pathways and finding the corresponding mechanisms – transcription factors and signals, and help with applying the knowledge generated in the course of plant sulfur research. Many traits are connected with sulfur, many sulfur compounds are important for various applications, e.g. the health promoting glucosinolates. The knowledge about control of sulfur fluxes will enable smart genetic engineering to create plants with tailored contents of diverse sulfur compounds.

What are the molecular mechanisms of the regulation? In the course of past investigations, much has been learned about the regulation of sulfur metabolism, and this trend will certainly continue. The number of known transcription factors, signalling compounds, and other regulatory mechanisms is limited, and it is crucial to find the full complement of transcription factors regulating the pathway, and to understand their functions. The bigger picture of transcriptional regulation has also to be considered, such as epigenetics and the role of transcriptional complexes, e.g. the Mediator. The redox regulation of APS kinase and APS reductase and possibly other enzymes has to be assessed in plants and the contribution to the control of sulfur fluxes quantified. We expect that further use of quantitative genetics will identify new genes and alleles controlling the variation in sulfur related traits in natural populations.

What are the biochemical properties of the new enzyme isoforms from algae? The new forms of ATP sulfurylase and APS reductase and their numerous fusions await characterization and determination of their biochemical properties. This will enable better understanding of sulfur metabolism in these organisms, that have often very high activity of these enzymes, but also generate sources of new more efficient enzymes for engineering of sulfur assimilation in plants or synthetic microbes.

What is the identity of the unknown sulfur compounds? In an untargeted metabolomics approach, a large number of unknown sulfur containing compounds have been detected in Arabidopsis (Glaser et al. 2014). This shows that the catalogue of sulfur compounds is far from complete even in model plants, and therefore new enzymes have to be connected to the sulfur network, e.g., a large number of sulfotransferase isoforms have unknown substrate specificity. Identification of these compounds will enable a better assessment of sulfur pools and may lead to discovery of new signalling and regulatory compounds.

How is sulfur metabolism integrated in the whole plant metabolism? Although the connections between sulfur and nitrogen nutrition, or sulfur and carbohydrates have frequently been described, the mechanisms and signals are unknown. The sensing of sulfur status and its transduction to the general response is part of this question and is also completely unclear, again a large contribution of mathematical tools will be necessary to answer this question.

What is the role of sulfur metabolism in evolution and adaptation of plants? Large variation in many sulfur related traits have been observed within Arabidopsis or Brassica species. There are also differences in localization of the pathway in C3 and C4 plants, which highlight the links between C4 photosynthesis and sulfur reduction. So, how conserved are the regulatory mechanisms in different plant families? With reduced costs for sequencing and genotyping, this may be answered for some plant species. However, research shows that variation in glucosinolates has a large role in underpinning variation of Arabidopsis accessions in resistance to insects (Kliebenstein et al. 2002). Therefore, the ecological significance of varied sulfate and sulfur levels is of immense interest.

How can we apply the knowledge on sulfur metabolism for improvement of plants and/or added value? Adequate sulfur nutrition is essential for high yields and quality of crops. Due to reduced atmospheric sulfur depositions yields can be sustained only through sulfur fertilization. Therefore, understanding sulfur homeostasis will underpin approaches to breed low input crop varieties allowing reduction of the environmental costs of intensive agriculture. In addition, sulfur compounds are often linked with resistance to pests or abiotic stress, so the improvement of synthesis of these compounds might be beneficial. Many sulfur compounds, however, are important beyond the plant; the best example is the glucosinolate from broccoli, glucoraphanine, which is a precursor of sulforaphane, a plant derived metabolite with a plethora of functions in cancer prevention (Clarke et al. 2008). Broccoli varieties with increased content of glucoraphanine have been created and shown to possess an increased capacity to prevent prostate cancer (Traka and Mithen 2009). A better knowledge of the control of plant sulfur metabolism will enable the generation of plants accumulating other beneficial compounds.

Altogether, we can surely look forward to many new exciting stories about sulfur and hope that the next 25 years of plant sulfur research will be as successful as the past 25 years.

Acknowledgements The research in SK's lab is funded by the Deutsche Forschungsgemeinschaft (EXC 1028). I apologize to all whose excellent papers I could not cite for the lack of space. I dedicate this overview to Christian Brunold, the pioneer of sulfur research, who formulated the first set of questions discussed here and who introduced me into the fascinating world of sulfur.

References

Aubry S, Smith-Unna RD, Boursnell CM, Kopriva S, Hibberd JM (2014) Transcript residency on ribosomes reveals a key role for the *Arabidopsis thaliana* bundle sheath in sulfur and glucosinolate metabolism. Plant J 78:659–673

Bhave DP, Hong JA, Keller RL, Krebs C, Carroll KS (2012) Iron-sulfur cluster engineering provides insight into the evolution of substrate specificity among sulfonucleotide reductases. ACS Chem Biol 7:306–315

Bick JA, Aslund F, Chen Y, Leustek T (1998) Glutaredoxin function for the carboxyl-terminal domain of the plant-type 5′-adenylylsulfate reductase. Proc Natl Acad Sci U S A 95:8404–8409

Bick JA, Dennis JJ, Zylstra GJ, Nowack J, Leustek T (2000) Identification of a new class of 5′-adenylylsulfate (APS) reductases from sulfate-assimilating bacteria. J Bacteriol 182:135–142

Bick JA, Setterdahl AT, Knaff DB, Chen Y, Pitcher LH, Zilinskas BA, Leustek T (2001) Regulation of the plant-type 5′-adenylyl sulfate reductase by oxidative stress. Biochemistry 40:9040–9048

Bochenek M, Etherington GJ, Koprivova A, Mugford ST, Bell TG, Malin G, Kopriva S (2013) Transcriptome analysis of the sulfate deficiency response in the marine microalga *Emiliania huxleyi*. New Phytol 199:650–662

Brunold C (1990) Reduction of sulfate to sulfide. In: Rennenberg H, Brunold C, De Kok LJ, Stulen I (eds) Sulfur nutrition and sulfur assimilation in higher plants. SPB Academic Publishing, The Hague, pp 13–31

Brunold C, Suter M (1984) Regulation of sulfate assimilation by nitrogen nutrition in the duckweed *Lemna minor* L. Plant Physiol 76:579–583

Brychkova G, Xia Z, Yang G, Yesbergenova Z, Zhang Z, Davydov O, Fluhr R, Sagi M (2007) Sulfite oxidase protects plants against sulfur dioxide toxicity. Plant J 50:696–709

Brychkova G, Grishkevich V, Fluhr R, Sagi M (2013) An essential role for tomato sulfite oxidase and enzymes of the sulfite network in maintaining leaf sulfite homeostasis. Plant Physiol 161:148–164

Calderwood A, Morris RJ, Kopriva S (2014) Predictive sulfur metabolism – a field in flux. Front Plant Sci 5:646

Carroll KS, Gao H, Chen H, Stout CD, Leary JA, Bertozzi CR (2005) A conserved mechanism for sulfonucleotide reduction. PLoS Biol 3(8):e250

Chan EK, Rowe HC, Corwin JA, Joseph B, Kliebenstein DJ (2011) Combining genome-wide association mapping and transcriptional networks to identify novel genes controlling glucosinolates in *Arabidopsis thaliana*. PLoS Biol 9(8):e1001125

Chan KX, Wirtz M, Phua SY, Estavillo GM, Pogson BJ (2013) Balancing metabolites in drought: the sulfur assimilation conundrum. Trends Plant Sci 18:18–29

Chao DY, Baraniecka P, Danku J, Koprivova A, Lahner B, Luo H, Yakubova E, Dilkes B, Kopriva S, Salt DE (2014) Variation in sulfur and selenium accumulation is controlled by naturally occurring isoforms of the key sulfur assimilation enzyme ADENOSINE 5′-PHOSPHOSULFATE REDUCTASE2 across the Arabidopsis species range. Plant Physiol 166:1593–1608

Clarke JD, Dashwood RH, Ho E (2008) Multi-targeted prevention of cancer by sulforaphane. Cancer Lett 269:291–304

Davidian JC, Kopriva S (2010) Regulation of sulfate uptake and assimilation-the same or not the same? Mol Plant 3:314–325

Eilers T, Schwarz G, Brinkmann H, Witt C, Richter T, Nieder J, Koch B, Hille R, Hansch R, Mendel RR (2001) Identification and biochemical characterization of *Arabidopsis thaliana* sulfite oxidase. A new player in plant sulfur metabolism. J Biol Chem 276:46989–46994

Estavillo GM, Crisp PA, Pornsiriwong W, Wirtz M, Collinge D, Carrie C, Giraud E, Whelan J, David P, Javot H, Brearley C, Hell R, Marin E, Pogson BJ (2011) Evidence for a SAL1-PAP chloroplast retrograde pathway that functions in drought and high light signaling in Arabidopsis. Plant Cell 23:3992–4012

Frerigmann H, Gigolashvili T (2014) MYB34, MYB51, and MYB122 distinctly regulate indolic glucosinolate biosynthesis in *Arabidopsis thaliana*. Mol Plant 7:814–828

Frigaard NU, Dahl C (2009) Sulfur metabolism in phototrophic sulfur bacteria. Adv Microb Physiol 54:103–200

Glaser K, Kanawati B, Kubo T, Schmitt-Kopplin P, Grill E (2014) Exploring the Arabidopsis sulfur metabolome. Plant J 77:31–45

Gutierrez-Marcos JF, Roberts MA, Campbell EI, Wray JL (1996) Three members of a novel small gene-family from *Arabidopsis thaliana* able to complement functionally an *Escherichia coli* mutant defective in PAPS reductase activity encode proteins with a thioredoxin-like domain and "APS reductase" activity. Proc Natl Acad Sci U S A 93:13377–13382

Haas FH, Heeg C, Queiroz R, Bauer A, Wirtz M, Hell R (2008) Mitochondrial serine acetyltransferase functions as a pacemaker of cysteine synthesis in plant cells. Plant Physiol 148:1055–1067

Hell R (1997) Molecular physiology of plant sulfur metabolism. Planta 202(2):138–148

Hirai MY, Fujiwara T, Awazuhara M, Kimura T, Noji M, Saito K (2003) Global expression profiling of sulfur-starved Arabidopsis by DNA macroarray reveals the role of *O*-acetyl-l-serine as a general regulator of gene expression in response to sulfur nutrition. Plant J 33:651–663

Hirai MY, Klein M, Fujikawa Y, Yano M, Goodenowe DB, Yamazaki Y, Kanaya S, Nakamura Y, Kitayama M, Suzuki H, Sakurai N, Shibata D, Tokuhisa J, Reichelt M, Gershenzon J, Papenbrock J, Saito K (2005) Elucidation of gene-to-gene and metabolite-to-gene networks in Arabidopsis by integration of metabolomics and transcriptomics. J Biol Chem 280:25590–25595

Hothorn M, Wachter A, Gromes R, Stuwe T, Rausch T, Scheffzek K (2006) Structural basis for the redox control of plant glutamate cysteine ligase. J Biol Chem 281:27557–27565

Hubberten HM, Drozd A, Tran BV, Hesse H, Hoefgen R (2012a) Local and systemic regulation of sulfur homeostasis in roots of *Arabidopsis thaliana*. Plant J 72:625–635

Hubberten HM, Klie S, Caldana C, Degenkolbe T, Willmitzer L, Hoefgen R (2012b) Additional role of *O*-acetylserine as a sulfur status-independent regulator during plant growth. Plant J 70:666–677

Jost R, Altschmied L, Bloem E, Bogs J, Gershenzon J, Hahnel U, Hansch R, Hartmann T, Kopriva S, Kruse C, Mendel RR, Papenbrock J, Reichelt M, Rennenberg H, Schnug E, Schmidt A, Textor S, Tokuhisa J, Wachter A, Wirtz M, Rausch T, Hell R (2005) Expression profiling of metabolic genes in response to methyl jasmonate reveals regulation of genes of primary and secondary sulfur-related pathways in *Arabidopsis thaliana*. Photosynth Res 86:491–508

Kawashima CG, Yoshimoto N, Maruyama-Nakashita A, Tsuchiya YN, Saito K, Takahashi H, Dalmay T (2009) Sulphur starvation induces the expression of microRNA-395 and one of its target genes but in different cell types. Plant J 57:313–321

Kawashima CG, Matthewman CA, Huang SQ, Lee BR, Yoshimoto N, Koprivova A, Rubio-Somoza I, Todesco M, Rathjen T, Saito K, Takahashi H, Dalmay T, Kopriva S (2011) Interplay of SLIM1 and miR395 in the regulation of sulfate assimilation in Arabidopsis. Plant J 66:863–876

Khan MS, Haas FH, Samami AA, Gholami AM, Bauer A, Fellenberg K, Reichelt M, Hansch R, Mendel RR, Meyer AJ, Wirtz M, Hell R (2010) Sulfite reductase defines a newly discovered bottleneck for assimilatory sulfate reduction and is essential for growth and development in *Arabidopsis thaliana*. Plant Cell 22:1216–1231

Kliebenstein DJ, Gershenzon J, Mitchell-Olds T (2001a) Comparative quantitative trait loci mapping of aliphatic, indolic and benzylic glucosinolate production in *Arabidopsis thaliana* leaves and seeds. Genetics 159:359–370

Kliebenstein DJ, Kroymann J, Brown P, Figuth A, Pedersen D, Gershenzon J, Mitchell-Olds T (2001b) Genetic control of natural variation in Arabidopsis glucosinolate accumulation. Plant Physiol 126:811–825

Kliebenstein DJ, Lambrix VM, Reichelt M, Gershenzon J, Mitchell-Olds T (2001c) Gene duplication in the diversification of secondary metabolism: tandem 2-oxoglutarate-dependent dioxygenases control glucosinolate biosynthesis in Arabidopsis. Plant Cell 13:681–693

Kliebenstein D, Pedersen D, Barker B, Mitchell-Olds T (2002) Comparative analysis of quantitative trait loci controlling glucosinolates, myrosinase and insect resistance in *Arabidopsis thaliana*. Genetics 161:325–332
Kopriva S (2006) Regulation of sulfate assimilation in Arabidopsis and beyond. Ann Bot-London 97:479–495
Kopriva S (2015) Plant sulfur nutrition: from Sachs to Big Data. Plant Signal Behav 10(9): e1055436
Kopriva S, Koprivova A (2004) Plant adenosine 5′-phosphosulphate reductase: the past, the present, and the future. J Exp Bot 55:1775–1783
Kopriva S, Rennenberg H (2004) Control of sulphate assimilation and glutathione synthesis: interaction with N and C metabolism. J Exp Bot 55:1831–1842
Kopriva S, Muheim R, Koprivova A, Trachsel N, Catalano C, Suter M, Brunold C (1999) Light regulation of assimilatory sulphate reduction in *Arabidopsis thaliana*. Plant J 20:37–44
Kopriva S, Buchert T, Fritz G, Suter M, Weber M, Benda R, Schaller J, Feller U, Schurmann P, Schunemann V, Trautwein AX, Kroneck PMH, Brunold C (2001) Plant adenosine 5′-phospho sulfate reductase is a novel iron-sulfur protein. J Biol Chem 276:42881–42886
Kopriva S, Buchert T, Fritz G, Suter M, Benda RD, Schunemann V, Koprivova A, Schurmann P, Trautwein AX, Kroneck PMH, Brunold C (2002) The presence of an iron-sulfur cluster in adenosine 5′-phosphosulfate reductase separates organisms utilizing adenosine 5′-phosphosulfate and phosphoadenosine 5′-phosphosulfate for sulfate assimilation. J Biol Chem 277:21786–21791
Kopriva S, Fritzemeier K, Wiedemann G, Reski R (2007) The putative moss 3′-phosphoadenosine-5′-phosphosulfate reductase is a novel form of adenosine-5′-phosphosulfate reductase without an iron-sulfur cluster. J Biol Chem 282:22930–22938
Koprivova A, Kopriva S (2014) Molecular mechanisms of regulation of sulfate assimilation: first steps on a long road. Front Plant Sci 5:589
Koprivova A, Suter M, Op den Camp R, Brunold C, Kopriva S (2000) Regulation of sulfate assimilation by nitrogen in Arabidopsis. Plant Physiol 122:737–746
Koprivova A, Giovannetti M, Baraniecka P, Lee BR, Grondin C, Loudet O, Kopriva S (2013) Natural variation in the ATPS1 Isoform of ATP sulfurylase contributes to the control of sulfate levels in Arabidopsis. Plant Physiol 163:1133–1141
Kroymann J, Textor S, Tokuhisa JG, Falk KL, Bartram S, Gershenzon J, Mitchell-Olds T (2001) A gene controlling variation in Arabidopsis glucosinolate composition is part of the methionine chain elongation pathway. Plant Physiol 127:1077–1088
Lang C, Popko J, Wirtz M, Hell R, Herschbach C, Kreuzwieser J, Rennenberg H, Mendel RR, Hansch R (2007) Sulphite oxidase as key enzyme for protecting plants against sulphur dioxide. Plant Cell Environ 30:447–455
Lappartient AG, Touraine B (1996) Demand-driven control of root ATP sulfurylase activity and SO_4^{2-} uptake in intact canola (the role of phloem-translocated glutathione). Plant Physiol 111:147–157
Lappartient AG, Vidmar JJ, Leustek T, Glass AD, Touraine B (1999) Inter-organ signaling in plants: regulation of ATP sulfurylase and sulfate transporter genes expression in roots mediated by phloem-translocated compound. Plant J 18:89–95
Lee BR, Koprivova A, Kopriva S (2011) The key enzyme of sulfate assimilation, adenosine 5′-phosphosulfate reductase, is regulated by HY5 in Arabidopsis. Plant J 67:1042–1054
Lee BR, Huseby S, Koprivova A, Chetelat A, Wirtz M, Mugford ST, Navid E, Brearley C, Saha S, Mithen R, Hell R, Farmer EE, Kopriva S (2012) Effects of fou8/fry1 mutation on sulfur metabolism: is decreased internal sulfate the trigger of sulfate starvation response? PLoS One 7(6):e39425
Leustek T, Martin MN, Bick JA, Davies JP (2000) Pathways and regulation of sulfur metabolism revealed through molecular and genetic studies. Annu Rev Plant Physiol Plant Mol Biol 51:141–165

Liang G, Yang F, Yu D (2010) MicroRNA395 mediates regulation of sulfate accumulation and allocation in *Arabidopsis thaliana*. Plant J 62:1046–1057

Loudet O, Saliba-Colombani V, Camilleri C, Calenge F, Gaudon V, Koprivova A, North KA, Kopriva S, Daniel-Vedele F (2007) Natural variation for sulfate content in *Arabidopsis thaliana* is highly controlled by APR2. Nat Genet 39:896–900

Malitsky S, Blum E, Less H, Venger I, Elbaz M, Morin S, Eshed Y, Aharoni A (2008) The transcript and metabolite networks affected by the two clades of Arabidopsis glucosinolate biosynthesis regulators. Plant Physiol 148:2021–2049

Maruyama-Nakashita A, Inoue E, Watanabe-Takahashi A, Yamaya T, Takahashi H (2003) Transcriptome profiling of sulfur-responsive genes in Arabidopsis reveals global effects of sulfur nutrition on multiple metabolic pathways. Plant Physiol 132:597–605

Maruyama-Nakashita A, Nakamura Y, Yamaya T, Takahashi H (2004) A novel regulatory pathway of sulfate uptake in Arabidopsis roots: implication of CRE1/WOL/AHK4-mediated cytokinin-dependent regulation. Plant J 38:779–789

Maruyama-Nakashita A, Nakamura Y, Tohge T, Saito K, Takahashi H (2006) Arabidopsis SLIM1 is a central transcriptional regulator of plant sulfur response and metabolism. Plant Cell 18:3235–3251

Mugford SG, Yoshimoto N, Reichelt M, Wirtz M, Hill L, Mugford ST, Nakazato Y, Noji M, Takahashi H, Kramell R, Gigolashvili T, Flugge UI, Wasternack C, Gershenzon J, Hell R, Saito K, Kopriva S (2009) Disruption of adenosine-5′-phosphosulfate kinase in Arabidopsis reduces levels of sulfated secondary metabolites. Plant Cell 21:910–927

Mugford SG, Matthewman CA, Hill L, Kopriva S (2010) Adenosine-5′-phosphosulfate kinase is essential for Arabidopsis viability. FEBS Lett 584:119–123

Mugford SG, Lee BR, Koprivova A, Matthewman C, Kopriva S (2011) Control of sulfur partitioning between primary and secondary metabolism. Plant J 65:96–105

Neuenschwander U, Suter M, Brunold C (1991) Regulation of sulfate assimilation by light and *O*-acetyl-l-serine in *Lemna minor* L. Plant Physiol 97:253–258

Nikiforova V, Freitag J, Kempa S, Adamik M, Hesse H, Hoefgen R (2003) Transcriptome analysis of sulfur depletion in *Arabidopsis thaliana*: interlacing of biosynthetic pathways provides response specificity. Plant J 33:633–650

Nikiforova VJ, Daub CO, Hesse H, Willmitzer L, Hoefgen R (2005) Integrative gene-metabolite network with implemented causality deciphers informational fluxes of sulphur stress response. J Exp Bot 56:1887–1896

Patron NJ, Durnford DG, Kopriva S (2008) Sulfate assimilation in eukaryotes: fusions, relocations and lateral transfers. BMC Evol Biol 8:39

Rausch T, Wachter A (2005) Sulfur metabolism: a versatile platform for launching defence operations. Trends Plant Sci 10:503–509

Ravilious GE, Nguyen A, Francois JA, Jez JM (2012) Structural basis and evolution of redox regulation in plant adenosine-5′-phosphosulfate kinase. Proc Natl Acad Sci U S A 109:309–314

Robles P, Fleury D, Candela H, Cnops G, Alonso-Peral MM, Anami S, Falcone A, Caldana C, Willmitzer L, Ponce MR, Van Lijsebettens M, Micol JL (2010) The RON1/FRY1/SAL1 gene is required for leaf morphogenesis and venation patterning in Arabidopsis. Plant Physiol 152:1357–1372

Schmidt A (1972) On the mechanism of photosynthetic sulfate reduction. An APS-sulfotransferase from Chlorella. Arch Mikrobiol 84:77–86

Schmidt A (1973) Sulfate reduction in a cell-free system of Chlorella. The ferredoxin dependent reduction of a protein-bound intermediate by a thiosulfonate reductase. Arch Mikrobiol 93:29–52

Schmidt A, Jäger K (1992) Open questions about sulfur metabolism in plants. Annu Rev Plant Physiol Plant Mol Biol 43:325–349

Schweizer F, Fernandez-Calvo P, Zander M, Diez-Diaz M, Fonseca S, Glauser G, Lewsey MG, Ecker JR, Solano R, Reymond P (2013) Arabidopsis basic helix-loop-helix transcription factors MYC2, MYC3, and MYC4 regulate glucosinolate biosynthesis, insect performance, and feeding behavior. Plant Cell 25:3117–3132

Setya A, Murillo M, Leustek T (1996) Sulfate reduction in higher plants: molecular evidence for a novel 5′-adenylylsulfate reductase. Proc Natl Acad Sci U S A 93:13383–13388

Sonderby IE, Burow M, Rowe HC, Kliebenstein DJ, Halkier BA (2010) A complex interplay of three R2R3 MYB transcription factors determines the profile of aliphatic glucosinolates in Arabidopsis. Plant Physiol 153:348–363

Suter M, von Ballmoos P, Kopriva S, den Camp RO, Schaller J, Kuhlemeier C, Schurmann P, Brunold C (2000) Adenosine 5′-phosphosulfate sulfotransferase and adenosine 5′-phosphosulfate reductase are identical enzymes. J Biol Chem 275:930–936

Takahashi H, Yamazaki M, Sasakura N, Watanabe A, Leustek T, Engler JA, Engler G, Van Montagu M, Saito K (1997) Regulation of sulfur assimilation in higher plants: a sulfate transporter induced in sulfate-starved roots plays a central role in *Arabidopsis thaliana*. Proc Natl Acad Sci U S A 94:11102–11107

Takahashi H, Kopriva S, Giordano M, Saito K, Hell R (2011) Sulfur assimilation in photosynthetic organisms: molecular functions and regulations of transporters and assimilatory enzymes. Annu Rev Plant Biol 62:157–184

Thompson JF (1967) Sulfur metabolism in plants. Annu Rev Plant Physiol 18:59–84

Traka M, Mithen R (2009) Glucosinolates, isothiocyanates and human health. Phytochem Rev 8:269–282

Vauclare P, Kopriva S, Fell D, Suter M, Sticher L, von Ballmoos P, Krahenbuhl U, den Camp RO, Brunold C (2002) Flux control of sulphate assimilation in *Arabidopsis thaliana*: adenosine 5′-phosphosulphate reductase is more susceptible than ATP sulphurylase to negative control by thiols. Plant J 31:729–740

Wawrzynska A, Sirko A (2014) To control and to be controlled: understanding the Arabidopsis SLIM1 function in sulfur deficiency through comprehensive investigation of the EIL protein family. Front Plant Sci 5:575

Weber M, Suter M, Brunold C, Kopriva S (2000) Sulfate assimilation in higher plants – characterization of a stable intermediate in the adenosine 5′-phosphosulfate reductase reaction. Eur J Biochem 267:3647–3653

Williams SJ, Senaratne RH, Mougous JD, Riley LW, Bertozzi CR (2002) 5′-adenosinephosphosulfate lies at a metabolic branch point in mycobacteria. J Biol Chem 277:32606–32615

Wirtz M, Hell R (2006) Functional analysis of the cysteine synthase protein complex from plants: structural, biochemical and regulatory properties. J Plant Physiol 163:273–286

Yatusevich R, Mugford SG, Matthewman C, Gigolashvili T, Frerigmann H, Delaney S, Koprivova A, Flugge UI, Kopriva S (2010) Genes of primary sulfate assimilation are part of the glucosinolate biosynthetic network in *Arabidopsis thaliana*. Plant J 62:1–11

Expression Profile of the Serine Acetyltransferase (*SERAT*) and *O*-Acetylserine (thiol)lyase (*OASTL*) Gene Families in Arabidopsis

Mutsumi Watanabe and Rainer Hoefgen

Abstract Cysteine synthesis in plants constitutes the entry of reduced sulfur from assimilatory sulfate reduction into metabolism. Cysteine synthesis is catalyzed by the sequential action of serine acetyltransferase (SERAT) and *O*-acetylserine (thiol) lyase (OASTL). In the Arabidopsis genome, there are five *SERAT* and three *OASTL* genes. Analysis of the expression data obtained from micro array data in database such as Arabidopsis eFP browser and publications indicates that the *SERAT* genes and *OASTL* genes show distinct expression patterns during development and under diurnal regulation as well as under stress conditions, suggesting the specific function/regulation of SERATs and OASTLs in different subcellular compartments.

Introduction

Serine acetyltransferase (SERAT), which catalyzes the formation of *O*-acetylserine (OAS) from serine and acetyl coenzyme A (acetyl-CoA), links serine metabolism to cysteine biosynthesis. OAS reacts with sulfide to yield cysteine catalyzed by *O*-acetylserine(thiol)-lyase (OASTL). In the Arabidopsis genome, there are five *SERAT* genes, *SERAT1;1* (At5g56760), *SERAT2;1* (At1g55920), *SERAT2;2* (At3g13110), *SERAT3;1* (At2g17640) and *SERAT3;2* (At4g35640) and three *OASTL* genes, *OASTL1;1* (At4g14880), *OASTL2;1* (At2g43750) and *OASTL2;2* (At3g59760) (Table 1). OASTLs belong to the beta-substituted alanine synthase (BSAS) family in the large superfamily of pyridoxal 5′-phosphate-dependent enzymes, which comprises beta-cyanoalanine synthase (CAS) (Hatzfeld et al. 2000; Yamaguchi et al. 2000), L-cysteine desulfhydrase (DES) (Alvarez et al. 2010) and *S*-sulfocysteine synthase (SSCS) (Bermudez et al. 2010) in Arabidopsis. The OASTL1;1, OASTL2;1 and OASTL2;2 are also called BSAS1;1, BSAS2;1 and BSAS2;2, respectively (Watanabe et al. 2008a). Decades research has provided

M. Watanabe • R. Hoefgen (✉)
Max Planck Institute of Molecular Plant Physiology, 14476 Potsdam-Golm, Germany
e-mail: hoefgen@mpimp-golm.mpg.de

L.J. De Kok et al. (eds.), *Sulfur Metabolism in Higher Plants - Fundamental, Environmental and Agricultural Aspects*, Proceedings of the International Plant Sulfur Workshop, DOI 10.1007/978-3-319-56526-2_3

Table 1 SERAT and OASTL gene families

Family	Group	Gene name	AGI code	Localization
SERAT	Group I	SERAT1;1	At5g56760	Cytosol
	Group II	SERAT2;1	At1g55920	Plastid
		SERAT2;2	At3g13110	Mitochondrion
	Group III	SERAT3;1	At2g17640	Cytosol
		SERAT3;2	At4g35640	Cytosol
OASTL	Group I	OASTL1;1	At4g14880	Cytosol
	Group II	OASTL2;1	At2g43750	Plastid
		OASTL2;2	At3g59760	Mitochondrion

cumulative insights into the function of SERATs and OASTLs by biochemical characterization *in vitro*, subcellular localization studies and *in vivo* functionality studies using the T-DNA knockout mutants. SERATs and OASTLs are present in three compartments; cytosol, plastid and mitochondrion in Arabidopsis (Noji et al. 1998; reviewed in Hell et al. 2002; Kawashima et al. 2005) (Table 1). Among five SERAT isoforms, three isoforms (SERAT1;1, SERAT2;1 and SERAT2;2) are biochemically more active than other two isoforms in group III (Noji et al. 1998; Kawashima et al. 2005). Cytosolic SERAT isoforms (SERAT1;1 and SERAT3;2) are feedback sensitive by cysteine (Noji et al. 1998; Kawashima et al. 2005). Among nine BSAS isoforms in Arabidopsis, three isoforms (OASTL1;1, OASTL2;1 and OASTL2;2) are biochemically most active in terms of OASTL activity (Jost et al. 2000). The fact that the single knockout mutants of *SERATs* and *OASTLs* did not reveal lethal phenotype indicated that the individual SERAT and OASTL isoforms are functionally redundant and OAS and cysteine are efficiently transported between cytosol and organelles (Heeg et al. 2008; Watanabe et al. 2008a, b). Although they display some functional redundancy, it appears that their contributions of OAS and cysteine synthesis are different in plant tissues and growth conditions. The analysis of *serat* and *oastl* mutants revealed that mitochondrial SERAT2;2 and cytosolic OASTL1;1 predominantly contributes to OAS and cysteine synthesis, respectively, in leaves of Arabidopsis grown on agar plate (Watanabe et al. 2008a, b) while the predominant isoforms change in other tissues. For example, in case of SERAT, cytosolic forms SERAT1;1, SERAT3;1 and SERAT3;2 seem responsible for OAS formation rather than mitochondrial SERAT2;2 in the Arabidopsis siliques (Watanabe et al. 2008b). This is consistent with their gene expression patterns in the public transcriptome database such as Arabidopsis eFP browser (Winter et al. 2007; Fig. 1) and gene expression analysis by qRT-PCR (Kawashima et al. 2005); the decrease of gene expression level of *SERAT2;2*, increase of *SERAT3;1* and *SERAT3;2* and constant level of *SERAT1;1* during silique and seed development. As in this example, the detailed gene expression analysis of *SERATs* and *OASTLs* with the experimental knowledge could be more useful for predicting their contributions and specific functions.

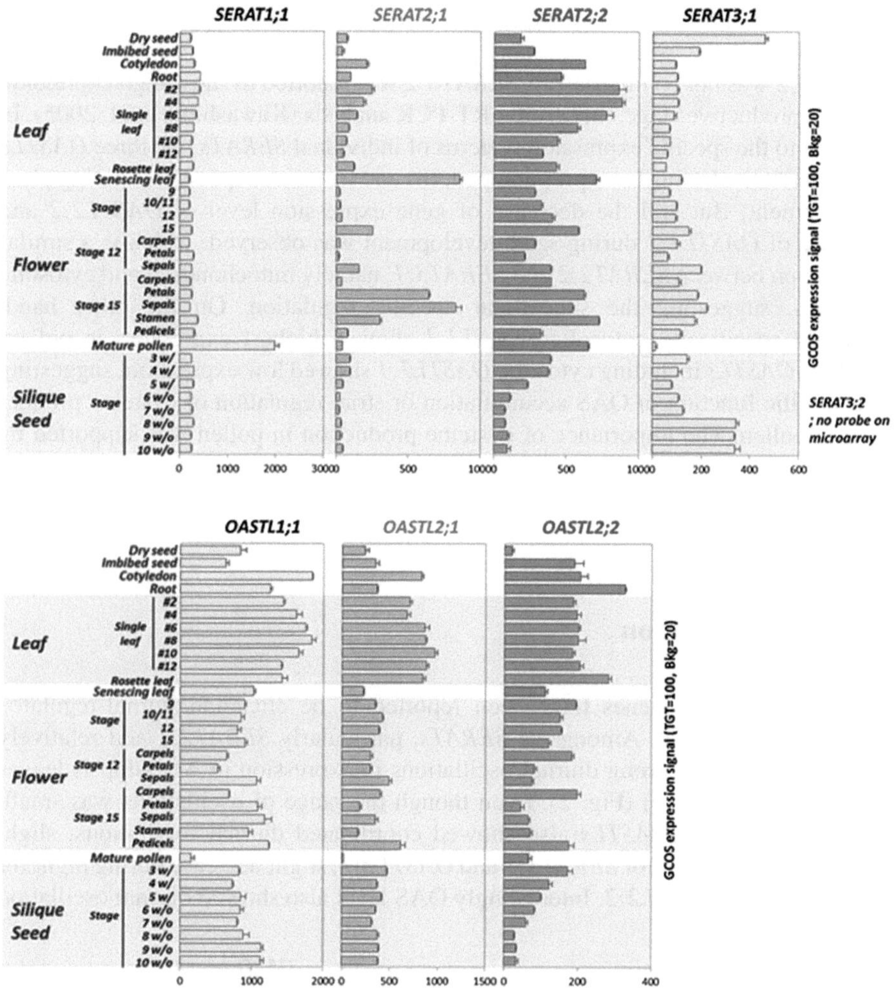

Fig. 1 Developmental and tissue-specific regulation. Gene expression patters in Arabidopsis tissues (Arabidopsis eFP Browser; Winter et al. 2007). The expression data was normalized with the GCOS (Gene Chip Operating Software) method and Target intensity (TGT) value of 100. Error bars represent standard deviation. Most tissues were sampled in triplicate. The specific probe for *SERAT3;2* gene is not found in ATH1 Affymetrix microarray

Developmental and Tissue Specific Regulation

During plant development, the *SERAT* genes showed tissue-specific expression patterns (Fig. 1). High gene expression levels were observed in mature pollen for *SERAT1;1*, in senescent leaves and late stage of flower for *SERAT2;1* and in silique, especially seed for *SERAT3;1*. *SERAT2;2* showed high expression at the vegetative stage and in flower, but low expression during seed development in a negative correlation to the gene expression pattern of *SERAT3;1*. Since the probe of *SERAT3;2* was not on microarray, *SERAT3;2* was reported to show high expression at the reproductive stage only from qRT-PCR analysis (Kawashima et al. 2005). In contrast to the specific expression patterns of individual *SERATs*, the three *OASTLs* showed relatively similar expression patterns and less tissue specificity during plant development. But still the decrease of gene expression level of *OASTL2;2* and increase of *OASTL1;1* during seed development was observed, which is a similar correlation between *SERAT2;2* and *SERAT3;1*, namely mitochondrial and cytosolic isoforms, suggesting the subcellular specific regulation. On the other hand, although specifically cytosolic *SERAT1;1* showed highest expression in pollen, all three *OASTLs* including cytosolic *OASTL1;1* showed low expression, suggesting the specific function of OAS accumulation or strict regulation of cysteine production in pollen. The importance of cysteine production in pollen was supported by the report that at least a functional one of three OASTLs in the pollen is required for the successful fertilization (Birke et al. 2013).

Diurnal Regulation

Sulfur assimilation genes have been reported to be circadian/diurnal-regulated (Harmer et al. 2000). Among the *SERATs*, particularly *SERAT2;1* and relatively *SERAT2;2* showed strong diurnal oscillations in expression in Arabidopsis leaves (Espinoza et al. 2010) (Fig. 2). Even though the range of oscillations was small, other *SERATs* and *OASTLs* also showed coordinated diurnal oscillations; slight increase during night for *SERAT3;1* and *OASTL2;1*, slight decrease during night for *OASTL1;1* and *OASTL2;2*. Interestingly OAS level also showed diurnal oscillation

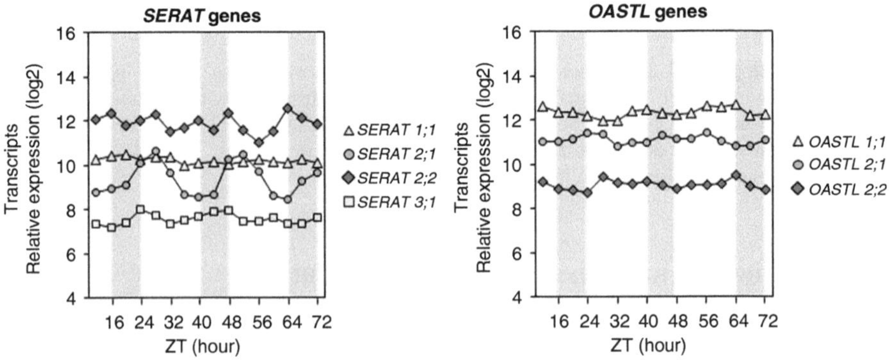

Fig. 2 Diurnal regulation. Gene expression patters in diurnal cycles (Espinoza et al. 2010). Plants were grown under long day condition (16 h light/8 h dark) at 20 °C. Relative expression (log2) is indicated. *White* and *grey bars* indicate the day and night periods, respectively. *ZT* zeitgeber time. The specific probe for *SERAT3;2* gene is not found in ATH1 Affymetrix microarray

with a peak at middle of night whilst cysteine level was not changed (Espinoza et al. 2010). Although any *SERATs* did not display the peak expression at the same phase of the OAS peak, *SERAT2;2*, which were increased in early onset of night, may contribute to the OAS peak considering the time lag between changes in levels of transcripts, proteins and metabolites. This also fits to the predominant role of SERAT2;2 in Arabidopsis leaves (Watanabe et al. 2008b). However, it cannot be excluded that several regulatory mechanisms such as protein-protein interactions including SERAT-OASTL complex and phosphorylation, which have been described for affecting the activities of SERAT and OASTL and sensitivity of SERAT to the feedback inhibition by cysteine (Liu et al. 2006; reviewed in Wirtz and Hell 2006), might be involved in the diurnal cycles in OAS level. Actually transient increase in OAS levels was also observed in Arabidopsis leaves 5–10 min after transfer to darkness (Caldana et al. 2011), which seems not regulated at the transcriptional level.

Regulation Under Stress Conditions

Nutrient starvation stresses such as sulfate, nitrate and phosphate starvations have been known to share several responsive genes and phenotypes, but specific gene inductions were observed in *SERATs* and *OASTLs* under nutrient starvations. Under sulfur starvation *SERAT3;2* was highly induced in leaf and root tissues and *SERAT3;1* was slightly induced in roots and seedling, but the expression levels of the other three *SERATs* and *OASTLs* were not significantly altered except for >2-fold change of *SERAT2;1* and *OASTL2;1* in seedling (Fig. 3). In contrast to the specific induction of *SERATs* in the group III under sulfur starvation, under other nutrient starvations, for example, under phosphate starvation *SERAT2;1* and *SERAT2;2* in the group II were induced (Fig. 3). The transcript levels of the group II *SERATs*, particularly *SERAT2;1*, were also increased under oxidative stress conditions caused by treatments of menadione and H_2O_2 (Fig. 3). By contrast, the three *OASTLs* and *SERAT3;1* showed decreases except for slight increase of

Experiment	Tissue	*SERAT 1;1*	*SERAT 2;1*	*SERAT 2;2*	*SERAT 3;1*	*SERAT 3;2*	*OASTL 1;1*	*OASTL 2;1*	*OASTL 2;2*	References
-Sulfate	shoot	0.9	0.8	0.9	1.0	40.0	nd	nd	nd	Kawashima et al. (2005)
-Sulfate	root	1.2	0.8	0.9	2.0	45.0	nd	nd	nd	Kawashima et al. (2005)
-Sulfate	root	1.2	1.3	1.1	2.6	nd	1.1	1.3	1.0	Maruyama-Nakashita et al. (2006)
-Sulfate	seedling	1.2	2.1	1.2	2.5	nd	1.0	2.0	0.8	Bielecka et al. (2015)
-Nitrate	seedling	1.1	1.7	1.3	0.7	nd	0.7	0.6	1.2	Scheible et al. (2004)
-Phosphate	seedling	1.6	2.2	4.5	0.6	nd	1.0	0.7	1.4	Morcuende et al. (2007)
-Potassium	seedling	1.2	1.4	1.3	0.6	nd	1.0	0.6	0.9	Armengaud et al. (2004)
+Menadione	root	2.7	10.3	8.8	0.5	nd	0.7	nd	0.6	Lehmann et al. (2009)
$+H_2O_2$	seedling	0.8	17.0	2.3	0.8	nd	0.9	1.0	1.0	Davletova et al. (2005)

0.125 (2^{-3}) 1 8 (2^3)

Fig. 3 Response to several stress conditions. The fold change in each stress condition relative to the control condition is shown. *Blue* and *red* colors represent decreases and increases, respectively, as compared with the control plants. *nd* not determined. The specific probe for *SERAT3;2* gene is not found in ATH1 Affymetrix microarray

OASTL2;2 under nitrate and phosphate starvations. It is noteworthy that the pairs *SERAT2;1*/*SERAT2;2* and *SERAT3;1*/*SERAT3;2* were generated by a gene duplication event in Arabidopsis, respectively (Watanabe et al. 2008b), suggesting that they have kept similar upstream elements and enhancers for the stress inducibility. The high level of *SERAT2;1* transcript was also observed in senescing leaves (Fig. 1), where oxidative stress arises with the accumulation of reactive oxygen species (ROS). A key role of SERAT2;1 under oxidative stress was supported by the report that SERAT2;1 interacts with cyclophilin CYP20-3 in plastid, which has peptidyl prolyl isomerase activity for protein folding and assembly (Dominguez-Solis et al. 2009; Park et al. 2013). The formation of SERAT2;1-CYP20-3 complex resulted in activation or stabilization of the SERAT-OASTL complex to produce more OAS, sequentially cysteine and GSH to protect the cell from oxidative stress. Actually the Arabidopsis *cyp20-3* mutant with less SERAT activity and thiols was hypersensitive to oxidative stress conditions caused by high light, rose bengal, high salt, and osmotic stress (Dominguez-Solis et al. 2009). The response of *SERAT2;1* to oxidative/ROS stress is consistent with the strong diurnal oscillations of *SERAT2;1* (Fig. 2) since ROS production is also diurnal-regulated with a peak at midday (Lai et al. 2012). As in the case of the specific role of SERAT2;1 under oxidative stress, the specific induction of *SERAT3;1* and *SERAT3;2* under sulfate starvations (Fig. 3) and the sensitivity of SERAT3;2 to feedback inhibition by Cys (Kawashima et al. 2005) suggests their distinct roles in sulfur metabolism, especially under sulfur starvation although both SERATs in group III had less capacity to supply OAS *in vivo* than other isoforms.

The variety of gene expression pattern of *SERATs* and *OASTLs* in plant tissues and environmental conditions, which causes the difference in the ratio of SERAT and OASTL activities, sequentially the balance of OAS and cysteine productions in different subcellular compartments, might be benefit for using substrates such as serine and sulfide efficiently, releasing substrates and products such as sulfide, OAS and cysteine to other compartments and responding to the immediate needs or overcoming transport limitation of OAS and cysteine in specific compartments under specific conditions. Furthermore, the unbalancing in the ratio of SERAT and OASTL (e.g. high SERAT activity and low OASTL activity) might cause a specific OAS accumulation, which induces a specific gene set, the OAS cluster genes (Hubberten et al. 2012).

Acknowledgements We thank the Max Planck Society (MPG) for funding.

References

Alvarez C, Calo L, Romero LC, Garcia I, Gotor C (2010) An *O*-acetylserine(thiol)lyase homolog with L-cysteine desulfhydrase activity regulates cysteine homeostasis in Arabidopsis. Plant Physiol 152:656–669

Angeles Bermudez M, Angeles Paez-Ochoa M, Gotor C, Romero LC (2010) Arabidopsis *S*-sulfocysteine synthase activity is essential for chloroplast function and long-day light-dependent redox control. Plant Cell 22:403–416

Armengaud P, Breitling R, Amtmann A (2004) The potassium-dependent transcriptome of Arabidopsis reveals a prominent role of jasmonic acid in nutrient signaling. Plant Physiol 136:2556–2576

Bielecka M, Watanabe M, Morcuende R, Scheible W-R, Hawkesford MJ, Hesse H, Hoefgen R (2015) Transcriptome and metabolome analysis of plant sulfate starvation and resupply provides novel information on transcriptional regulation of metabolism associated with sulfur nitrogen and phosphorus nutritional responses in Arabidopsis. Front Plant Sci 5:805

Birke H, Heeg C, Wirtz M, Hell R (2013) Successful fertilization requires the presence of at least one major *O*-acetylserine(thiol)lyase for cysteine synthesis in pollen of Arabidopsis. Plant Physiol 163:959–972

Caldana C, Degenkolbe T, Cuadros-Inostroza A, Klie S, Sulpice R, Leisse A, Steinhauser D, Fernie AR, Willmitzer L, Hannah MA (2011) High-density kinetic analysis of the metabolomic and transcriptomic response of Arabidopsis to eight environmental conditions. Plant J 67:869–884

Davletova S, Schlauch K, Coutu J, Mittler R (2005) The zinc-finger protein Zat12 plays a central role in reactive oxygen and abiotic stress signaling in Arabidopsis. Plant Physiol 139:847–856

Dominguez-Solis JR, He Z, Lima A, Ting J, Buchanan BB, Luan S (2009) A cyclophilin links redox and light signals to cysteine biosynthesis and stress responses in chloroplasts. Proc Natl Acad Sci U S A 106:1292–1292

Espinoza C, Degenkolbe T, Caldana C, Zuther E, Leisse A, Willmitzer L, Hincha DK, Hannah MA (2010) Interaction with diurnal and circadian regulation results in dynamic metabolic and transcriptional changes during cold acclimation in Arabidopsis. PLoS One 5:e14102

Harmer SL, Hogenesch LB, Straume M, Chang HS, Han B, Zhu T, Wang X, Kreps JA, Kay SA (2000) Orchestrated transcription of key pathways in Arabidopsis by the circadian clock. Science 290:2110–2113

Hatzfeld Y, Maruyama A, Schmidt A, Noji M, Ishizawa K, Saito K (2000) β-Cyanoalanine synthase is a mitochondrial cysteine synthase-like protein in spinach and Arabidopsis. Plant Physiol 123:1163–1171

Heeg C, Kruse C, Jost R, Gutensohn M, Ruppert T, Wirtz M, Hell R (2008) Analysis of the Arabidopsis *O*-acetylserine(thiol)lyase gene family demonstrates compartment-specific differences in the regulation of cysteine synthesis. Plant Cell 20:168–185

Hell R, Jost R, Berkowitz O, Wirtz M (2002) Molecular and biochemical analysis of the enzymes of cysteine biosynthesis in the plant *Arabidopsis thaliana*. Amino Acids 22:245–257

Hubberten H-M, Klie S, Caldana C, Degenkolbe T, Willmitzer L, Hoefgen R (2012) Additional role of *O*-acetylserine as a sulfur status-independent regulator during plant growth. Plant J 70:666–677

Jost R, Berkowitz O, Wirtz M, Hopkins L, Hawkesford MJ, Hell R (2000) Genomic and functional characterization of the oas gene family encoding *O*-acetylserine (thiol) lyases enzymes catalyzing the final step in cysteine biosynthesis in *Arabidopsis thaliana*. Gene 253:237–247

Kawashima CG, Berkowitz O, Hell R, Noji M, Saito K (2005) Characterization and expression analysis of a serine acetyltransferase gene family involved in a key step of the sulfur assimilation pathway in Arabidopsis. Plant Physiol 137:220–230

Lai AG, Doherty CJ, Mueller-Roeber B, Kay SA, Schippers JH, Dijkwel PP (2012) CIRCADIAN CLOCK-ASSOCIATED 1 regulates ROS homeostasis and oxidative stress responses. Proc Natl Acad Sci U S A 109:17129–17134

Lehmann M, Schwarzlaender M, Obata T, Sirikantaramas S, Burow M, Olsen CE, Tohge T, Fricker MD, Moller BL, Fernie AR, Sweetlove LJ, Laxa M (2009) The metabolic response of Arabidopsis roots to oxidative stress is distinct from that of heterotrophic cells in culture and highlights a complex relationship between the levels of transcripts metabolites and flux. Mol Plant 2:390–406

Liu F, Yoo B-C, Lee J-Y, Pan W, Harmon AC (2006) Calcium-regulated phosphorylation of soybean serine acetyltransferase in response to oxidative stress. J Biol Chem 281:27405–27415

Maruyama-Nakashita A, Nakamura Y, Tohge T, Saito K, Takahashi H (2006) Arabidopsis SLIM1 is a central transcriptional regulator of plant sulfur response and metabolism. Plant Cell 18:3235–3251

Morcuende R, Bari R, Gibon Y, Zheng W, Pant BD, Blasing O, Usadel B, Czechowski T, Udvardi MK, Stitt M, Scheible W-R (2007) Genome-wide reprogramming of metabolism and regulatory networks of Arabidopsis in response to phosphorus. Plant Cell Environ 30:85–112

Noji M, Inoue K, Kimura N, Gouda A, Saito K (1998) Isoform-dependent differences in feedback regulation and subcellular localization of serine acetyltransferase involved in cysteine biosynthesis from *Arabidopsis thaliana*. J Biol Chem 273:32739–32745

Park S-W, Li W, Viehhauser A, He B, Kim S, Nilsson AK, Andersson MX, Kittle JD, Ambavaram MMR, Luan S, Esker AR, Tholl D, Cimini D, Ellerstrom M, Coaker G, Mitchell TK, Pereira A, Dietz K-J, Lawrence CB (2013) Cyclophilin 20-3 relays a 12-oxo-phytodienoic acid signal during stress responsive regulation of cellular redox homeostasis. Proc Natl Acad Sci U S A 110:9559–9564

Scheible WR, Morcuende R, Czechowski T, Fritz C, Osuna D, Palacios-Rojas N, Schindelasch D, Thimm O, Udvardi MK, Stitt M (2004) Genome-wide reprogramming of primary and secondary metabolism protein synthesis cellular growth processes and the regulatory infrastructure of Arabidopsis in response to nitrogen. Plant Physiol 136:2483–2499

Watanabe M, Kusano M, Oikawa A, Fukushima A, Noji M, Saito K (2008a) Physiological roles of the beta-substituted alanine synthase gene family in Arabidopsis. Plant Physiol 146:310–320

Watanabe M, Mochida K, Kato T, Tabata S, Yoshimoto N, Noji M, Saito K (2008b) Comparative genomics and reverse genetics analysis reveal indispensable functions of the serine acetyltransferase gene family in Arabidopsis. Plant Cell 20:2484–2496

Winter D, Vinegar B, Nahal H, Ammar R, Wilson GV, Provart NJ (2007) An "Electronic Fluorescent Pictograph" browser for exploring and analyzing large-scale biological data sets. PLoS One 2:e718

Wirtz M, Hell R (2006) Functional analysis of the cysteine synthase protein complex from plants: structural biochemical and regulatory properties. J Plant Physiol 163:273–286

Yamaguchi Y, Nakamura T, Kusano T, Sano H (2000) Three Arabidopsis genes encoding proteins with differential activities for cysteine synthase and beta-cyanoalanine synthase. Plant Cell Physiol 41:465–476

Elucidating the Effects of Higher Expression Level of Cystathionine γ-Synthase on Methionine Contents in Transgenic Arabidopsis, Soybean and Tobacco Seeds

Hagai Cohen, Yael Hacham, Ifat Matityahu, and Rachel Amir

Abstract Plant seeds accumulate low contents of methionine in their seeds, limiting their nutritional values as a source of proteins. Previous conventional and molecular attempts to increase methionine levels in seeds by classical breeding, selection of mutants or creating 'additional protein sinks' for soluble methionine by expressing methionine-rich seed-storage proteins, have yielded only limited success. Here, we summarize our efforts to increase methionine contents in transgenic Arabidopsis, soybean and tobacco seeds by seed-specific expression of feedback-insensitive mutated forms of the *Arabidopsis thaliana* cystathionine γ-synthase (AtCGS), methionine main regulatory enzyme. Each of these species represents different plant families (Solanaceae, Brassicaceae and Fabaceae). The effects of the manipulations on the levels of soluble and total methionine and on the accumulation of storage compounds in these transgenic seeds are discussed.

Introduction

Legumes and cereals supplied as grains and/or forage are among the most important nutritional sources of protein for humans and livestock, but contain limited levels of several essential amino acids. Legume grains are mainly deficient in the sulfur-containing amino acids, methionine and cysteine, while cereals are deficient mainly in lysine, threonine and tryptophan. Combinations of legumes and cereals in diets do not reach the recommended nutritional levels of methionine, which supposed to be 3.5% out of the total consumed amino acids, according to the World Health Organization (WHO 2007). Thus, improving the nutritional quality of seeds by increasing the content of methionine is of critical importance.

The attempts to increase methionine levels in seeds by conventional breeding and selection of mutants have yielded only limited success and were usually

H. Cohen • Y. Hacham • I. Matityahu • R. Amir (✉)
Laboratory of Plant Science, Migal Galilee Technology Center, Kiryat Shmona 12100, Israel
e-mail: rachel@migal.org.il

L.J. De Kok et al. (eds.), *Sulfur Metabolism in Higher Plants - Fundamental, Environmental and Agricultural Aspects*, Proceedings of the International Plant Sulfur Workshop, DOI 10.1007/978-3-319-56526-2_4

associated with a significant reduction in yield (Imsande 2001). Numerous molecular approaches were taken to create 'additional protein sinks' for soluble methionine by expressing methionine-rich seed-storage proteins, such as the 2S-albumin proteins from Brazil nut, sunflower and sesame (Altenbach and Simpson 1990; Altenbach et al. 1992; Muntz 1997; Tabe and Higgins 1998; Lee et al. 2003), and the maize zein proteins (Chui and Falco 1995; Hoffman et al. 1987; Kim and Krishnan 2004; Anthony et al. 1997). These transgenic seeds showed no net increases or only slightly higher levels of total methionine indicating that the levels of soluble methionine may limit the synthesis of seed proteins. This assumption encouraged further approaches to elevate the levels of soluble methionine in seeds. One of them is the manipulation of methionine biosynthetic pathway in seeds by expressing cystathionine γ-synthase (CGS), methionine main regulatory enzyme. Mutated forms of the *Arabidopsis thaliana* CGS, which are insensitive to methionine/SAM were used (Hacham et al. 2006; Hanafy et al. 2013).

Here, we summarize our efforts to increase methionine contents in transgenic Arabidopsis, soybean and tobacco seeds expressing one of the mutated forms of AtCGS (Hacham et al. 2006) in a seed-specific manner (Cohen et al. 2014; Song et al. 2013; Matityahu et al. 2013). Each of these species represents different plant families (Solanaceae, Brassicaceae and Fabaceae). The effects of the manipulation on the levels of soluble and total methionine and on the accumulation of storage compounds in these transgenic seeds are discussed.

Natural Differences in Soluble and Total Methionine Contents in Arabidopsis, Soybean and Tobacco Wild Type Seeds

The analyses highlighted natural differences in soluble and total methionine contents in WT seeds of Soybean (*Glycine max* (L.) Merr. cv Zigongdongdou), tobacco (*Nicotiana tabacum* cv Samsun NN) and Arabidopsis (*Arabidopsis thaliana* ecotype Colombia-0). Soybean and tobacco WT seeds accumulated around 10-fold higher soluble methionine than Arabidopsis seeds (based on nmol per g^{-1} seed fresh weight; Fig. 1a); possibly since *Glycine* and *Nicotiana* genera belong to the Fabaceae and Solanaceae families, respectively, that accumulate more storage proteins in their seeds, while the *Arabidopsis* genus belongs to the Brassicaceae family that accumulate more oil. Accordingly, the Arabidopsis WT seeds accumulated 56- and 26-fold less total methionine than soybean and tobacco WT seeds, respectively (Fig. 1b), emphasizing natural differences in seed metabolism existing between Arabidopsis and the other two species.

To further understand the natural differences observed in methionine contents in WT seeds from the three species, we also calculated their total/soluble methionine ratios. This highlighted the notion that soybean seeds incorporated most of their methionine into proteins as they accumulate 72-fold higher total methionine than soluble methionine. Tobacco seeds exhibited a 36-fold ratio, while Arabidopsis

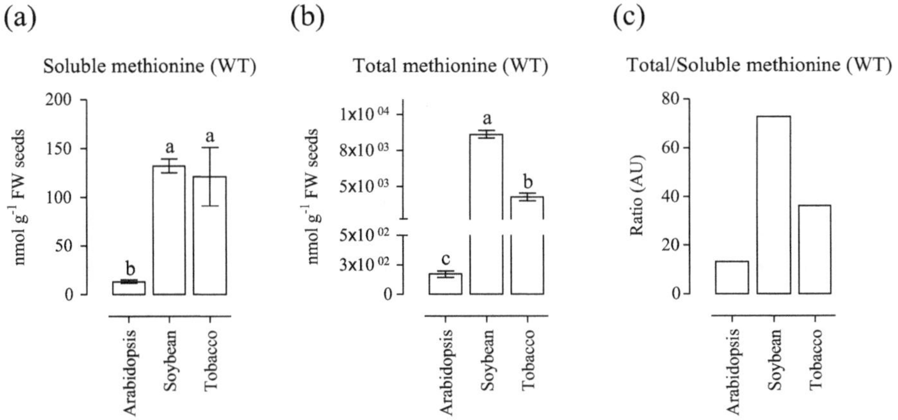

Fig. 1 Natural differences in soluble and total methionine contents in Arabidopsis, soybean and tobacco WT seeds. (**a**) Soluble methionine contents and (**b**) total methionine contents following protein hydrolysis, measured by GC-MS. (**c**) Calculated total/soluble methionine ratios. Significance was calculated by the ANOVA one-way test of $P < 0.05$ and identified by *letters*

seeds accumulate only 13-fold more total methionine than soluble methionine. These differences might results from different regulation of methionine metabolism, higher usage of soluble methionine into its diverse metabolism, and different incorporation capacity of methionine into seed storage proteins due to different methionine content in their codons (Cohen et al. 2016).

Seed-Specific Expression of Mutated Forms of AtCGS Leads to Higher Soluble Methionine Content in Transgenic Arabidopsis and Soybean Seeds, but Not in Transgenic Tobacco Seeds

To assess the regulatory roles of CGS in methionine synthesis in seeds of various plant species, we have generated transgenic Arabidopsis, soybean and tobacco seeds, expressing mutated forms of the AtCGS under the control of the seed-specific promoters of phaseolin or legumin B4 (Sundaram et al. 2013; Zakharov et al. 2004). These promoters belong to the most abundant seed-storage protein in the common bean (*Phaseolus vulgaris*) and *Vicia faba* which are stringently turned off during all vegetative stages of plant development (Sundaram et al. 2013). It was also demonstrated that these promoters are induced constitutively during maturation and desiccation stages of seed development (Fait et al. 2011; Zakharov et al. 2004).

Soluble methionine levels were evaluated in the transgenic seeds from the three species, and compared to their respective WT seeds. Transgenic Arabidopsis and soybean seeds accumulated 6- and 1.6-fold higher levels of soluble methionine,

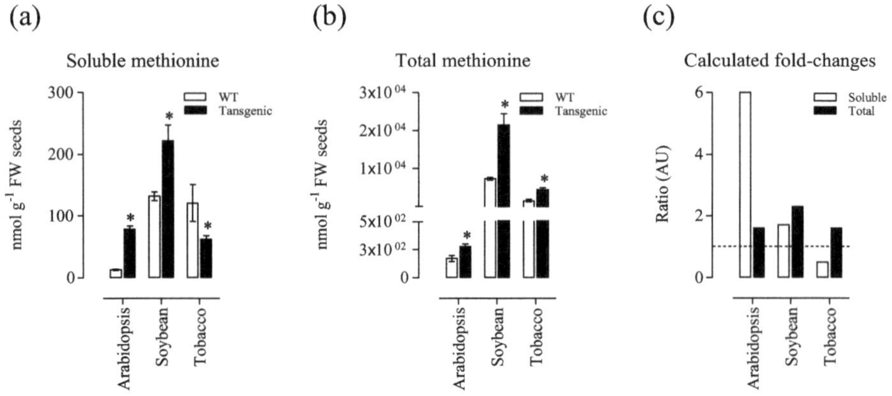

Fig. 2 Seed-specific expression of AtD-CGS leads to differential accumulation of soluble and total methionine contents in transgenic Arabidopsis, soybean and tobacco seeds. (**a**) Soluble methionine contents and (**b**) total methionine contents following protein hydrolysis, measured by GC-MS. (**c**) Calculated fold-changes of soluble and total methionine contents in transgenic seeds compared to their respective WT. *Dashed line* represents the relative levels in the respective WT seeds. Significance was calculated by the Student's *t*-test of $P < 0.05$ and identified by *asterisks*

respectively, compared to their respective WT seeds (Fig. 2a). These observations stand in line with previous reports demonstrating that transgenic seeds of *Misuzudaizu* and *Bert* soybean cultivars and azuki beans expressing another mutated form of AtCGS accumulated about two-fold higher levels of soluble methionine (Hanafy et al. 2013).

However, the levels of soluble methionine in the transgenic tobacco seeds did not increase beyond the levels of their respective WT seeds, but was even significantly reduced by about two-fold (Fig. 2a). One possible explanation for the lower soluble methionine found in these seeds is higher catabolism of methionine, as previously reported in transgenic tobacco seeds expressing a key-enzyme of lysine synthesis. These seeds accumulated significantly higher lysine content compared to WT during their development; however, this triggered the expression of lysine catabolic enzyme at later stages of seed development (Karchi et al. 1994). Another possible explanation might be that the manipulation in tobacco seeds triggered the synthesis of additional storage proteins, thus, creating a higher demand to the incorporation of soluble methionine and other amino acids.

Transgenic Seeds of All Three Species Accumulate Higher Levels of Total Methionine in Their Proteins

To reveal whether higher levels of soluble methionine affected total methionine contents in the transgenic seeds, including methionine incorporated into seed proteins, we measured methionine levels after protein hydrolysis. Generally, transgenic seeds from all three species accumulated significantly higher levels of total methionine, where the Arabidopsis and tobacco seeds exhibited 1.6-fold increases and the soybean seeds a 2.3-fold increase (Fig. 2b). The increases in soluble and total methionine contents detected in transgenic soybean seeds were of relatively similar magnitudes (Fig. 2c), suggesting that the excess of soluble methionine due to AtD-CGS expression most probably incorporated into the seed storage proteins. The transgenic Arabidopsis seeds, however, accumulated 6-fold more soluble methionine but only 1.6-fold more total methionine (Fig. 2c). This might be related to the observation that Arabidopsis seeds naturally accumulate less protein, and thus, are characterized by a lower demand for the excess of soluble methionine. Yet, previous observations showed that higher methionine in transgenic Arabidopsis seeds altered the contents of several 12-globulin subunits (Cohen et al. 2016). The transgenic tobacco seeds accumulated 1.6-fold more total methionine even though they exhibited less soluble methionine compared to their respective WT (Fig. 2c). Based on these results, we assume that the levels of soluble methionine in the transgenic tobacco seeds are lower due to demand for protein synthesis, as previously suggested for transgenic seeds with higher levels of soluble lysine, tryptophan and cysteine (Falco et al. 1995; Kita et al. 2010; Kim et al. 2012).

The Effects of Higher Methionine Levels on the Accumulation of Other Soluble and Total Amino Acids

The total soluble amino acid contents in soybean and tobacco WT seeds are around 10-fold higher than those detected in Arabidopsis WT seeds (Fig. 3a). The highest levels of total amino acid contents were observed in soybean seeds, the tobacco exhibited mid-levels, and the lowest levels were detected in the Arabidopsis seeds (Fig. 3b).

The elevations of soluble methionine contents in transgenic Arabidopsis and soybean seeds were associated with 3.6- and 1.1-fold, respectively significant increases in soluble amino acid contents (Fig. 3c). These findings suggest a yet unknown tight connection between the metabolism of methionine and the accumulation of other amino acids. Such elevations were not observed in transgenic seeds with elevated levels of other amino acids. For example, transgenic soybean seeds with higher levels of total tryptophan and Arabidopsis seeds having higher lysine levels did not show increases in total amino acid contents (Kita et al. 2010; Zhu and

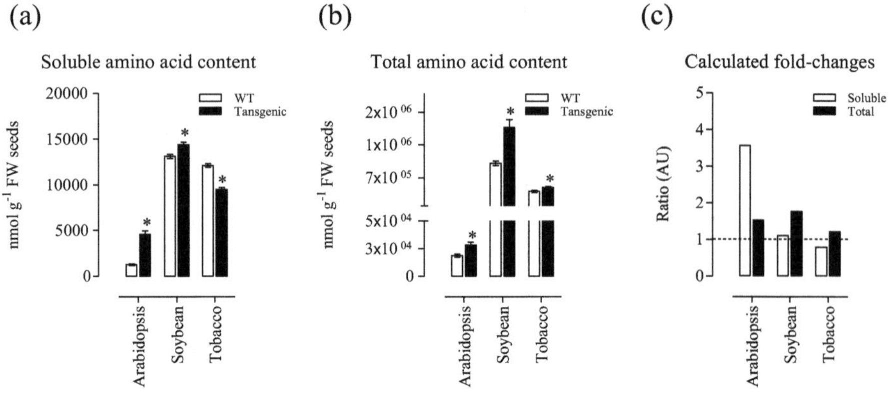

Fig. 3 The effects of higher methionine levels on the accumulation of other soluble and total amino acids. (**a**) Soluble amino acid contents and (**b**) total amino acid contents following protein hydrolysis, measured by GC-MS. (**c**) Calculated fold-changes of soluble and total amino acid contents in transgenic seeds compared to their respective WT. *Dashed line* represents the relative levels in the respective WT seeds. Significance was calculated by the Student's t-test of $P < 0.05$ and identified by *asterisks*

Galili 2003). This kind of association between the metabolism of methionine and other amino acids is further emphasized in the transgenic tobacco seeds that accumulated less soluble methionine, but also lower soluble amino acid contents compared to their respective WT seeds (Fig. 3a). Higher levels of soluble amino acids can be incorporated into proteins leading to an overall increase in total amino acid contents. Indeed, both transgenic Arabidopsis and soybean seeds exhibited 1.5- and 1.8-fold increases of total amino acid contents, respectively (Fig. 3b, c). The transgenic tobacco seeds also exhibited a significant 1.2-fold increased content of total amino acids, providing a reliable explanation for the lower soluble amino acid contents found in these seeds (Fig. 3b, c).

Altogether, the results suggest that methionine level is a limiting factor of protein synthesis in seeds. Additionally, we observed that when methionine synthesis is enhanced, the levels of most other soluble amino acids increased, enabling them to incorporate into proteins. Yet, further studies are required to define the mechanisms by which methionine metabolism is associated with the accumulation of other amino acids, and to determine if such phenomenon is indeed unique to methionine, or the elevation of other amino acids can also yield higher contents of other soluble and total amino acids.

Increased Methionine Levels in Transgenic Seeds Lead to Differential Accumulation of Total Protein and Lipid Contents

The increased amounts of total amino acids incorporated in proteins suggest that the transgenic seeds have higher protein contents. To test this possibility, the levels of total protein contents in seeds were determined according to the Kjeldahl method. As expected, transgenic seeds from all three species exhibited significantly higher protein contents compared to their respective WT seeds (Fig. 4a). Interestingly, even though the transgenic tobacco seeds exhibited the lowest increases total amino acids compared to the transgenic Arabidopsis and soybean seeds, they exhibited the highest net increases in total protein contents, 4.8% (Fig. 4c). The transgenic Arabidopsis and soybean seeds contained 2.2% and 2.1% more protein than their respective WT seeds (Fig. 4c). The observation that no correlation exists between the increases in total amino acid and total protein contents in transgenic tobacco seeds, might be explained by the fact that the first parameter is determined by protein hydrolysis that extracts only the water-soluble protein fractions, while the second by the Kjeldahl method hydrolyses all proteins existing in seeds. Yet the results imply that the regulatory mechanisms underlying the accumulation of protein in tobacco seeds differ than those operate in Arabidopsis and soybean seeds.

Higher protein levels may affect the levels of total lipid contents as another major reserve component in seeds, as reported previously (Hernandez-Sebastia et al. 2005). Indeed, measurements of total lipid contents according to the Soxhlet method revealed a slight significant 2.2% and 1.0% decreases in both transgenic Arabidopsis and soybean seeds, respectively (Fig. 4b, c). The transgenic tobacco seeds, however, exhibited a significant 4% increase in their total lipid contents (Fig. 4b, c), suggesting that the regulatory relationships exist between the metabolism of these reserve compounds in tobacco seeds differ than those in Arabidopsis and soybean seeds.

Conclusions and Future Prospective

The results presented here suggest a tight link between higher methionine contents during seed development and the accumulation of other soluble amino acids through yet-unknown mechanisms, urging for further research to elucidate the regulation behind these observations. As a result, more soluble amino acids can be incorporated into storage proteins during seed maturation, and the levels of total proteins increase in seeds, affecting the accumulation of total lipids. As the level of soluble methionine in the transgenic seeds remained high, most of it was incorporated into proteins. We assume that other amino acids become rate-limiting for protein synthesis, and thus, methionine cannot be incorporated further into proteins. This might explain the higher soluble methionine levels found in transgenic

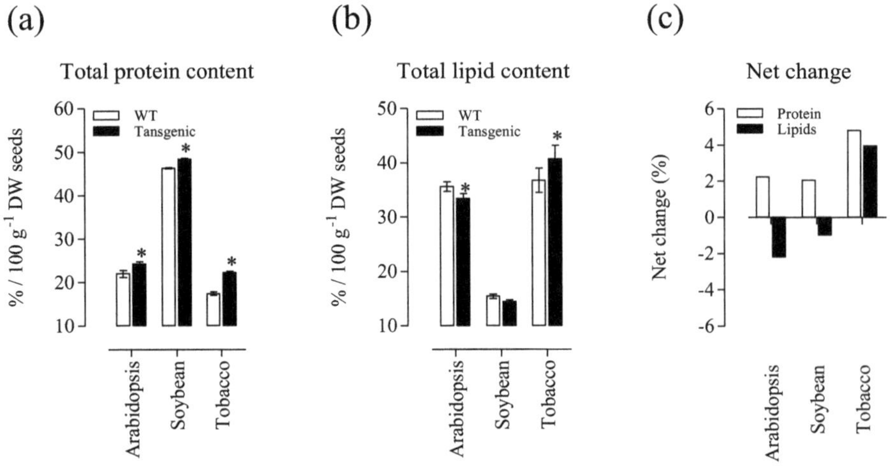

Fig. 4 Increased methionine levels in transgenic seeds lead to differential accumulation of total protein and lipid contents. (**a**) Total protein contents measured according to the Kjeldahl method. (**b**) Total lipid contents measured according to the Soxhlet method. (**c**) Net change in total protein and lipid contents in transgenic seeds compared to their respective WT seeds. Significance was calculated by the Student's *t*-test of $P < 0.05$ and identified by *asterisks*

Arabidopsis and soybean seeds. Another explanation for this phenomenon is that the capacity of methionine codons in WT seed storage proteins limits the incorporation of the overproduced methionine into proteins. This option was suggested previously for cysteine in lupin plants (Tabe et al. 2010).

WT seeds from the three species exhibit natural differences in the amounts of soluble and total methionine, soluble and total amino acid contents, and the accumulation patterns of total proteins and lipids. This is expected as the three species belong to evolutionary-separated families. For example, soybean seeds as a protein-accumulating legume contain 46% total protein compared to Arabidopsis and tobacco seeds that accumulate 17–22%. On the other hand, these two species contain 36% of total lipids while soybean only 15%. Despite several similarities, when examining the changes occurred due to the genetic manipulation we could clearly distinguish the transgenic tobacco seeds from the Arabidopsis and soybean seeds. It seems that the regulatory mechanisms mediating the effects of AtD-CGS and/or higher methionine in tobacco seeds significantly differ than the mechanisms operating in Arabidopsis and soybean seeds.

Altogether, the results indicate a way to improve the nutritional quality of crop plants by elevating the levels of methionine and total protein. The data provide new insights into the factors participating in the regulation of methionine and the mechanisms mediating the effects of elevated methionine levels on seed metabolism and composition.

References

Altenbach SB, Simpson RB (1990) Manipulation of methionine-rich protein genes in plant seeds. Trends Biotechnol 8:156–160

Altenbach SB, Kuo CC, Staraci LC, Pearson KW, Wainwright C, Georgescu A, Townsend J (1992) Accumulation of Brazil nut albumin in seeds of transgenic canola results in enhanced levels of seed protein methionine. Plant Mol Biol 18:235–245

Anthony J, Brown W, Buhr D, Ronhovde G, Genovesi D, Lane T, Yingling R, Aves K, Rosato M, Anderson P (1997) Transgenic maize with elevated 10 kD zein and methionine. In: Cram H, Stulen I, Brunold C, Rennenberg H (eds) Sulfur metabolism in higher plants: molecular, ecophysiological and nutritional aspects. Backhuys Publishers, Leiden, pp 295–297

Chui CF, Falco S (1995) A new methionine-rich seed storage protein from maize. Plant Physiol 107:291

Cohen H, Israeli H, Matityahu I, Amir R (2014) Seed-specific expression of a feedback-insensitive form of cystathionine γ-synthase in Arabidopsis stimulates metabolic and transcriptomic responses associated with desiccation stress. Plant Physiol 166:1575–1592

Cohen H, Pajak A, Pandurangan S, Amir R, Marsolais F (2016) Higher endogenous methionine in transgenic Arabidopsis seeds affects the composition of storage proteins and lipids. Amino Acids. doi:10.1007/s00726-016-2193-4

Fait A, Nesi AN, Angelovici R, Lehmann M, Pham PA, Song L, Haslam RP, Napier JA, Galili G, Fernie AR (2011) Targeted enhancement of glutamate-to-γ-aminobutyrate conversion in Arabidopsis seeds affects carbon nitrogen balance and storage reserves in a development-dependent manner. Plant Physiol 157:1026–1042

Falco SC, Guida T, Locke M et al (1995) Transgenic canola and soybean seeds with increased lysine. Biotechnology 13:577–582

Hacham Y, Schuster G, Amir R (2006) An *in vivo* internal deletion in the n-terminus of cystathionine c-synthase in Arabidopsis results with decreased modulation of expression by methionine. Plant J 45:955–967

Hanafy MS, Rahman SM, Nakamoto Y, Fujiwara T, Naito S, Wakasa K, Ishimoto M (2013) Differential response of methionine metabolism in two grain legumes, soybean and azuki bean, expressing a mutated form of Arabidopsis cystathionine γ-synthase. J Plant Physiol 170:338–345

Hernandez-Sebastia C, Marsolais F, Saravitz C et al (2005) Free amino acid profiles suggest a possible role for asparagine in the control of storage-product accumulation in developing seeds of low- and high-protein soybean lines. J Exp Bot 56:1951–1963

Hoffman LM, Donaldson DD, Bookland R, Rashka K, Herman EM (1987) Synthesis and protein body deposition of maize 15-KD zein in transgenic tobacco seeds. EMBO J 6:3213–3221

Imsande J (2001) Selection of soybean mutants with increased concentrations of seed methionine and cysteine. Crop Sci 41:510–515

Karchi H, Shaul O, Galili G (1994) Lysine synthesis and catabolism are coordinately regulated during tobacco seed development. Proc Natl Acad Sci U S A 91:2577–2581

Kim WS, Krishnan H (2004) Expression of an 11 kDa methionine-rich delta-zein in transgenic soybean results in the formation of two types of novel protein bodies in transitional cells situated between the vascular tissue and storage parenchyma cells. Plant Biotechnol J 2:199–210

Kim WS, Chronis D, Juergens M, Schroeder AC, Hyun SW, Jez JM, Krishnan HB (2012) Transgenic soybean plants overexpressing *O*-acetylserine sulfhydrylase accumulate enhanced levels of cysteine and Bowman-Birk protease inhibitor in seeds. Planta 235:13–23

Kita Y, Nakamoto Y, Takahashi M, Kitamura K, Wasaka K, Ishimoto M (2010) Manipulation of amino acid composition in soybean seeds by the combination of deregulated tryptophan biosynthesis and storage protein deficiency. Plant Cell Rep 29:87–95

Lee TT, Hou RC, Chen LJ, Su RC, Wang CS, Tzen JT (2003) Enhanced methionine and cysteine levels in transgenic rice seeds by the accumulation of sesame 2S albumin. Biosci Biotechnol Biochem 67:1699–1705

Matityahu I, Godo I, Hacham Y, Amir R (2013) Tobacco seeds expressing a feedback-insensitive cystathionine gamma-synthase exhibit elevated content of methionine and altered primary metabolic profile. BMC Plant Biol 13:206

Muntz K (1997) How does the seed's sulphur metabolism react on high level formation of foreign methionine rich proteins in transgenic narbon bean (*Vicia narbonensis* L.)? In: The 39th NIBB conference: dynamic aspects of seed maturation and germination. National Institute for Basic Biology, Okazaki, p 444

Song S, Hou W, Godo I et al (2013) Soybean seeds expressing feedback-insensitive cystathionine γ-synthase exhibit a higher content of methionine. J Exp Bot 64:1917–1926

Sundaram S, Kertbundit S, Shakirov EV et al (2013) Gene networks and chromatin and transcriptional regulation of the phaseolin promoter in Arabidopsis. Plant Cell 25:2601–2617

Tabe LM, Higgins T (1998) Engineering plant protein composition for improved nutrition. Trends Plant Sci 3:282–286

Tabe L, Molvig L, Droux M et al (2010) Overexpression of serine acetyltransferase produced large increases in *O*-acetylserine and free cysteine in developing seeds of a grain legume. J Exp Bot 61:721–733

World Health Organization (2007) Protein and amino acid requirements in human nutrition. Report of a joint WHO/FAO/UNU expert consultation (WHO Technical Report Serious 935). Switzerland WHO Press, Geneva

Zakharov A, Giersberg M, Hosein F, Melzer M, Muntz K, Saalbach I (2004) Seed-specific promoters direct gene expression in non-seed tissues. J Exp Bot 55:1463–1471

Zhu X, Galili G (2003) Increased lysine synthesis coupled with a knockout of its catabolism synergistically boosts lysine content and also transregulates the metabolism of other amino acids in Arabidopsis seeds. Plant Cell 15:845–853

Biosynthesis of *S*-Alk(en)yl-L-Cysteine Sulfoxides in *Allium*: Retro Perspective

Naoko Yoshimoto and Kazuki Saito

Abstract The biosynthesis of *S*-alk(en)yl-L-cysteine sulfoxides is one of the most characteristic feature of plants that belong to the genus *Allium*. Upon tissue damage, these compounds are hydrolyzed by the enzyme alliinase to generate their respective sulfenic acids, which are spontaneously converted to a series of volatile sulfur-containing compounds with a range of health-beneficial activities. Therefore, the molecular understanding of the mechanism for the biosynthesis of *S*-alk(en)yl-L-cysteine sulfoxides is important for both basic and applied pharmaceutical researches. Information from chemical analysis and radiolabeling experiments, conducted in the latter half of last century, has suggested that *S*-alk(en)yl-L-cysteine sulfoxides are biosynthesized from glutathione via γ-glutamyl-*S*-alk(en)yl-L-cysteines; however, the molecular components that contribute to this biosynthetic pathway and their exact reaction order have long been unclear. Very recently, some genes encoding enzymes involved in the biosynthesis of *S*-alk(en)yl-L-cysteine sulfoxides have been identified through transcriptome-based approaches, and the characterization of these genes and the encoded enzymes has provided insights into this biosynthetic "black box". Here we briefly summarize the current knowledge on the molecular basis of the generation of bioactive sulfur-containing compounds and the biosynthesis of *S*-alk(en)yl-L-cysteine sulfoxides in *Allium* plants.

N. Yoshimoto (✉)
Graduate School of Pharmaceutical Sciences, Chiba University, 1-8-1 Inohana, Chuo-ku, Chiba 260-8675, Japan
e-mail: naokoy@faculty.chiba-u.jp

K. Saito
Graduate School of Pharmaceutical Sciences, Chiba University, 1-8-1 Inohana, Chuo-ku, Chiba 260-8675, Japan

RIKEN Center for Sustainable Resource Science, 1-7-22 Suehiro-cho, Tsurumi-ku, Yokohama 230-0045, Japan

L.J. De Kok et al. (eds.), *Sulfur Metabolism in Higher Plants - Fundamental, Environmental and Agricultural Aspects*, Proceedings of the International Plant Sulfur Workshop, DOI 10.1007/978-3-319-56526-2_5

Introduction

The genus *Allium* is one of the largest plant genera including approximately 700 species (Block 2010). Traditionally, *Allium* plants, especially garlic, have been cultivated worldwide, not only as flavory crops but also as herbal medicines that are effective in prevention and treatment of various human diseases (Rivlin 2001). The oldest reports of medicinal effects of garlic are found in the Ebers Papyrus, which is an Egyptian medical document of herbal knowledge dating back to the sixteenth century BC. Modern researches have confirmed the medicinal properties of garlic, such as antibacterial, antifungal, antivirus, immunostimulating, antioxidant, cholesterol- and triglyceride-lowering, anti-aggregatory, and hypotensive effects (Iciek et al. 2009). Both the distinctive flavors and the medicinal effects of *Allium* plants are attributed to their organosulfur compounds. Generally, intact *Allium* plants are odorless, and it is only when the tissue is damaged that the volatile organosulfur compounds are generated. The origin of these volatile compounds is *S*-alk(en)yl-L-cysteine sulfoxides, which are non-volatile organosulfur compounds accumulated in the cell. When the tissue is damaged, *S*-alk(en)yl-L-cysteine sulfoxides are enzymatically hydrolyzed to generate corresponding sulfenic acid, which is spontaneously converted into a series of sulfur-containing compounds with various medicinal properties (Jones et al. 2004; Rose et al. 2005). Because of the chemical diversity and the medicinal properties of sulfur-containing compounds generated from *S*-alk(en)yl-L-cysteine sulfoxides, *Allium* plants have attracted considerable interest from both chemists and biologists. In this chapter, we describe the current molecular understanding of the production of bioactive sulfur-containing compounds from *S*-alk(en)yl-L-cysteine sulfoxides and the biosynthesis of *S*-alk(en)yl-L-cysteine sulfoxides in *Allium* plants.

S-Alk(en)yl-L-Cysteine Sulfoxides and Their Decomposition by Alliinase

The origin of flavor and medicinal compounds in *Allium* plants has been investigated from the middle of the twentieth century. The first identified flavor precursor from *Allium* plants was *S*-allyl-L-cysteine sulfoxide (alliin) isolated from garlic (Stoll and Seebeck 1948, 1951). Further studies revealed that *Allium* plants generally contain high concentrations of *S*-alk(en)yl-L-cysteine sulfoxides as flavor precursors. To date, four major *S*-alk(en)yl-L-cysteine sulfoxides, alliin, *S*-*trans*-1-propenyl-L-cysteine sulfoxide (isoalliin), *S*-propyl-L-cysteine sulfoxide (propiin), and *S*-methyl-L-cysteine sulfoxide (methiin), have been identified from *Allium* plants (Fig. 1a; Jones et al. 2004; Rose et al. 2005). The abundance ratio of these four sulfoxides is different among *Allium* species (Block 2010; Fritsch and Keusgen 2006). For example, alliin is a major component in garlic (*Allium sativum*), isoalliin

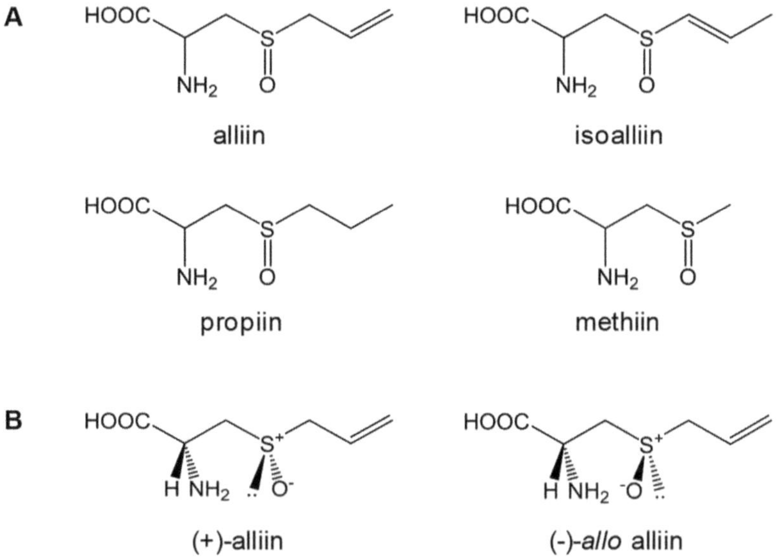

Fig. 1 Chemical structures of *S*-alk(en)yl-L-cysteine sulfoxides in *Allium* plants. (**a**) The four major *S*-alk(en)yl-L-cysteine sulfoxides. (**b**) The stereochemistry of (+)-alliin [($R_C S_S$)-*S*-allylcysteine sulfoxide] and (−)-*allo* alliin [($R_C R_S$)-*S*-allylcysteine sulfoxide]

is characteristic of onion (*Allium cepa*), methiin is found in most *Allium* plants, and propiin is a minor component in most *Allium* plants.

In intact *Allium* plants, *S*-alk(en)yl-L-cysteine sulfoxides are physically separated from the enzyme alliinase (EC. 4.4.1.4; Stoll and Seebeck 1949a,b), which is a pyridoxal 5′-phosphate-dependent C-S lyase that can hydrolyze *S*-alk(en)yl-L-cysteine sulfoxides. While *S*-alk(en)yl-L-cysteine sulfoxides are accumulated in the cytosol of the storage mesophyll cells, alliinase is confined to the vacuole of the vascular bundle sheath cells at least in garlic, onion, and *Allium tuberosum*, as revealed by the subcellular fractionation and the immunostaining analyses (Ellmore and Feldberg 1994; Lancaster and Collin 1981; Yamazaki et al. 2002). When their tissue is damaged by cutting or crushing, alliinase is released from the bundle sheath cells, and comes into contact with *S*-alk(en)yl-L-cysteine sulfoxides. Alliinase immediately cleaves the C-S bond of *S*-alk(en)yl-L-cysteine sulfoxides to generate their respective sulfenic acids, which are rapidly converted to thiosulfinates, such as allicin (diallyl thiosulfinate), by the non-enzymatic self-condensation. Thiosulfinates, which are chemically unstable, undergo further spontaneous reactions, yielding a variety of sulfur-containing bioactive compounds (Jones et al. 2004; Rose et al. 2005). Representative sulfur-containing compounds produced from alliin and the plausible mechanism of production of them are shown in Fig. 2 (El-Aasr et al. 2011; Nohara et al. 2012, 2013, 2014, 2015).

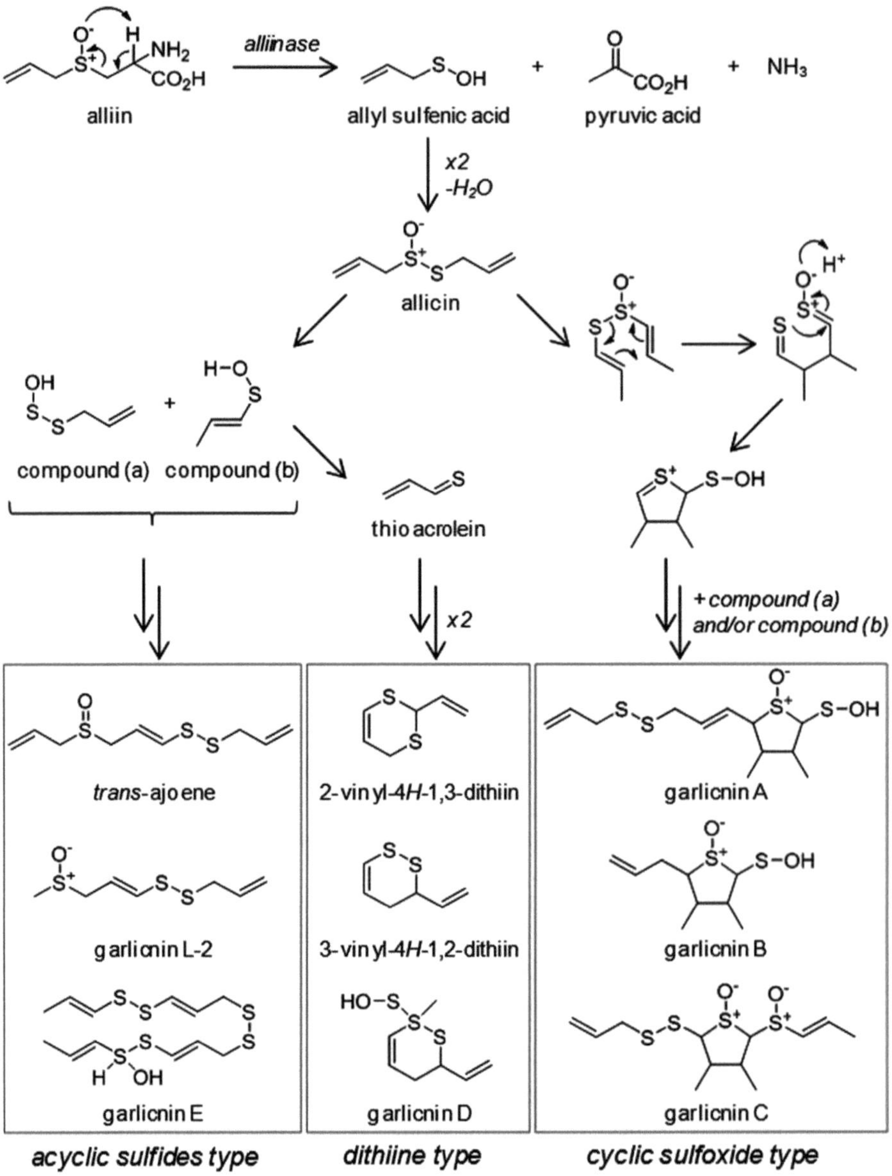

Fig. 2 Alliinase-mediated hydrolysis of alliin and plausible route for the production of various sulfur-containing compounds

The Biosynthetic Pathway for *S*-Alk(en)yl-L-Cysteine Sulfoxides

The biosynthetic pathway for *S*-alk(en)yl-L-cysteine sulfoxides in *Allium* plants has been proposed based on chemical analysis and radiotracer studies that fed [^{35}S] sulfate or [^{14}C] amino acids, in the latter part of the twentieth century (Ettala and Virtanen 1962; Lancaster and Shaw 1989; Suzuki et al. 1961, 1962; Turnbull et al. 1980). The origin of sulfur atom in *S*-alk(en)yl-L-cysteine sulfoxides is mainly sulfate, as [^{35}S] was found in *S*-alk(en)yl-L-cysteine sulfoxides in onion, garlic, and *Allium siculum*, which were supplied with [^{35}S] sulfate (Suzuki et al. 1961; Lancaster and Shaw 1989). This was further supported by the finding that the application of $CaSO_4$ to soil as a sulfur fertilizer resulted in the increase in the concentration of alliin in garlic (Arnault et al. 2003). As in other plant species, *Allium* plants take up sulfate from the soil solution by the functions of sulfate transporters, and biosynthesize cysteine through the reductive sulfate assimilation pathway. Cysteine residue in glutathione or γ-glutamylcysteine is *S*-alk(en)ylated, and the biosynthetic intermediate γ-glutamyl-*S*-alk(en)yl-L-cysteine undergoes deglutamylation and *S*-oxygenation to give *S*-alk(en)yl-L-cysteine sulfoxide (Fig. 3; Lancaster and Shaw 1989). The order of deglutamylation and *S*-oxygenation might be different among *Allium* species and/or among four *S*-alk(en)yl-L-cysteine sulfoxides; the pulse-labeling experiment suggested that *S*-oxygenation likely occur before deglutamylation in isoalliin biosynthesis in onion (Lancaster and Shaw 1989), while *in vitro* determination of catalytic activities of deglutamylation enzymes and *S*-oxygenation enzyme suggested that deglutamylation likely occur prior to *S*-oxygenation in alliin biosynthesis in garlic, as described later in this chapter (Yoshimoto et al. 2015a, b).

By contrast, the origin of *S*-alk(en)yl groups remains obscure. From garlic and onion, *S*-(2-carboxypropyl)glutathione has been isolated and is suggested as an intermediate for the biosynthesis of alliin, isoalliin, and propiin. Feeding of uniformly labeled [^{14}C] valine to garlic resulted in the production of [^{14}C]-labeled *S*-(2-carboxypropyl)glutathione and its derivative, *S*-(2-carboxypropyl)cysteine (Suzuki et al. 1962). Similarly, onion that was exposed to [^{14}C] valine produced [^{14}C]-labeled *S*-(2-carboxypropyl)cysteine (Turnbull et al. 1980). These findings strongly suggested that compound synthesized from valine *in vivo* is the origin of *S*-alk(en)yl group. Based on the chemical structure of *S*-(2-carboxypropyl)glutathione, methacrylyl-CoA, which is known to be synthesized from valine in animal (Shimomura et al. 2004), is suggested to be a plausible candidate compound for the origin of *S*-alk(en)yl group of alliin, isoalliin, and propiin (Suzuki et al. 1962). However, to the best of our knowledge, the presence of methacrylyl-CoA in *Allium* plants has not been confirmed. The origin of *S*-methyl group for methiin is also unclear and needs to be elucidated.

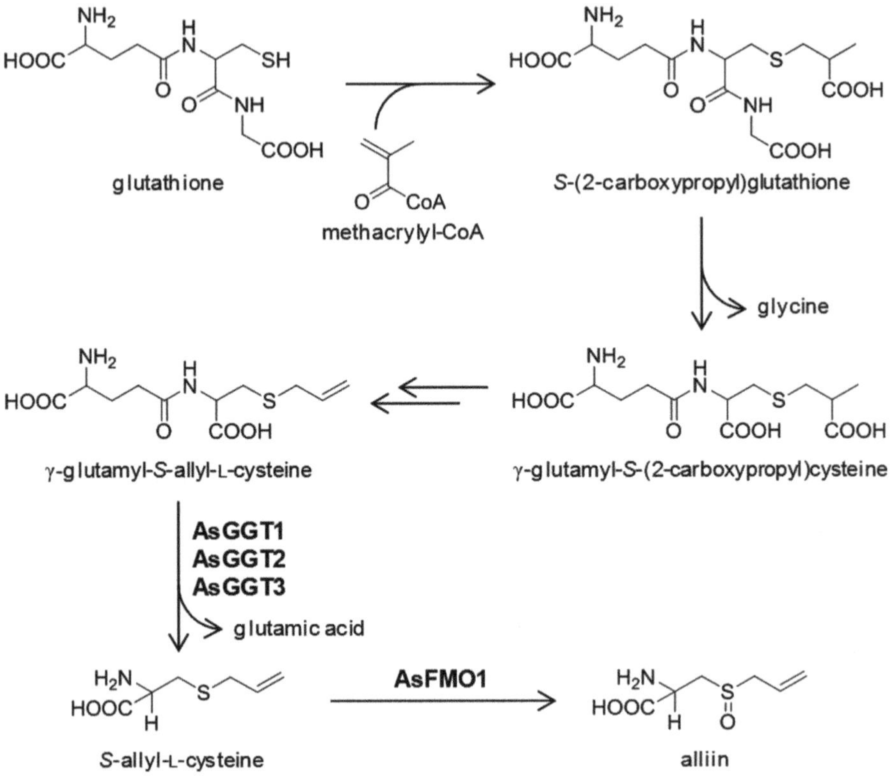

Fig. 3 Putative alliin biosynthetic pathway in garlic

Enzymes Involved in the Biosynthesis of *S*-Alk(en)yl-L-Cysteine Sulfoxides

In the proposed pathway for the biosynthesis of *S*-alk(en)yl-L-cysteine sulfoxides from glutathione, enzymes responsible for the following four steps, respectively, are suggested to catalyze: (1) *S*-alk(en)ylation, (2) removal of glycyl group, (3) removal of γ-glutamyl group, and (4) *S*-oxygenation (Fig. 3). Recent studies identified the enzymes catalyzing the removal of γ-glutamyl group and the enzyme catalyzing *S*-oxygenation.

γ-Glutamyl Transpeptidases Catalyze Deglutamylation Reaction

The most extensively investigated enzymatic step among these four steps is the removal of γ-glutamyl group from the biosynthetic intermediates. γ-Glutamyl transpeptidase (GGT; EC 2.3.2.2) is the enzyme widespread both in prokaryotes and eukaryotes, which catalyzes the transfer of the γ-glutamyl group from γ-glutamyl compounds to the acceptor molecules such as water, amino acids, and small peptides (Tate and Meister 1981). In Arabidopsis, GGT functions in the breakdown of oxidized glutathione in the extracellular space, in the breakdown of glutathione *S*-conjugates of xenobiotics, and presumably in camalexin biosynthesis that requires the removal of γ-glutamyl group from the precursor molecule (Ohkama-Ohtsu et al. 2007a, b, 2011; Su et al. 2011).

Because of its catalytic properties and the potential contribution to the biosynthesis of natural products in non-*Allium* plants, involvement of GGT as the deglutamylation enzyme in the biosynthesis of *S*-alk(en)yl-L-cysteine sulfoxides in *Allium* plants has been proposed. In fact, a GGT partially purified from sprouting onion bulbs showed high substrate specificity for the biosynthetic intermediates of *S*-alk(en)yl-L-cysteine sulfoxides (Lancaster and Shaw 1994), supporting the idea that GGT catalyzes deglutamylation in the biosynthesis of *S*-alk(en)yl-L-cysteine sulfoxides. However, another onion GGT, which was purified to homogeneity from sprouting bulbs, showed a high-affinity to glutathione and glutathione *S*-conjugates, but utilized γ-glutamyl-*S*-propenyl-L-cysteine sulfoxide as a poor substrate (Shaw et al. 2005), indicating that onion has at least two distinct GGT proteins with different *in vivo* functions.

More recently, three genes encoding GGTs, *AsGGT1*, *AsGGT2*, and *AsGGT3*, have been identified from garlic, by homology searching of *Allium* EST database and by using degenerate primers designed based on the conserved regions of known plant GGTs (Yoshimoto et al. 2015a). In the proposed pathway for alliin biosynthesis, a biosynthetic intermediate γ-glutamyl-*S*-allyl-L-cysteine is deglutamylated and *S*-oxygenated to yield alliin (Fig. 3), although the order of deglutamylation and *S*-oxygenation has long been unclear. Recombinant proteins of AsGGT1, AsGGT2, and AsGGT3 catalyzed deglutamylation of γ-glutamyl-*S*-allyl-L-cysteine to yield *S*-allyl-L-cysteine, with the apparent K_m values for γ-glutamyl-*S*-allyl-L-cysteine of 86 μM, 1.1 mM, and 9.4 mM, respectively. By contrast, these GGT proteins showed very low deglutamylation activity toward the corresponding sulfoxide, γ-glutamyl-*S*-allyl-L-cysteine sulfoxide, which is another possible biosynthetic intermediate of alliin, suggesting that deglutamylation occurs prior to *S*-oxygenation in the biosynthesis of alliin in garlic.

Flavin-Containing Monooxygenase Catalyzes Highly Stereoselective S-*Oxygenation Reaction*

The other enzymatic step that has been investigated at the molecular level is *S*-oxygenation, particularly in the biosynthesis of alliin in garlic. This step occurs in a highly stereoselective manner, since almost all *S*-alk(en)yl-L-cysteine sulfoxides contained in *Allium* plants have (*S*)-stereochemistry at their sulfur atoms. For example, in case of alliin in garlic, (+)-alliin [($R_C S_S$)-*S*-allylcysteine sulfoxide] is the major natural form, while (−)-*allo* alliin [($R_C R_S$)-*S*-allylcysteine sulfoxide], one of the diastereomers of (+)-alliin, can be detected only at trace levels (Fig. 1b; Block 2010).

Previous findings that some animal flavin-containing monooxygenase (FMO; EC 1.14.13.8) can catalyze the *S*-oxygenation of *S*-allyl-L-cysteine to yield alliin (Krause et al. 2002; Novick and Elfarra 2008; Ripp et al. 1997), and the fact, that five Arabidopsis FMOs classified into plant clade III FMO are responsible for *S*-oxygenation of *S*-methylthioalkyl glucosinolates to yield their respective sulfoxides (Hansen et al. 2007; Li et al. 2008), suggested the possibility that an FMO is responsible for the *S*-oxygenation reaction in the biosynthesis of *S*-alk(en)yl-L-cysteine sulfoxides in *Allium* plants. Based on homology searching of *Allium* EST database, *AsFMO1*, which encodes a clade III FMO, has been identified from garlic (Yoshimoto et al. 2015b). Recombinant AsFMO1 catalyzed highly stereoselective *S*-oxygenation of *S*-allyl-L-cysteine to nearly exclusively yield (+)-alliin, with the apparent K_m value of 0.25 mM. The *S*-oxygenation activity of AsFMO1 was much higher against *S*-allyl-L-cysteine than against γ-glutamyl-*S*-allyl-L-cysteine, which is another possible biosynthetic intermediate. Together with the substrate specificity of AsGGT1, AsGGT2, and AsGGT3, the substrate preference of AsFMO1 strongly suggested that, in garlic, the biosynthetic intermediate γ-glutamyl-*S*-allyl-L-cysteine is deglutamylated by GGTs to yield *S*-allyl-L-cysteine, which is further *S*-oxygenated by AsFMO1 to yield (+)-alliin (Fig. 3).

Expression Profiles of AsGGT1, AsGGT2, AsGGT3, and AsFMO1, and Their Possible In Vivo *Roles in Garlic*

In growing garlic plants with green foliage leaves, alliin is suggested to be actively biosynthesized from glutathione in green foliage leaves (Bloem et al. 2004; Koch and Lawson 1996; Ueda et al. 1991). During bulb formation, alliin in senescing foliage leaves is presumably translocated to developing bulbs (Bloem et al. 2004). Consequently, mature bulbs contain high amounts of alliin (Ichikawa et al. 2006a). In mature bulbs, the biosynthetic intermediate γ-glutamyl-*S*-allyl-L-cysteine is also accumulated at high levels (ca. 26 mM), as a storage peptide, especially in storage leaves (Matsuura et al. 1996; Ichikawa et al. 2006a, b; Yoshimoto et al. 2015a). γ-Glutamyl-*S*-allyl-L-cysteine in storage leaves of bulbs is stored during dormancy, while it is immediately converted to alliin via two enzymatic reactions, deglutamylation and *S*-oxygenation, during sprouting (Ichikawa et al. 2006b). Alliin produced in storage leaves during sprouting is likely to be translocated to white foliage leaves inside the bulbs (Yoshimoto et al. 2015b). The two-step enzymatic production of alliin from γ-glutamyl-*S*-allyl-L-cysteine may enable rapid production of high amounts of alliin, which is effective in protecting sprouts from microorganisms and animals.

Gene expression profiles of *AsGGT1*, *AsGGT2*, *AsGGT3*, and *AsFMO1* suggested their possible *in vivo* roles in garlic plants. In pre-emergent nearly sprouting bulbs, the mRNA levels of *AsGGT1*, *AsGGT2*, *AsGGT3*, and *AsFMO1* were higher in storage leaves than in white foliage leaves inside the bulbs (Yoshimoto et al. 2015b). This fact suggested the important roles of AsGGT1, AsGGT2, AsGGT3, and AsFMO1 in synthesizing alliin from γ-glutamyl-*S*-allyl-L-cysteine stored at high levels in storage leaves. Chemical analysis suggested that alliin produced in this process is presumably translocated from storage leaves to white foliage leaves inside the bulbs.

By contrast, in sprouted bulbs, *AsGGT1* mRNA accumulated at high levels in green foliage leaves, *AsGGT2* mRNA accumulated in roots, *AsGGT3* mRNA was abundant in storage leaves, and *AsFMO1* mRNA accumulated in various tissues at similar levels (Yoshimoto et al. 2015b). These findings suggested that the contribution of the three GGTs to alliin biosynthesis is different among tissues, while AsFMO1 functions in alliin biosynthesis in various tissues, at this growth stage. Interestingly, the expression patterns of the three GGTs seem to be closely associated with their affinity for γ-glutamyl-*S*-allyl-L-cysteine and the concentration of γ-glutamyl-*S*-allyl-L-cysteine in tissues. In green foliage leaves, the concentration of γ-glutamyl-*S*-allyl-L-cysteine is approximately 38 μM (Matsuura et al. 1996; Yoshimoto et al. 2015a). AsGGT1, which has highest affinity for γ-glutamyl-*S*-allyl-L-cysteine (K_m = 86 μM) among the three GGTs, is suggested to be the most important GGT responsible for the biosynthesis of alliin from glutathione in green foliage leaves. In storage leaves of sprouted bulbs, AsGGT3, which has lowest affinity for γ-glutamyl-*S*-allyl-L-cysteine (K_m = 9.4 mM) among the three GGTs, is suggested to be responsible for deglutamylating γ-glutamyl-*S*-allyl-L-cysteine present at high levels even after sprouting. AsGGT2, which has intermediate affinity for γ-glutamyl-*S*-allyl-L-cysteine (K_m = 1.1 mM), may contribute to both the biosynthesis of alliin from glutathione in green foliage leaves and the deglutamylation of γ-glutamyl-*S*-allyl-L-cysteine in storage leaves. The different organ-specificity of the three GGTs with different kinetic properties may enable the production of optimized levels of *S*-allyl-L-cysteine in each tissue, which is further *S*-oxygenated by constitutively-expressing AsFMO1 to yield alliin. Future investigations of transgenic plants with altered expression levels of AsGGT1, AsGGT2, AsGGT3, and AsFMO1, respectively, will provide a better understanding of their *in vivo* functions.

Conclusions and Future Perspectives

In the past few years, significant advances have been made in our understanding of the molecular basis for the biosynthesis of *S*-alk(en)yl-L-cysteine sulfoxides in *Allium* plants, especially for the biosynthesis of alliin in garlic. Three GGT genes encoding deglutamylation enzymes and one FMO gene encoding *S*-oxygenation enzyme have been identified based on the mining of transcriptome data, and the substrate specificities of their encoded proteins suggested the reaction order of deglutamylation and *S*-oxygenation in alliin biosynthesis in garlic. However, enzymes catalyzing the removal of glycyl group or *S*-alk(en)ylation have not been identified and await discovery. Although the quantification analysis of *S*-alk (en)yl-L-cysteine sulfoxides in garlic and onion suggested that *S*-alk(en)yl-L-cysteine sulfoxides are translocated among organs (Bloem et al. 2004; Lancaster et al. 1986; Yoshimoto et al. 2015b), the transport proteins mediating these transport processes have not been identified. Regulatory factors controlling the rate of biosynthesis also remain to be determined. Recent advances in omics technologies and computational systems biology will accelerate the identification of additional members contributing to the production and the accumulation of *S*-alk(en)yl-L-cysteine sulfoxides in *Allium* plants. Elucidation of the difference in the biosynthesis among *Allium* species is the area also requiring further investigation. Application of genome editing technologies to *Allium* plants will facilitate not only a better understanding of *in vivo* functions of genes involved in the biosynthesis of *S*-alk(en) yl-L-cysteine sulfoxides, but also the molecular breeding of plants beneficial for human health.

Acknowledgements This study was supported in part by JSPS KAKENHI Grant Number 26460118 (to N.Y.) of Japan.

References

Arnault I, Christidès JP, Mandon N, Haffner T, Kahane R, Auger J (2003) High-performance ion-pair chromatography method for simultaneous analysis of alliin, deoxyalliin, allicin and dipeptide precursors in garlic products using multiple mass spectrometry and UV detection. J Chromatogr A 991:69–75

Block E (2010) Garlic and other Alliums: the Lore and the science. The Royal Society of Chemistry, Cambridge

Bloem E, Haneklaus S, Schnug E (2004) Influence of nitrogen and sulfur fertilization on the alliin content of onions and garlic. J Plant Nutr 27:1827–1839

El-Aasr M, Fujiwara Y, Takeya M, Ono M, Nakano D, Okawa M, Kinjo J, Ikeda T, Miyashita H, Yoshimitsu H, Nohara T (2011) Garlicnin A from the fraction regulating macrophage activation of *Allium sativum*. Chem Pharm Bull 59:1340–1343

Ellmore GS, Feldberg RS (1994) Alliin lyase localization in bundle sheaths of the garlic clove (*Allium sativum*). Am J Bot 81:89–94

Ettala T, Virtanen AI (1962) Labeling of sulfur-containing amino acids and γ-glutamylpeptides after injection of labeled sulfate into onion (*Allium cepa*). Acta Chem Scand 16:2061–2063

Fritsch RM, Keusgen M (2006) Occurrence and taxonomic significance of cysteine sulphoxides in the genus *Allium* L (Alliaceae). Phytochemistry 67:1127–1135

Hansen BG, Kliebenstein DJ, Halkier BA (2007) Identification of a flavin-monooxygenase as the *S*-oxygenating enzyme in aliphatic glucosinolate biosynthesis in Arabidopsis. Plant J 50:902–910

Ichikawa M, Ide N, Yoshida J, Yamaguchi H, Ono K (2006a) Determination of seven organosulfur compounds in garlic by high-performance liquid chromatography. J Agric Food Chem 54:1535–1540

Ichikawa M, Ide N, Ono K (2006b) Changes in organosulfur compounds in garlic cloves during storage. J Agric Food Chem 54:4849–4854

Iciek M, Kwiecień I, Włodek L (2009) Biological properties of garlic and garlic-derived organosulfur compounds. Environ Mol Mutagen 50:247–265

Jones MG, Hughes J, Tregova A, Milne J, Tomsett AB, Collin HA (2004) Biosynthesis of the flavour precursors of onion and garlic. J Exp Bot 55:1903–1918

Koch HP, Lawson LD (1996) Garlic: the science and therapeutic application of *Allium sativum* L. and related species, 2nd edn. Williams & Wilkins, Baltimore

Krause RJ, Glocke SC, Elfarra AA (2002) Sulfoxides as urinary metabolites of *S*-allyl-L-cysteine in rats: evidence for the involvement of flavin-containing monooxygenases. Drug Metab Dispos 30:1137–1142

Lancaster JE, Collin HA (1981) Presence of alliinase in isolated vacuoles and of alkyl cysteine sulphoxides in the cytoplasm of bulbs of onion (*Allium cepa*). Plant Sci Lett 22:169–176

Lancaster JE, Shaw ML (1989) γ-Glutamyl peptides in the biosynthesis of *S*-alk(en)yl-L-cysteine sulphoxides (flavour precursors) in *Allium*. Phytochemistry 28:455–460

Lancaster JE, Shaw ML (1994) Characterization of purified γ-glutamyl transpeptidase in onions: evidence for *in vivo* role as a peptidase. Phytochemistry 36:1351–1358

Lancaster JE, McCallion BJ, Shaw ML (1986) The dynamics of the flavour precursors, the *S*-alk (en)yl-L-cysteine sulphoxides, during leaf blade and scale development in the onion (*Allium cepa*). Physiol Plant 66:293–297

Li J, Hansen BG, Ober JA, Kliebenstein DJ, Halkier BA (2008) Subclade of flavin-monooxygenases involved in aliphatic glucosinolate biosynthesis. Plant Physiol 148:1721–1733

Matsuura H, Inagaki M, Maeshige K, Ide N, Kajimura Y, Itakura Y (1996) Changes in contents of γ-glutamyl peptides and fructan during growth of *Allium sativum*. Planta Med 62:70–71

Nohara T, Kiyota Y, Sakamoto T, Manabe H, Ono M, Ikeda T, Fujiwara Y, Nakano D, Kinjo J (2012) Garlicnins B_1, C_1, and D, from the fraction regulating macrophage activation of *Allium sativum*. Chem Pharm Bull 60:747–751

Nohara T, Fujiwara Y, Ikeda T, Murakami K, Ono M, Nakano D, Kinjo J (2013) Cyclic sulfoxides garlicnins B_2, B_3, B_4, C_2, and C_3 from *Allium sativum*. Chem Pharm Bull 61:695–699

Nohara T, Fujiwara Y, Ikeda T, Yamaguchi K, Manabe H, Murakami K, Ono M, Nakano D, Kinjo J (2014) Acyclic sulfoxides, garlicnins L-1–L-4, E, and F, from *Allium sativum*. Chem Pharm Bull 62:477–482

Nohara T, Fujiwara Y, Komota Y, Kondo Y, Saku T, Yamaguchi K, Komohara Y, Takeya M (2015) Cyclic sulfoxides-garlicnins K_1, K_2, and H_1-extracted from *Allium sativum*. Chem Pharm Bull 63:117–121

Novick RM, Elfarra AA (2008) Purification and characterization of flavin-containing monooxygenase isoform 3 from rat kidney microsomes. Drug Metab Dispos 36:2468–2474

Ohkama-Ohtsu N, Radwan S, Peterson A, Zhao P, Badr AF, Xiang C, Oliver DJ (2007a) Characterization of the extracellular γ-glutamyl transpeptidases, GGT1 and GGT2, in Arabidopsis. Plant J 49:865–877

Ohkama-Ohtsu N, Zhao P, Xiang C, Oliver DJ (2007b) Glutathione conjugates in the vacuole are degraded by γ-glutamyl transpeptidase GGT3 in Arabidopsis. Plant J 49:878–888

Ohkama-Ohtsu N, Sasaki-Sekimoto Y, Oikawa A, Jikumaru Y, Shinoda S, Inoue E, Kakude Y, Yokoyama T, Hirai MY, Shirasu K, Kamiya Y, Oliver DJ, Saito K (2011) 12-Oxo-phytodienoic acid-glutathione conjugate is transported into the vacuole in Arabidopsis. Plant Cell Physiol 52:205–209

Ripp SL, Overby LH, Philpot RM, Elfarra AA (1997) Oxidation of cysteine *S*-conjugates by rabbit liver microsomes and cDNA-expressed flavin-containing mono-oxygenases: studies with *S*-(1,2-dichlorovinyl)-L-cysteine, *S*-(1,2,2-trichlorovinyl)-L-cysteine, *S*-allyl-L-cysteine, and *S*-benzyl-L-cysteine. Mol Pharmacol 51:507–515

Rivlin RS (2001) Historical perspective on the use of garlic. J Nutr 131:951S–954S

Rose P, Whiteman M, Moore PK, Zhu YZ (2005) Bioactive *S*-alk(en)yl cysteine sulfoxide metabolites in the genus *Allium*: the chemistry of potential therapeutic agents. Nat Prod Rep 22:351–368

Shaw ML, Pither-Joyce MD, McCallum JA (2005) Purification and cloning of a γ-glutamyl transpeptidase from onion (*Allium cepa*). Phytochemistry 66:515–522

Shimomura Y, Honda T, Goto H, Nonami T, Kurokawa T, Nagasaki M, Murakami T (2004) Effects of liver failure on the enzymes in the branched-chain amino acid catabolic pathway. Biochem Biophys Res Commun 313:381–385

Stoll A, Seebeck E (1948) Allium compounds. I. *Alliin*, the true mother compound of garlic oil. Helv Chim Acta 31:189–210

Stoll A, Seebeck E (1949a) Allium compounds. II. Enzymic degradation of *alliin* and the properties of alliinase. Helv Chim Acta 32:197–205

Stoll A, Seebeck E (1949b) Allium compounds. III. Specificity of alliinase and synthesis of compounds related to alliin. Helv Chim Acta 32:866–876

Stoll A, Seebeck E (1951) Chemical investigations on alliin, the specific principle of garlic. Adv Enzymol 11:377–400

Su T, Xu J, Li Y, Lei L, Zhao L, Yang H, Feng J, Liu G, Ren D (2011) Glutathione-indole-3-acetonitrile is required for camalexin biosynthesis in *Arabidopsis thaliana*. Plant Cell 23:364–380

Suzuki T, Sugii M, Kakimoto T (1961) New γ-glutamyl peptides in garlic. Chem Pharm Bull 9:77–78

Suzuki T, Sugii M, Kakimoto T (1962) Incorporation of L-valine-[^{14}C] into *S*-(2-carboxypropyl) glutathione and *S*-(2-carboxypropyl)cysteine in garlic. Chem Pharm Bull 10:328–331

Tate SS, Meister A (1981) γ-Glutamyl transpeptidase: catalytic, structural and functional aspects. Mol Cell Biochem 39:357–368

Turnbull A, Galpin IJ, Collin HA (1980) Comparison of the onion plant (*Allium cepa*) and onion tissue culture. III. Feeding of ^{14}C labeled precursors of the flavor precursor compounds. New Phytol 85:483–487

Ueda Y, Kawajiri H, Miyamura N, Miyajima R (1991) Content of some sulfur-containing components and free amino acids in various strains of garlic. Nippon Shokuhin Kogyo Gakkaishi 38:429–434

Yamazaki M, Sugiyama M, Saito K (2002) Intercellular localization of cysteine synthase and alliinase in bundle sheaths of *Allium* plants. Plant Biotechnol 19:7–10

Yoshimoto N, Yabe A, Sugino Y, Murakami S, Sai-ngam N, Sumi S, Tsuneyoshi T, Saito K (2015a) Garlic γ-glutamyl transpeptidases that catalyze deglutamylation of biosynthetic intermediate of alliin. Front Plant Sci 5:758

Yoshimoto N, Onuma M, Mizuno S, Sugino Y, Nakabayashi R, Imai S, Tsuneyoshi T, Sumi S, Saito K (2015b) Identification of a flavin-containing *S*-oxygenating monooxygenase involved in alliin biosynthesis in garlic. Plant J 83:941–951

The Effect of Sulfur Nutrition on Glucosinolate Patterns and Their Breakdown Products in Vegetable Crops

Britta Pitann, Carolin Heyer, and Karl H. Mühling

Abstract Glucosinolates (GSLs) are amino acid derived secondary metabolites naturally occurring in the order of Brassicales. They represent an important class of phytochemicals involved in plant–microbe, plant–insect, plant–animal and plant–human interactions. In *Brassica* vegetables GSL are known as the bioactive compounds giving the typical flavor and odor, being involved in natural pest control. Still, in high doses GSL remain highly toxic. Even though the GSL content in *Brassica* species is genetically fixed, breeding programs already aimed for reducing the GSL content, with the engineering of 00-varieties of rapeseed (*Brassica napus*) being the most prominent example. Contrary to their negative effects, GSLs are also discussed to have beneficial nutritional and health effects. But it is more their breakdown products, particularly isothiocyanates (ITCs) and nitriles, formed after hydrolysis within the glucosinolate-myrosinase-system, which the health-promoting effects can be ascribed to when taken up in low doses. Besides genetic approaches to influence GSL content and pattern and their breakdown products, little is yet known about how agronomic and particularly plant nutritional factors can alter the GSL content and pattern of their different hydrolysis products in the context of improving food quality. Therefore, the influence of the sulfur (S) supply on GSLs, ITCs and nitriles in various *Brassica* species, such as Indian mustard (*Brassica juncea*), kohlrabi (*Brassica oleracea*), and Chinese cabbage (*Brassica rapa* spp. *pekinensis*), are exemplarily discussed in relation to nitrogen nutrition.

Introduction

Glucosinolates (GSLs) have been part of human life for thousands of years, as being found in edible plants belonging to the order of Capparales, including the family of *Brassicaceae*. Within this family it is mainly the vegetables, such as broccoli, cauliflower, radish, and Brussels sprouts, which play an important role in today's

---

B. Pitann • C. Heyer • K.H. Mühling (✉)
Institute of Plant Nutrition and Soil Science, Kiel University, Hermann-Rodewald-Strasse 2, 24118 Kiel, Germany
e-mail: khmuehling@plantnutrition.uni-kiel.de

L.J. De Kok et al. (eds.), *Sulfur Metabolism in Higher Plants - Fundamental, Environmental and Agricultural Aspects*, Proceedings of the International Plant Sulfur Workshop, DOI 10.1007/978-3-319-56526-2_6

diet, not least due to their distinct flavors and tastes (Charron et al. 2005). In 2011, a worldwide production of 100 million tons of *Brassica* vegetables was recorded (FAOSTAT 2013), while the average daily intake of GSLs was estimated to be approx. 14–15 mg per person in Germany (Schonhof et al. 2004; Steinbrecher and Linseisen 2009). Though treated as unfavorable compound in the past because of their putative goitrogenic effect, GSLs have generated considerable pharmacological interest as components of human food not least since epidemiological studies confirmed the positive effect of vegetable consumption on human health.

Besides their effects in human metabolism, GSLs and their breakdown products are known as allelochemicals. Thus, they play an important role in plant defenses, inhibiting, for example, microbial growth (Glenn et al. 1988). Moreover, they are an intermediate in the biosynthesis of indole phytoalexins (Griffiths et al. 1994), or stimulate oviposition and feeding by insects (Mewis et al. 2005). Glucosinolates also act as feeding deterrents for polyphagous herbivores and feeding stimulants for specialist insects in cruciferous crops (Bartlet et al. 1999). The traditional use of crucifers as green manures with phytosanitary effects is partly based upon the toxic nature of the glucosinolate degradation products, which serve to reduce the soil inoculum of pathogens and pests for subsequent agricultural and horticultural crops (Mithen et al. 2000).

Glucosinolates, which belong to the mustard oil glucosides, are a large group of secondary metabolites in plants, containing abundant nitrogen (N) and sulfur (S). Not least because of their enrichment in sulfur, it is contradictorily discussed whether GSLs might play an important role as a S-reservoir in times of sulfur deficiency to maintain ordinary metabolism (Schonhof et al. 2007; Gerendás et al. 2009). Today, more than 120 different GSL have already been identified, which all comprise a ß-D-thioglucose group, a sulphonated oxime moiety (glucone) and a variable side-chain (aglycone) derived from one of eight amino acids (Fig. 1). Accordingly, GSLs are classified by their precursor amino acids and further modifications of the side chains. While GSLs derived from alanine, leucine, isoleucine, methionine or valine characterize aliphatic GSL, aromatic GSLs derive from phenylalanine and tyrosine, and tryptophane is part of indole GSLs (Mithen et al. 2000; Fahey et al. 2001). Additionally, most *R* groups are elongated by methylene moieties and are subjected to further transformations such as hydroxylation, methylation, desaturation or glycosylation (Halkier and Gershenzon 2006). Eventually, it is just this side chain modification that can have a significant impact on the biological activity of the GSLs and their various hydrolysis products (Brown and Morra 1995). However, there is only limited information on the specific effect of each compound and their breakdown products.

Glucosinolates and Their Breakdown Products

In plant tissues, GSLs remain chemically stable within the cytoplasm until they come into contact with the enzyme myrosinase, which is located in the vacuole. After physical damage of plant tissues, e.g., by cutting, chewing, milling or

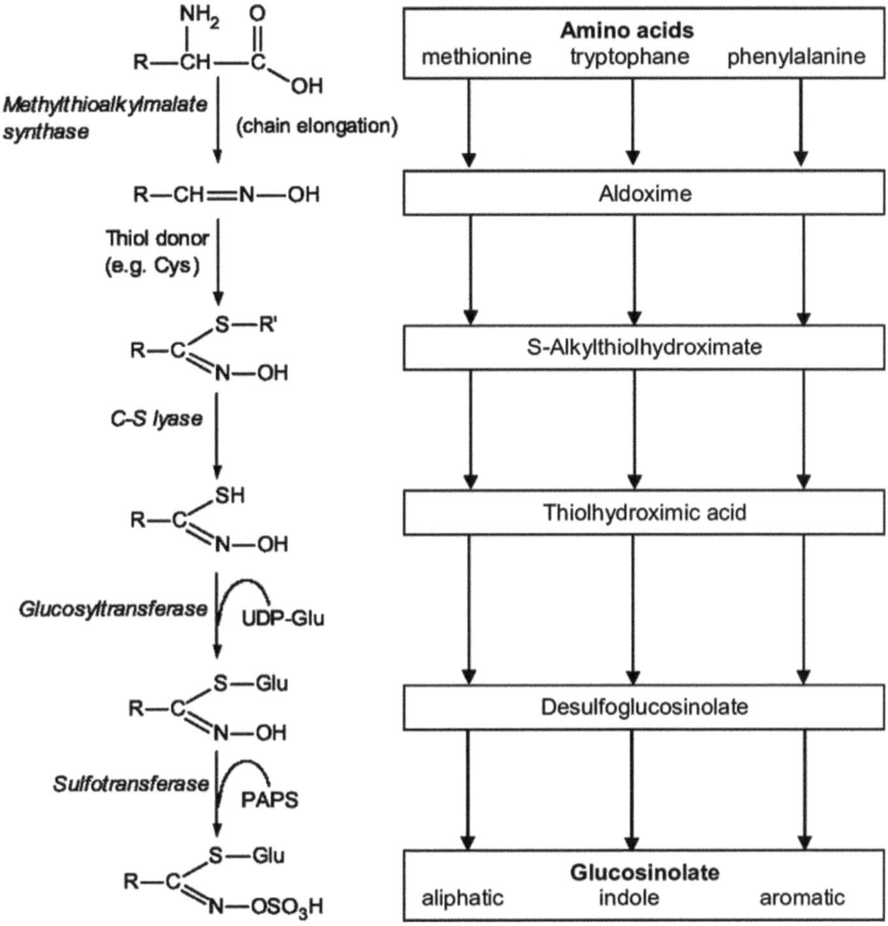

Fig. 1 Biosynthesis and general structure of glucosinolates. *PAPS* 3′-phosphoadenosine-5′-phosphosulfate

penetration, this effective compartmentation is broken down, thus releasing myrosinase, which subsequently leads to the hydrolytic breakdown of GSLs in an ascorbic acid activated manner (e.g., Bones and Rossiter 1996; Kleinwächter and Selmar 2004). This process, also referred to as myrosinase-glucosinolate-system, is regarded as a defense system against any herbivore attack. The myrosinase (EC 3.2.1.147), which belongs to the family of glucoside hydrolases, catalyzes the cleavage d-glucose from GSLs in the presence of water resulting in the formation of thiohydroximate-O-sulfonate (Fig. 2). In a next step, the aglycone undergoes a spontaneous Lossen-like rearrangement, releasing sulfate. Depending on various physiological conditions such as pH or the presence of cofactors, this molecule is further decomposed to form nitriles, epithionitriles, thiocyanates and isothiocyanates (ITCs), which all, but isothiocyanates in particular, contribute to

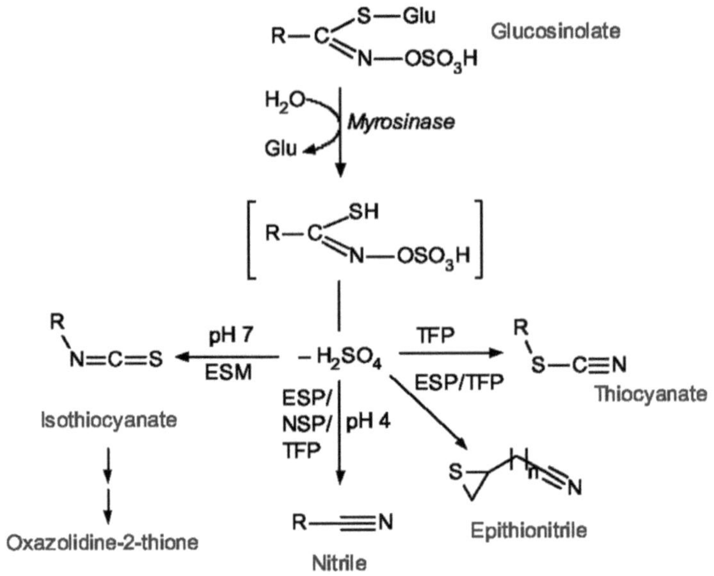

Fig. 2 Formation of enzymatic breakdown products within the glucosinolates–myrosinase system. *ESP* epithiospecifier protein, *TFP* thiocyanate-forming protein, *ESM* epithiospecifier modifier protein, *NSP* nitrile-specifier proteins

the typical smell and flavor of *Brassica*-vegetables (Kliebenstein et al. 2005; Krumbein et al. 2001).

In human nutrition, GSLs and their hydrolysis products were originally considered as antinutritives and being toxic (Mithen et al. 2000; Vermorel et al. 1988), with nitrile depressing growth and causing liver and kidney lesions, thiocyanates inhibiting iodine uptake by the thyroid, and isothiocyanates having a strong antithyroid effect (Holst and Williamson 2004; Bonnesen et al. 2001). Especially, effects of the latter can be traced back to the formation of oxazolidine-2-thiones (Fig. 2), goitrogenic compounds which are all products from chemically unstable isothiocyanates (Hanschen et al. 2014). However, actual human clinical studies confirm that critical threshold levels for uptake are hardly met in human nutrition (Fahey et al. 2001), which is mainly a result of enzymatic degradation of GSL breakdown products, protein cofactors, and processing conditions such as cooking prior to consumption (McMillan et al. 1986). Only in some parts of the world, where a low iodine supply is a problem, or in animal diets dominated by *Brassica*-originated fodder and feed, may these adverse effects still play a role (Tripathi and Mishra 2007). Today, GSLs and their breakdown products are even discussed among other bioactive substances to be linked, for example, with a reduced risk of developing several types of cancer and reduction of the cholesterol level, as well having an antimicrobial effect in humans (Fahey et al. 2001; Mithen et al. 2000; Watzl and Leitzmann 2005). As could be confirmed in several animal and human studies, the anticarcinogenic effect of isothiocyanates is closely related to the suppression and inhibition of phase-I enzymes together with a phase-II enzyme induction in the initiation process of carcinogenesis (Conaway et al. 1996; Boddupalli et al. 2012; James et al. 2012; Hecht 2000). To benefit from such health promoting effects, it is therefore not only desirable to foster an increased consumption of *Brassica*-vegetables, but also to improve the 'health quality' of plants. Increasing the content of bioactive compounds can, for example, be accomplished by including new plant breeding and management strategies, but an even more promising tool seems to be changing and adapting fertilization strategies for *Brassica* species.

Impact of Nitrogen and Sulfur on Glucosinolates and Their Breakdown Products

In general, the GSL content in plant tissue largely depends on pre-harvest variation (i.e., genotype selection, developmental stage, cultivation period; Schreiner 2005). This is always associated with ecophysiological factors, such as the climate parameters of irradiation and temperature, which exert an enormous impact on GSL accumulation in plants (Mithen et al. 2000; Fahey et al. 2001; Fallovo et al. 2009). Of all environmental factors nutrient management (e.g., nitrogen and sulfur status) has been shown to significantly affect the GSL content in plant tissue (De Pascale et al. 2007; Chen et al. 2006; Zhou et al. 2013). A strong interactive effect between N and S on the GSL content can be expected as their side chains are derived from S-containing or S-free, but in any, case N-containing amino acids, the thiol bridge to glucose from cysteine, and the sulfonate group from phospho-adenosin-phosphosulfate (Fahey et al. 2001). Further GSLs are still discussed as a transient S-reservoir in plants (Bones and Rossiter 1996; Schnug et al. 1995), assuming that under S deficiency GSL concentration may decrease leading to accumulation of their breakdown products (i.e., ITC), presumably resulting from less efficient compartmentalization of GSLs from myrosinase. This is in contrast to other studies, which is nowadays generally disproved (e.g., Gerendás et al. 2008b; 2009;). In addition, sulfate released by myrosinase activity (Fig. 2) is easily consumed by metabolic processes but any further conversion of low molecular weight ITCs may be restricted due to their inefficient compartmentalization. Although energy coupled uptake by protoplasts (Chen and Halkier 2000) and site-directed translocation of intact GSLs have been documented (Brudenell et al. 1999), this seems not to be the case for ITCs. The effect of S limitation may be particularly strong in plants grown with marginal S but ample N supply.

Indeed, while *Brassica nigra* deficient in S but sufficient in N showed lower levels of allyl isothiocyanates, thus leading to increased feeding by *Spodoptera eridania*, plants supplied with high S but low N showed a reduced infestation (Wolfson 1982). This is in line with studies on *Brassica napus*, which showed that high N and low S can minimize the GSL content of seeds (Zhao et al. 1993,

1994). It can be reasoned that the effect of N supply largely depends on the S supply. Whenever the soil was deficient in S, N fertilization decreased the GSL content, whereas N addition to adequately S-supplied soils resulted in either increased GSL concentration or had no effect (Zhao et al. 1993; Bilsborrow et al. 1993). Palaniswamy et al. (1995) also showed that when decreasing the N:S ratio in the growth medium, the phenylethyl isothiocyanate concentration increased in water cress. Similarly, glucoraphanin and glucobrassicin concentrations increased in broccoli florets upon S fertilization (Krumbein et al. 2001). These results are in contrast to that of Vallejo et al. (2003) who reported no effect of S supply on the GSL content of broccoli. However, the soil S status needs to be taken into account since already well S supplied soils will not respond to S fertilization, thus having no effect on GSL synthesis.

Effect of S Supply on GSL and ITC in Chinese Cabbage

Up to date, numerous studies investigated the general effect of the N and S supply on GSLs and their breakdown products in *Brassica* vegetables, but only few were designed to show effects of incremental increases of S fertilization (Schnug 1990; Gerendás et al. 2008a, b, 2009; Schnug and Haneklaus 2016). Thus, the minimum level of S supply after which a significant increase in GSL and GSL pattern becomes obvious was investigated. Furthermore, the effect of increasing S on GSL hydrolysis products, especially the ITCs, was investigated. Additionally, it aimed to clarify whether GSLs might serve as S source under limited S conditions.

As Chinese cabbage (*Brassica rapa* spp. *pekinensis*) is one of the most popular leafy-cabbage vegetables in mainland China and is increasingly being used in the western, Mediterranean and American cuisine, it seemed to be an adequate trial plant besides other *Brassica* vegetables. Chinese cabbage has a long history in being used in Traditional Chinese Medicine, not least due to its known high nutritional value and health promoting effects by thiocyanates and isothiocyanates amongst others (USDA 2015).

Being a S indigent plant, Chinese cabbage responded well to increasing levels of S ($MgSO_4$; S1: 0 g; S2: 0.01 g; S3: 0.03 g; S4: 0.06 g; S5: 0.1 g; S6: 0.3 g; per 6 kg quartz sand; all other nutrients in optimum), following the classical Law of the Optimum (data not shown, see also Gerendás et al. 2008a, 2009. Surprisingly, increasing S supply did not result in a similar steady incline of total S in plant tissue but more successive increase in the order: S1, S2, S3 $<$ S4, S5 $<$ S6 (Fig. 3a). As there are strong interactive effects of S and N in plant metabolism (e.g., nutrient uptake, protein synthesis), it was expected that N is also responding to increasing S, but which was only the case in the highest S treatments (Fig. 3b). The N/S ratio is frequently used to characterize the nutritional status of crops ((Reuter and Robinson 1988; Bergmann 1993) (Fig. 3c; Ahmad et al. 1999; Schonhof et al. 2007; Gerendás et al. 2009).

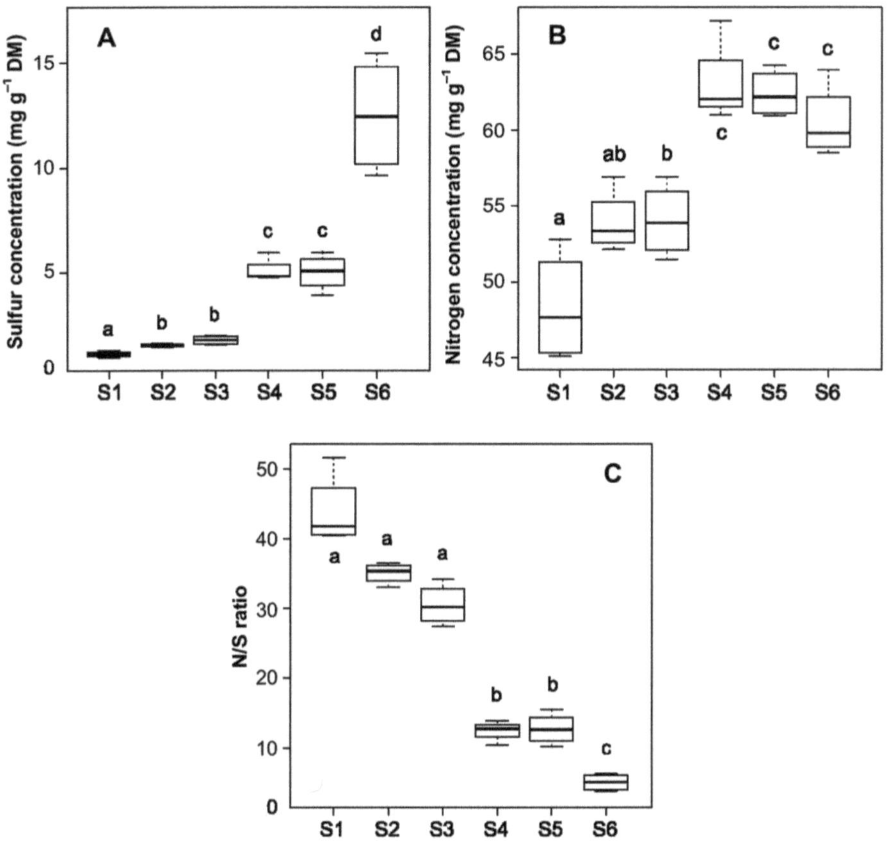

Fig. 3 Effect of sulfur (S) supply on sulfur concentration (**a**), nitrogen concentration (**b**), and N/S ratio (**c**) in *Brassica rapa* spp. *pekinensis* ($MgSO_4$; S1 = 0 g, S2 = 0.01 g, S3 = 0.03 g, S4 = 0.06 g, S5 = 0.1 g, S6 = 0.3 g)

Due to the dependence of GSLs on S, it is not surprising that GSL synthesis is attenuated under limited S conditions (Fig. 4a), but responded well to increasing supply (Mithen et al. 2000). Consequently, also the GSL showed the same successive increase as did the S (Fig. 3a) concentration. However, the GSL pattern varied with enhanced S supply. In total, eight GSLs could be detected: aromatic GSL–glucotropaeolin (GTL); aliphatic GSL–progoitrin (PRO), gluconapin (GNA), glucobrassicanapin (GBN); indole GSL–4-hydroxy glucobrassicin (4OH), glucobrassicin (GBC), 4-methoxy glucobrassicin (4MG), and neoglucobrassicin (NGBS), which all increased upon S supply. This is in line with Gerendás et al. (2008a, b, 2009) in which plants were supplied with moderate S, even though here (Fig. 3) plants were grown without S before the initial S-fertilization.

In this context, aromatic and aliphatic GSLs were more responsive to S (Fig. 4b), which might be attributable to the fact that more thiols are provided under optimum S conditions (Falk et al. 2007). In contrast, it is described that indole GSLs are

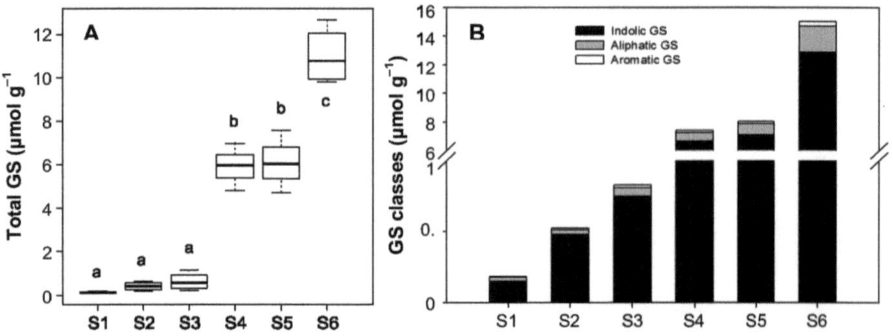

Fig. 4 Effect of increasing sulfur supply on concentration of (**a**) total GSL and (**b**) GSL classes in *Brassica rapa* spp. *pekinensis*. Aromatic GSL–glucotropaeolin; Aliphatic GSL–progoitrin + gluconapin + glucobrassicin; Indolic GSL–glucobrassicanapin + neoglucobrassicin + 4-hydroxyglucobrassicin + 4-methoxyglucobrassicin ($MgSO_4$; S1 = 0 g, S2 = 0.01 g, S3 = 0.03 g, S4 = 0.06 g, S5 = 0.1 g, S6 = 0.3 g)

positively affected by increasing N (Omirou et al. 2009) due to the nitrogenous origin of tryptophan, the precursor amino acid for indole GSLs (Fig. 1) (Kim et al. 2002; Falk et al. 2007). Thus, GBC, NGBS, and 4-MG dominated in S deficient plants, whereas with increasing S supply the pattern changed with a tendency to GBN, GNA, PRO, and GTL. Nonetheless, indole GSLs dominated even under high S supply, possibly due to the constant and sufficient N supply (Fig. 4b).

Based on the positive effect of S fertilization on the GSL content and with regard to possible health promoting effects, it was further questioned whether the ITC profile corresponds to the GSL profile. Like GSLs, the concentration of their breakdown products showed an increasing trend especially in the high S treatments (S4, S5, S6; Fig. 5a). This is in accordance with the results of Gerendás et al. (2008a) who demonstrated that in kohlrabi a low N/S ratio led to a high concentration of GSL breakdown products. However, total concentration of GSL hydrolysis products was lower than detected by Gerendás et al. (2008b) in cress. After all, differences between species are quite likely. Of all GSL breakdown products, a total of four ITCs and one nitrile could be identified, and the profile was dominated by 2-methylpropyl isothiocyanate (MePrITC), followed by 3-butenyl isothiocyanate (ButITC), 2-phenylethyl isothiocyanate (PEITC), and 4-pentenyl isothiocyanate (PentITC) (Fig. 5b–f) resulting from the hydrolysis of 2-methylpropyl GSL, gluconapin, gluconasturtiin, and glucobrassicanapin. All responded in a similar way, showing concentrations tending upwards upon increasing S supply (Fig. 5b–f). Again, relevant concentrations became only obvious under high S conditions (S5, S6).

All identified ITCs have in common that they represent highly volatile GSL breakdown products giving the typical odor of *Brassica* species. Besides antioxidative and herbicidal functions, it is further known that they also have a proven antibacterial activity against, for example, pathogenic bacteria, as well as being cytotoxic (Jang et al. 2010; Khemani et al. 2012). Under health

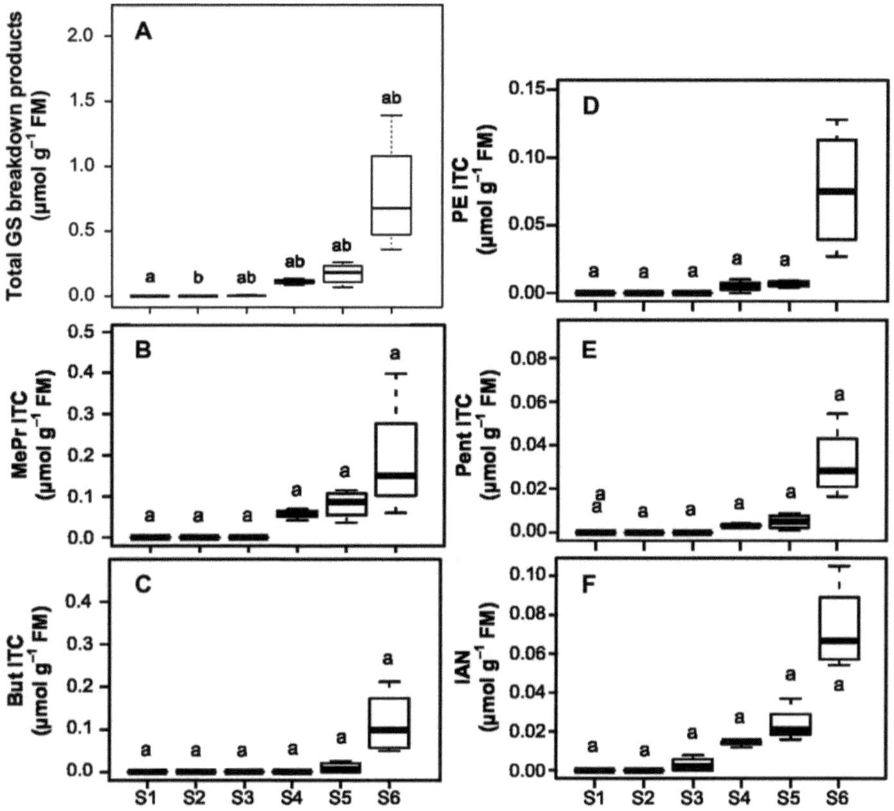

Fig. 5 Effect of sulfur supply on concentration of (**a**) total and (**b–f**) single GSL breakdown products in *Brassica rapa* spp. *pekinensis*. MePrITC–2-methylpropyl isothiocyanate; ButITC–3-butenyl isothiocyanate; PEITC–2-phenylethyl isothiocyanate; PentITC–4-pentenyl isothiocyanate; IAN–indole-3-acetonitrile ($MgSO_4$; S1 = 0 g, S2 = 0.01 g, S3 = 0.03 g, S4 = 0.06 g, S5 = 0.1 g, S6 = 0.3 g)

considerations, especially PEITC has been found to inhibit the lung tumorigenesis induced by a potent tobacco-specific carcinogenic nitrosamine in animals (Chung et al. 1992). It was further in humans found that the consumption of watercress led to a dose dependent urinary excretion of the N-acetylcysteine conjugate of PEITC indicative for a pulmonary carcinogen. In animal studies it is suggested that PEITC might inhibit carcinogen activation by CYP enzymes and further decreases the secretion of inflammatory signaling molecules by white blood cells (Hecht 2000; Conaway et al. 2002).

Despite high concentrations of indole GSL concentration (Fig. 4b), concentrations of their breakdown products were low. This is possibly due to the reaction conditions during hydrolysis. The formation of IAN, instead of indolylmethylisothiocyanate as the primary breakdown product, was promoted by ESP (Brown and Morra 1995) and resulted from the breakdown of indole GSLs such as GBC and NGBS, as well as indol-3-ylmethyl GSL, one of the predominant indole GSLs in *Arabidopsis* but also other *Brassica* species (Burow et al. 2008). This reaction usually runs at the expense of generally more toxic isothiocyanates (Lambrix et al. 2001), thus, even as nitrile IAN can be classified as less harmful compared to other ITCs. Further, GBC can be enzymatically hydrolyzed to indole-3-carbinol, which in turn reacts with L-ascorbic acid to ascorbigen, which can induce phase I and II enzymes that are centrally involved in the detoxification of xenobiotics (Wagner and Rimbach 2009).

Conclusions

As was shown in various studies, fertilization of *Brassica* species, especially with sulfur, results in a higher GSL and ITC content. But it is not only the total concentration which is affected but rather the GSL and ITC pattern that can be changed by targeted S supply. While indolic GSLs dominated under a low S status, aliphatic GSLs clearly prevailed under high S supply, resulting in a corresponding pattern of ITCs. However, indolic GSLs dominated under all conditions. Especially in Chinese cabbage a clear effect on particularly PEITC, a known ITC of anticarcinogenic activity, could be identified. It can be concluded that biofortification with S seems to be an appropriate tool to improve the S status and thus the health promoting effect of *Brassica* vegetables such as Chinese cabbage.

References

Ahmad A, Abraham G, Abdin MZ (1999) Physiological investigation of the impact of nitrogen and sulphur application on seed and oil yield of rapeseed (*Brassica campestris* L.) and mustard (*Brassica juncea* L. Czern. and Coss.) genotypes. J Agron Crop Sci 183:19–25

Bartlet E, Kiddle G, Williams I, Wallsgrove R (1999) Wound-induced increases in the glucosinolate content of oilseed rape and their effect on subsequent herbivory by a crucifer specialist. In: Simpson SJ, Mordue AJ, Hardie J (eds) Proceedings of the 10th international symposium on insect-plant relationships. Springer Netherlands, Dordrecht, pp 163–167

Bergmann W (1993) Ernährungsstörungen bei Kulturpflanzen, 3rd edn. Gustav Fischer Verlag, Jena

Bilsborrow PE, Evans EJ, Zhao FJ (1993) The influence of spring nitrogen on yield, yield components and glucosinolate content of autumn-sown oilseed rape (*Brassica napus*). J Agric Sci 120:219–224

Boddupalli SS, Mein JR, James DR, Lakkanna S (2012) Induction of phase 2 antioxidant enzymes by broccoli sulforaphane: perspectives in maintaining the antioxidant activity of vitamins A, C, and E. Front Genet 3:7

Bones AM, Rossiter JT (1996) The myrosinase-glucosinolate system, its organisation and biochemistry. Physiol Plant 97:194–208

Bonnesen C, Eggleston IM, Hayes JD (2001) Dietary indoles and isothiocyanates that are generated from cruciferous vegetables can both stimulate apoptosis and confer protection against DNA damage in human colon cell lines. Cancer Res 61:6120–6130

Brown PD, Morra MJ (1995) Glucosinolate-containing plant tissues as bioherbicides. J Agric Food Chem 43:3070–3074

Brudenell AJP, Griffiths H, Rossiter JT, Baker DA (1999) The phloem mobility of glucosinolates. J Exp Bot 50:745–756

Burow M, Zhang Z-Y, Ober JA, Lambrix VM, Wittstock U, Gershenzon J, Kliebenstein DJ (2008) ESP and ESM1 mediate indol-3-acetonitrile production from indol-3-ylmethyl glucosinolate in Arabidopsis. Phytochemistry 69:663–671

Charron CS, Saxton AM, Sams CE (2005) Relationship of climate and genotype to seasonal variation in the glucosinolate–myrosinase system. II. Myrosinase activity in ten cultivars of *Brassica oleracea* grown in fall and spring seasons. J Sci Food Agric 85:682–690

Chen SX, Halkier BA (2000) Characterization of glucosinolate uptake by leaf protoplasts of *Brassica napus*. J Biol Chem 275:22955–22960. doi:10.1074/jbc.M002768200

Chen X-J, Zhu Z-J, Ni X-L, Qian Q-Q (2006) Effect of nitrogen and sulfur supply on glucosinolates in *Brassica campestris* ssp. *chinensis*. Agric Sci China 5:603–608

Chung FL, Morse MA, Eklind KI, Lewis J (1992) Quantitation of human uptake of the anticarcinogen phenethyl isothiocyanate after a watercress meal. Cancer Epidemiol Biomark Prev 1:383–388

Conaway CC, Jiao D, Chung F-L (1996) Inhibition of rat liver cytochrome P450 isozymes by isothiocyanates and their conjugates: a structure-activity relationship study. Carcinogenesis 17:2423–2427

Conaway C, Yang Y, Chung F (2002) Isothiocyanates as cancer chemopreventive agents: their biological activities and metabolism in rodents and humans. Curr Drug Metab 3:233–255

De Pascale S, Maggio A, Pernice R, Fogliano V, Barbieri G (2007) Sulphur fertilization may improve the nutritional value of *Brassica rapa* L. subsp. sylvestris. Eur J Agron 26:418–424

Fahey JW, Zalcmann AT, Talalay P (2001) The chemical diversity and distribution of glucosinolates and isothiocyanates among plants. Phytochemistry 56:5–51

Falk KL, Tokuhisa JG, Gershenzon J (2007) The effect of sulfur nutrition on plant glucosinolate content: physiology and molecular mechanisms. Plant Biol 9:573–581

Fallovo C, Colla G, Schreiner M, Krumbein A, Schwarz D (2009) Effect of nitrogen form and radiation on growth and mineral concentration of two Brassica species. Sci Hortic 123:170–177

FAOSTAT (2013) Food and Agricultural Organization of the United Nations, Rome. http://faostat.fao.org/

Gerendás J, Breuning S, Stahl T, Mersch-Sundermann V, Mühling KH (2008a) Isothiocyanate concentration in kohlrabi (*Brassica oleracea* L. Var. *gongylodes*) plants as influenced by sulfur and nitrogen supply. J Agric Food Chem 56(18):8334–8342

Gerendás J, Sailer M, Fendrich M-L, Stahl T, Mersch-Sundermann V, Mühling KH (2008b) Influence of sulfur and nitrogen supply on growth, nutrient status and concentration of benzyl-isothiocyanate in cress (*Lepidium sativum* L.) J Sci Food Agric 88:2576–2580

Gerendás J, Podestát J, Stahl T, Kübler K, Brückner H, Mersch-Sundermann V, Mühling KH (2009) Interactive effects of sulfur and nitrogen supply on the concentration of sinigrin and allyl isothiocyanate in Indian mustard (*Brassica juncea* L.) J Agric Food Chem 57:3837–3844

Glenn MG, Chew FS, Williams PH (1988) Influence of glucosinolate content of Brassica (Cruciferae) roots on growth of vesicular–arbuscular mycorrhizal fungi. New Phytol 110:217–225. doi:10.1111/j.1469-8137.1988.tb00255.x

Griffiths DW, Birch ANE, Macfarlane-Smith WH (1994) Induced changes in the indole glucosinolate content of oilseed and forage rape (*Brassica napus*) plants in response to either turnip root fly (*Delia floralis*) larval feeding or artificial root damage. J Sci Food Agric 65:171–178

Halkier BA, Gershenzon J (2006) Biology and biochemistry of glucosinolates. Annu Rev Plant Biol 57:303–333

Hanschen FS, Lamy E, Schreiner M, Rohn S (2014) Reactivity and stability of glucosinolates and their breakdown products in foods. Angew Chem Int Ed 53:11430–11450

Hecht SS (2000) Inhibition of carcinogenesis by isothiocyanates. Drug Metab Rev 32:395–411
Holst B, Williamson G (2004) A critical review of the bioavailability of glucosinolates and related compounds. Nat Prod Rep 21:425–447
James D, Devaraj S, Bellur P, Lakkanna S, Vicini J, Boddupalli S (2012) Novel concepts of broccoli sulforaphanes and disease: induction of phase II antioxidant and detoxification enzymes by enhanced-glucoraphanin broccoli. Nutr Rev 70:654–665
Jang M, Hong E, Kim G-H (2010) Evaluation of antibacterial activity of 3-butenyl, 4-pentenyl, 2-phenylethyl, and benzyl isothiocyanate in *Brassica* vegetables. J Food Sci 75:M412–M416
Khemani L, Srivastava M, Srivastava S (eds) (2012) Chemistry of phytopotentials: health, energy and environmental perspectives, 1st edn. Springer, Berlin/Heidelberg
Kim S-J, Matsuo T, Watanabe M, Watanabe Y (2002) Effect of nitrogen and sulphur application on the glucosinolate content in vegetable turnip rape (*Brassica rapa* L.) Soil Sci Plant Nutr 48:43–49
Kleinwächter M, Selmar D (2004) A novel approach for reliable activity determination of ascorbic acid depending myrosinases. J Biochem Biophys Methods 59:253–265
Kliebenstein DJ, Kroymann J, Mitchell-Olds T (2005) The glucosinolate–myrosinase system in an ecological and evolutionary context. Curr Opin Plant Biol 8:264–271
Krumbein A, Schonhof I, Rühlmann J, Widell S (2001) Influence of sulphur and nitrogen supply on flavour and health-affecting compounds in Brassicaceae. In: Horst WJ, Schenk MK, Bürkert A et al (eds) Plant nutrition: food security and sustainability of agro-ecosystems through basic and applied research. Springer Netherlands, Dordrecht, pp 294–295
Lambrix V, Reichelt M, Mitchell-Olds T, Kliebenstein DJ, Gershenzon J (2001) The *Arabidopsis* epithiospecifier protein promotes the hydrolysis of glucosinolates to nitriles and influences *Trichoplusia ni* herbivory. Plant Cell 13:2793–2807
McMillan M, Spinks EA, Fenwick GR (1986) Preliminary observations on the effect of dietary brussels sprouts on thyroid function. Hum Exp Toxicol 5:15–19
Mewis I, Appel HM, Hom A, Raina R, Schultz JC (2005) Major signaling pathways modulate *Arabidopsis* glucosinolate accumulation and response to both phloem-feeding and chewing insects. Plant Physiol 138:1149–1162
Mithen RF, Dekker M, Verkerk R, Rabot S, Johnson IT (2000) The nutritional significance, biosynthesis and bioavailability of glucosinolates in human foods. J Sci Food Agric 80:967–984
Omirou MD, Papadopoulou KK, Papastylianou I, Constantinou M, Karpouzas DG, Asimakopoulos I, Ehaliotis C (2009) Impact of nitrogen and sulfur fertilization on the composition of glucosinolates in relation to sulfur assimilation in different plant organs of broccoli. J Agric Food Chem 57:9408–9417
Palaniswamy U, McAvoy R, Bible B, Singha S, Hill D (1995) Phenylethyl isothiocyanate concentration in watercress (*Nasturtium officinale* R.Br.) is altered by the nitrogen to sulfur ratio in hydroponic solution. In: Gustine DL, Flores HE (eds) Phytochemicals and health, vol 15. American Society of Plant Physiologists, Rockville, pp 280–283
Reuter DJ, Robinson JB (1988) Plant analysis – an interpretation manual, 2nd edn. Inkata Press, Sydney
Schnug E (1990) Sulphur nutrition and quality of vegetables. Sulphur Agric 14:3–7
Schnug E, Haneklaus S (2016) Chapter Ten – Glucosinolates – the agricultural story. In: Kopriva S (ed) Glucosinolates, Advances in botanical research, vol 80. Academic, London, pp 281–302
Schnug E, Haneklaus S, Borchers A, Polle A (1995) Relations between sulphur supply and glutathione and ascorbate concentrations in *Brassica napus*. Z Pflanzenernähr Bodenkd 158:67–69
Schonhof I, Krumbein A, Brückner B (2004) Genotypic effects on glucosinolates and sensory properties of broccoli and cauliflower. Food/Nahrung 48:25–33
Schonhof I, Blankenburg D, Müller S, Krumbein A (2007) Sulfur and nitrogen supply influence growth, product appearance, and glucosinolate concentration of broccoli. J Plant Nutr Soil Sci 170:65–72

Schreiner M (2005) Vegetable crop management strategies to increase the quantity of phytochemicals. Eur J Nutr 44:85–94

Steinbrecher A, Linseisen J (2009) Dietary intake of individual glucosinolates in participants of the EPIC-Heidelberg Cohort Study. Ann Nutr Metab 54:87–96

Tripathi MK, Mishra AS (2007) Glucosinolates in animal nutrition: a review. Anim Feed Sci Technol 132:1–27

USDA (2015) USDA nutritional database for standard reference. U.S. Department of Agriculture, Washington, DC

Vallejo F, Tomás-Barberán FA, Benavente-García AG, García-Viguera C (2003) Total and individual glucosinolate contents in inflorescences of eight broccoli cultivars grown under various climatic and fertilisation conditions. J Sci Food Agric 83:307–313

Vermorel M, Heaney RK, Fenwick GR (1988) Antinutritional effects of the rapeseed meals, darmor and jet neuf, and progoitrin together with myrosinase, in the growing rat. J Sci Food Agric 44:321–334

Wagner AE, Rimbach G (2009) Ascorbigen: chemistry, occurrence, and biologic properties. Clin Dermatol 27:217–224

Watzl B, Leitzmann C (2005) Bioaktive Substanzen in Lebensmitteln. Georg Thieme Verlag, Stuttgart

Wolfson JL (1982) Developmental responses of *Pieris rapae* and *Spodoptera eridania* to environmentally induced variation in *Brassica nigra*. Environ Entomol 11:207–213

Zhao F, Evans EJ, Bilsborrow PE, Syers JK (1993) Influence of sulphur and nitrogen on seed yield and quality of low glucosinolate oilseed rape (*Brassica napus* L). J Sci Food Agric 63:29–37

Zhao F, Evans EJ, Bilsborrow PE, Syers JK (1994) Influence of nitrogen and sulphur on the glucosinolate profile of rapeseed (*Brassica napus* L). J Sci Food Agric 64:295–304

Zhou C, Zhu Y, Luo Y (2013) Effects of sulfur fertilization on the accumulation of health-promoting phytochemicals in radish sprouts. J Agric Food Chem 61:7552–7559

Prospects for Agricultural Sulfur Research

Silvia H. Haneklaus, Elke Bloem, and Ewald Schnug

Abstract Only a small number of basic research studies find their way into applied treatments in agricultural production. Advances in plant sulfur research within the past 25 years are documented in the series of the proceedings of the International Sulfur Workshop. Here, key findings in different disciplines such as molecular biology, physiology, agronomy and environment have been published. Based on this substantial knowledge pool the potential significance for future practical applications on production fields, actual knowledge gaps and technology transfer projects for an adaptation to global change and scarcity of fossil fuels has been assessed. An example for each of these contemplations is given. In the first case, the concept of a functional bio-fertilizer, which consists of flexible, tailor-made mixtures of glucosinolate-containing plants against prevailing pathogens and pests is provided. Secondly, it will be shown that research concentrates on sulfur as a minimum factor, but an excessive supply to crops and an imbalanced dietary intake of sulfate by animals and humans unfolds obviously detrimental effects. Last, but not least a conception for the marketing of regional high-quality food products employing organic farming systems is presented which meets future demands of food production.

Introduction

Agricultural sulfur research experienced numerous highlights during the past 25 years. After clean air acts came into force in northern Europe during the 1980s the S supply of agricultural crops literally collapsed and caused yield losses in oilseed rape of more than 80% in remote areas. The contribution of soil organic matter to the overall S supply of crops has long been overestimated and finally in the late 1990s it was shown that soil physical and hydrological together with climatic conditions govern the actual sulfate supply in the root zone (Haneklaus et al. 2008). At the same time aspects of crop quality gained increasing interest

---

S.H. Haneklaus (✉) • E. Bloem • E. Schnug
Federal Research Centre for Cultivated Plants, Institute for Crop and Soil Science, Julius Kühn-Institut (JKI), Bundesallee 50, 38116 Braunschweig, Germany
e-mail: silvia.haneklaus@julius-kuehn.de

L.J. De Kok et al. (eds.), *Sulfur Metabolism in Higher Plants - Fundamental, Environmental and Agricultural Aspects*, Proceedings of the International Plant Sulfur Workshop, DOI 10.1007/978-3-319-56526-2_7

coinciding with a rising awareness and demand for nutritious and constitutional food products. Here, S containing S metabolites as for instance glucosinolates (GSLs) have been attributed a significant role in cancer prevention (Kumar et al. 2015) and S fertilization is a key instrument to increase the GSL content in vegetative and generative plant tissues (Haneklaus et al. 2006a). The significance of the S supply for improving plant health is depicted by the term Sulfur Induced Resistance (SIR) which denotes the reinforcement of the natural resistance of plants against fungal pathogens through triggering the stimulation of metabolic processes involving S by targeted soil-applied fertilizer strategies (Haneklaus et al. 2007). The interested reader may follow up the complete history of S research in the proceedings of the Internal Sulfur Workshop, which has been held every 3 years since 1989 (Anonymous 2015).

In general, the biological know-how is available in order to produce high-yielding crops of prime quality. This leaves the question for a justifiable development of agricultural S research that is promising to enhance food and feedstuff quality, and to reduce the input of pesticides.

From Cell to Crop or from Crop to Cell?

The usual procedure to implement new agro-technical measures is to quantify dose/effect data on lab, then greenhouse and finally field scale as pure effects become visible best if no other biotic and abiotic growth factor interferes with the result. Though the working hypothesis might be confirmed under greenhouse conditions, the outcome on production fields is often blurred and irreproducible. In comparison, agro-technical measures may cause distinct effects in the field, which can, however, not be triggered reliably. A prominent example is the phenomenon of biofumigation. The degradation of GSLs by the enzyme myrosinase delivers not only volatile isothiocyanates (ITCs), but also organic cyanides, nitriles, oxazolidinethiones and ionic ITCs all of which have allelopathic potential. GSLs may be released by root exudates of living plants and unfold their allelopathic effects. Another option is their degradation after decomposition of separated plant parts, and harvest residues. Their effect on soil-borne pathogens and pests is resumed by the term biofumigation. Biofumigation might advance to a promising and ecologically sound alternative for crop protection if its efficiency can be controlled.

GSL content and pattern vary in relation to plant species, plant part, growth stage and S supply (Haneklaus et al. 2006b). The efficiency of GSLs and/or ITCs against soil-borne fungal diseases, nematodes and weeds is related to the kind of pathogen and pathotype. ITCs differed in their nematicidal effect by factor 400; their efficacy was usually higher when exposure time was exalted (Lazzeri et al. 2004). Under laboratory conditions, a fungicidal/fungitoxic effect of ITCs lasted not longer than 6 days; only a continuous exposure reduced colony growth efficiently (Smolinska et al. 2003). Only 1–8% of the potential ITC concentration was found after incorporation of plant material into soil (Morra and Kirkegaard 2002).

The potency of ITCs was found to be distinctly higher under laboratory than field conditions if at all. Soils are open systems with a much higher volume than that of sealed containers in the lab, resulting putatively in a lower ITC concentration in the headspace of pathogens. Additional obstacles under field conditions are that the incorporation of the break crop is not homogenous; the GSL content is lower in plant residues than in younger plant material and degradation of GSLs is incomplete as it requires mechanical disruption to destroy cell structures and adequate water for a sufficiently high myrosinase activity. Under field conditions Smith et al. (2004) found no significant relationship between GSL content in roots of oilseed rape and phytosanitary effects. This is supposed to change with the development of a functional bio-fertilizer, which consists of a flexible, tailor-made blend of GSL-containing plants with view to the prevailing pathogen/nematode, a maximization of the GSL content through advanced cultivation, harvesting and preparation procedures and last but not least a mixed formulation of the fertilizer in terms of a short, medium and long-term release of bio-active compounds (Fig. 1; Bloem et al. 2007, 2013; Haneklaus et al. 2006b). A mixture with elemental S as its fungicidal effect on soil-borne pathogens has been shown to be promising (Yang et al. 2006).

Maximum Limit Values of S Supply in Plants, Animals and Humans

The classes of S supply are well documented for different crop plants in the deficiency and sufficiency range, while less attention has been paid to an excessive S supply. A similar situation prevails for the impact of excessive intake of sulfate on animal and human performance so that this angle of S research deserves intensified attention in future.

Plants

S is commonly evaluated as being highly bio-compliant in such way that excess S neither diminishes productivity, nor impairs quality of the plant products. There are, however, indications that overrated S fertilization may reduce crop yield and that this effect is related to crop type (Haneklaus et al. 2006b).

Important threshold markers for the S supply are: the symptomatological value, which reflects the S concentration below which deficiency symptoms become visible; the critical nutrient value, which stands for the S concentration above which the plant is sufficiently supplied with S for achieving the maximum potential yield or yield reduced by 5%, 10% and 20%; and the toxicological value, which

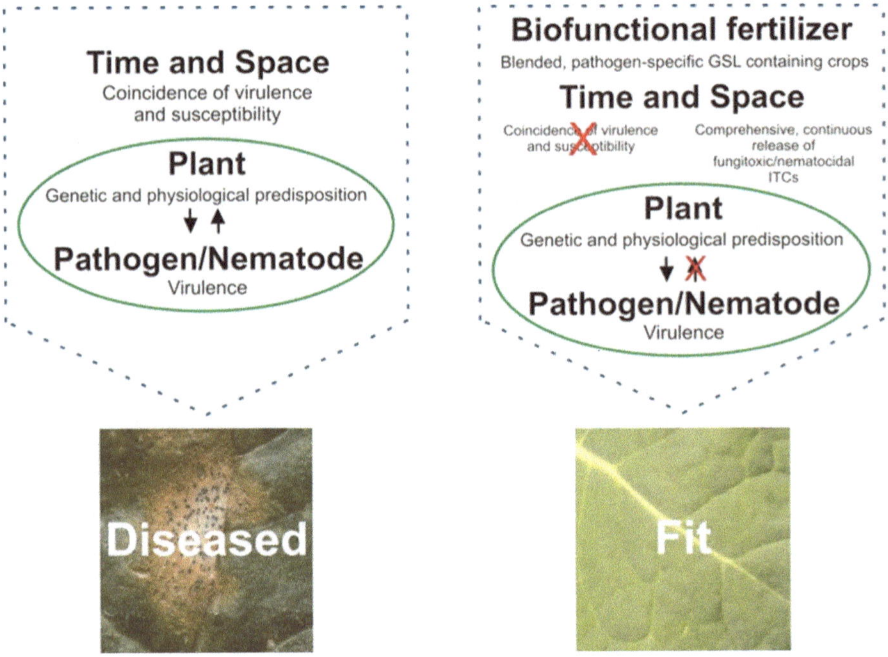

Fig. 1 Mode of action of a GSL-based biofunctional fertilizer

indicates the S concentration above which toxicity symptoms can be observed (Haneklaus et al. 2006a). In literature, toxicological threshold values for S are rare for the reasons mentioned earlier. An excessive S supply can be expected if plants contain more than 2800 mg/kg d.w. sulfate-S; for fodder crops total S concentrations of only 3.2 mg/g S d.w. may be already disproportionate, while the corresponding value for non-*Brassica* vegetables would equal 10 mg/g S d.w. (Haneklaus et al. 2006b). In general, it can be expected that yield depressions occur at lower S concentrations in plants when green matter is harvested such as forage grasses and cabbage. Upper critical S concentrations in cereals, oilseed rape and sugar beet, which result in yield depressions of 10% have been calculated by Haneklaus et al. (2006b, Table 1).

Animals

The dry matter composition of plant products is an important quality parameter of animal feed and foodstuffs. Prominent examples of adverse effects of high S intake on ruminants are polioencephalomalacia (PEM), a neurological disorder and haemolytic anaemia (Gould et al. 2002; Stoewsand 1995). The risk of PEM exists when grass is taken up by animals, which contains more than 0.38% S d.w. (Gould

Table 1 Threshold values for total S concentrations (mg/g S, d.w.) in younger leaves of oilseed rape and sugar beet, and whole above-ground biomass of cereals at start of stem extension and canopy closing

	Deficiency	Sufficiency		Excess
Crop	Symptomatological threshold	Lower critical value (−5% yield)	Maximum yield[a]	Upper critical value (−10% yield)
Cereals	<1.2	3.2	4.0	>7.5
Rape	<2.8[b] and <3.5[c]	5.5	6.5	>14.0
Sugar beet	<1.7	3.0	3.5	>4.5

Adapted from Haneklaus et al. (2006b)
[a]Seed (oilseed rape), grain (cereals), root and sugar (sugar beet) yield
[b]Single low
[c]Double low varieties

et al. 2002). According to Kamphues et al. (2015) noxious H_2S synthesis in the rumen is a possible cause of the disease.

An excessive intake of S and sulfate is associated with loss of appetite, watery feces, secondary trace element deficiency and signs of PEM (Kamphues et al. 2015). In an extended survey the total S content of grass and grass silage did not exceed 8.97 mg/g S d.w., while it was as high as 4.8, 7.6 and 11.8 in extracted soybean meal, oilseed rape meal and dried distillers grains (DDGS; Kamphues et al. 2015). The authors claim the GSL content in rapeseed meal and the sulfuric acid production from DDGS as sources for intolerance. Kamphues et al. (2015) calculated that in extracted rapeseed meal about 60% of the total S content is bound in amino acids, 11% in glucosinolates and 32% as sulfates. Additional S and sulfate sources are drinking water, whey powder, feed additives such as $(NH_2)_2SO_4$ and methionine hydroxy analogue, and acid salts such as sulfates for the prevention of milk fever. In Table 2 the impact of dietary S levels on animal health is summarized (Kamphues et al. 2015).

From viewpoint of plant nutrition the minimum total S and sulfate concentration for maximum crop production in roughage and corn silage are as follows 2.1 and 1.7 mg/g S d.w., and 500 and 150 mg/kg sulfate d.w., respectively (Haneklaus et al. 2006a). Thus it may be concluded that an adequate S supply of these fodder plants will not interfere with animal health while excessive S fertilization may yield adverse effects on animal health.

Humans

The S nutritional status of crops has a significant influence on the nutritive value and sensory features of plant products (Haneklaus et al. 2006b). Methionine is an essential amino acid for monogastric animals and humans; glucosinolates/ITCs have been attributed an anti-carcinogenic potential (Kumar et al. 2015) and S-alk

Table 2 Impact of the S content of feedstuff on health of cattle

Sulfur content (g/kg d.w.)	Significance
≤1	Deficiency in ruminal flora: degrading activity, synthesis of protein and vitamins is reduced
1.5–2	S requirements is covered; in general, positive S-balance in high producing cattle
>2	Linear, reduced feed intake in feed lot cattle with increasing S supply up to 4 g/kg S
>2.5	Absorption/utilization of copper and selenium are impaired (secondary deficiency)
>3	Risks for development of PEM at low fiber supply and high levels of concentrates
>4	Risk for PEM, altered metabolism at moderate fiber supply
>5	Risk for PEM and further disturbances/diseases even at high fiber supply

Extracted from Kamphues et al. (2015)

(en)yl-L-cysteine sulfoxide is active against a broad range of medical disorders. Glutathione (GSH) is another metabolite which content is closely related to the S supply of crop plants and which unfolds a key role in protection against disease in plants, animals and humans because of its protection against oxidative stress and promoting immune functions (Haneklaus et al. 2006a; Grimble 2009). If the dietary intake of S amino acids is insufficient this has serious negative effects on infections and injury (Grimble 2009).

Antinutritives are, for instance, substances, which interfere with the metabolic utilization of minerals (Berdanier 2002). The degradation of GSLs yields for example thiocyanates, isothiocyanates, cyclic S compounds and nitriles, which are goitrogenic. So-called cabbage goiter or struma inhibits the iodine uptake of the thyroid gland in humans (Berdanier 2002).

Less known is presumably the fact that dietary sulfate which is not absorbed in the small intestines is a substrate for sulfate reducing bacteria in the colonic lumen which dissimilate sulfate to H_2S which again is potentially toxic to the colonic mucosa (Florin et al. 1993). At the moment little is known about the formation of H_2S and occurrence of ulcerative colitis and colorectal cancer. Florin et al. (1993) identified beer, bread, sausage and cabbage as the main contributors to a sulfate-rich diet. The same authors suggest that excessive beer consumption may result in rectal carcinoma because of an over-proportionally high sulfate intake. A dose/response relationship exists between luminal H_2S of 0.2 μmol/g perfused into vascularized rat colon and damage to the colonic epithelium (Florin et al. 1993). Such concentrations are found in human feces, too (Florin et al. 1993). Schnug (1988) showed that a close relationship exists between total S and sulfate S content in the vegetative tissue of *Brassica* crops, which increases both, GSL and sulfate content. Fertilizer strategies for the optimization of the nutritional quality of agricultural crops should take possible contrary effects into account.

Regional Organic Products: Tasty and Wholesome

Peak models for the availability of oil and natural gas indicate that shortages are likely in the near future and, combined with extreme weather events, may threaten the productivity of agricultural systems. The production of mineral fertilizer and pesticides, and fuel for agricultural machinery will be affected as will the transport of agricultural commodities from producer to processor to consumer. Lee et al. (2008) concluded that food production will depend upon more localized farming that can operate effectively with reduced inputs and is adaptable to likely increases in weed, pest and disease problems. Consumer awareness about food quality including that of processed food is increasing and the demand accordingly growing. An expansion of organic farming in the outskirts of bigger cities will fully comply with these demands and is the basis of a holistic concept from cultivation to harvest, storage and consumption.

Favorable energy ratios in organic farming can largely be attributed to lower indirect fossil energy needs, since manufactured fertilizers and synthetic pesticides are omitted (Lee et al. 2008). Based upon an exhaustive understanding of biological and physiological processes, it should be possible to replace chemical aid by a scientific appreciation of these processes. An example of how research can provide substantial advances in crop protection is the concept of SIR (see above). Wider crop rotations and cultivation of catch crops enable the growth of herbal plants which are used as basic material for the production of functional fertilizers (see above).

Fresh organic vegetables had significantly better sensory features (taste, scent) than conventional products (Rembialkowska 2000) and Marckmann (2000) expects a higher intake of organic food because of favorable sensory features, which in return may positively affect diseases such type 2 diabetes and obesity. Sensory features of vegetables are linked to the S supply as for instance the pungency of mustard, radish, onion and garlic and S-containing secondary metabolites exert a health-promoting effect. Field trials have shown that the daily intake of only one onion which was fertilized with S would be sufficient to cover a demand of 12 mg (iso)alliin which corresponds to the recommended daily intake of alliin (Bloem et al. 2004). In comparison, it would be necessary to eat two average bulbs from fresh onions grown at a low S supply for a similar dose. Important will be a nutrient regime that fully satisfies the plant demand, while higher S doses which could further improve qualitative parameters should be set aside in order to avoid nutrient imbalances, enrichment of detrimental components such as sulfate and yield losses (see above).

Besides fresh products, consumers want high-quality products to be processed in restaurants. So, new restaurant chains such as "Hans im Glück" (Hans in Luck, http://hansimglueck-burgergrill.de/) promise fresh regional products of prime quality and address nutrition-conscious people. Here, organic products complete requests for sustainability and adaptation to climate change. The practical implementation of the concept seems affordable if organic farmers are not only paid for their products but also for their service with respect to precautionary flood protection by maintaining a high infiltration capacity of soils (Rogasik et al. 2007). The spatial extension of organic farming provides an efficient counter-measure against the adverse effects of the anthropogenic sealing of soils resulting from conurbation expansion.

References

Anonymous (2015) www.plantsulfur.org

Berdanier CD (2002) Food constituents. In: Berdanier D, Feldman EB, Flatt WP, Jeor ATS (eds) Handbook of nutrition and food. CRC Press, Boca Raton, pp 3–98

Bloem E, Haneklaus S, Schnug E (2004) Influence of nitrogen and sulphur fertilisation on the alliin content of onions (*Allium cepa* L.) and garlic (*Allium sativum* L.) J Plant Nutr 27:1827–1839

Bloem E, Haneklaus S, Schnug E (2007) Comparative effects of sulfur and nitrogen fertilization and post-harvest processing parameters on the glucotropaeolin content of *Tropaeolum majus* L. J Sci Food Agric 87:1576–1585

Bloem E, Haneklaus S, Schnug E (2013) *Tropaeolum majus* L life cycle and optimum harvest time for highest glucotropaeolin contents. Phyton 52:305–319

Florin THJ, Neale G, Goretski S, Cummings JH (1993) The sulfate content of foods and beverages. J Food Compos Anal 6:140–151

Gould DH, Dargatz DA, Garry FB, Hamar DH, Ross PF (2002) Potentially hazardous sulfur conditions on beef cattle ranches in the United States. JAVMA 221:673–677

Grimble R (2009) The impact of dietary sulfur amino acid intake in immune function in health and disease. In: Sirko A, DeKok LJ, Haneklaus S, Hawkesford MJ, Rennenberg H, Saito K, Schnug E, Stulen I (eds) Sulfur metabolism in plants, regulatory aspects, significance of sulfur in the food chain, agriculture and environment. Margraf Publisher, Weikersheim, pp 21–34

Haneklaus S, Bloem E, Schnug E, De Kok L, Stulen I (2006a) Sulphur. In: Barker AV, Pilbeam DJ (eds) Handbook of plant nutrition. CRC Press, Boca Raton, pp 183–238

Haneklaus S, Bloem E, Schnug E (2006b) Sulphur interactions in crop ecosystems. In: Hawkesford MJ, DeKok LJ (eds) Sulfur in plants – an ecological perspective. Springer, Dordrecht, pp 17–58

Haneklaus S, Bloem E, Schnug E (2007) Sulfur and plant disease. In: Datnoff L, Elmer W, Huber D (eds) Mineral nutrition and plant diseases. APS Press, St. Paul, pp 101–118

Haneklaus S, Bloem E, Schnug E (2008) History of sulfur deficiency in crops. In: Jez J (ed) Agronomy monograph: sulfur: a missing link between soils, crops, and nutrition, vol 50. ASA-CSSA-SSSA, Madison, pp 45–58

Kamphues J, Dohm A, Zimmermann J, Wolf P (2015) The dietary intake of sulfur in cattle – is there a need for setting upper levels? Lecture at 10th Jubilee plant sulfur workshop, Goslar, Germany, September 1–4, 2015

Kumar G, Tuli HS, Mittal S, Shandilya JK, Tiwari A, Sandhu SS (2015) Isothiocyanates: a class of bioactive metabolites with chemopreventive potential. Tumor Biol 36:4005–4016

Lazzeri L, Leoni O, Manici LM (2004) Biocidal plant dried pellets for biofumigation. Ind Crop Prod 20:59–65

Lee HC, Walker R, Haneklaus S, Philips L, Rahmann G, Schnug E (2008) Organic farming in Europe: a potential major contribution to food security in a scenario of climate change and fossil fuel depletion. VTI Agric For Res 58:145–152

Marckmann P (2000) Organic foods and allergies, cancers and other common diseases – present knowledge and future research. In: Proceeding 13th international IFOAM science conference, p 312

Morra MJ, Kirkegaard JA (2002) Isothiocyanate release from soil-incorporated *Brassica* tissues. Soil Biol Biochem 34:1683–1690

Rembialkowska E (2000) The nutritive and sensory quality of carrots and white cabbage from organic and conventional farms. In: Proceeding 13th international IFOAM science conference, p 297

Rogasik J, Panten K, Schnug E, Rogasik H (2007) Infiltration management factors. In: Lal R (ed) Encyclopedia of soil science, vol 1. Taylor & Francis, New York, pp 867–870

Schnug E (1988) Quantitative und qualitative Aspekte der Diagnose und Therapie der Schwefelversorgung von Raps (*Brassica napus* L.) unter besonderer Berücksichtigung glucosinolatarmer Sorten. Dsc thesis, Agrarwiss Fakultaet, Christian-Albrechts-University, Kiel, p 256

Smith BJ, Kirkegaard JA, Howe GN (2004) Impacts of *Brassica* break-crops on soil biology and yield of following wheat crops. Austr J Agric Res 55:1–11

Smolinska U, Morra MJ, Knudsen GR, James RL (2003) Isothiocyanates produced by *Brassicae* species as inhibitors of Fusarium oxysporum. Plant Dis 87:407–412

Stoewsand GS (1995) Bioactive organosulfur phytochemicals in *Brassica oleracea* vegetables – a review. Food Chem Toxicol 33:537–543

Yang Z, Haneklaus S, De KoK LJ, Schnug E, Singh BR (2006) Plant and soil effect of H_2S and DMS on growth and enzymatic activities of *Rhizoctonia solani* and its implications for sulphur induced resistance (SIR) of agricultural crops. Phyton 46:55–70

Part II
Research Papers

DMSP: Occurrence in Plants and Response to Salinity in *Zea mays*

Ties Ausma, Marko Kebert, Jacqueline Stefels, and Luit J. De Kok

Abstract Dimethylsulfoniopropionate (DMSP) is a secondary sulfur compound that is present in high levels in several marine algal species and some higher plant species. DMSP has also been detected in low levels in some other plant species, though its overall occurrence within the plant kingdom remains unclear. The physiological function of DMSP in saline algae and plants is largely unsolved. It is presumed that the compound can function as a constitutive osmolyte and/or antioxidant. In the current study the occurrence of DMSP within the plant kingdom was analyzed. DMSP was present in nanomolar concentrations in all investigated plant species, indicating that it appears to occur widespread within the plant kingdom. Maize (*Zea mays*) appeared also to be a low-DMSP containing species. Shoot DMSP content in this crop increased upon exposure to salinity, also in the presence of atmospheric H_2S and anoxia. This showed that the DMSP content in maize responded in a similar pattern to salinity exposure which has been observed in DMSP-accumulating plants.

Several marine algal species and some higher plant species contain high levels of the secondary sulfur compound dimethylsulfoniopropionate (DMSP; Stefels 2000, 2007). For instance, in the salt marsh halophyte *Spartina anglica* DMSP accounts for a substantial proportion of the total organic sulfur fraction, ranging from 4 to 50 μmol g FW^{-1} (Otte et al. 2004). Moreover, low levels of DMSP have been detected in some other plant species (Paquet et al. 1994, 1995), though the overall occurrence of the compound within the plant kingdom remains unclear. DMSP can – amongst others – function as a constitutive osmolyte and/or antioxidant (Stefels 2000; Otte et al. 2004). Similar to observations with marine algae, it has been reported that in *Spartina* species DMSP content may increase upon salinity exposure (Dacey et al. 1987; Stefels 2000; Mulholland and Otte 2001). However, in

T. Ausma • J. Stefels • L.J. De Kok (✉)
Laboratory of Plant Physiology, Groningen Institute for Evolutionary Life Sciences, University of Groningen, P.O. Box 11103, 9700 CC Groningen, The Netherlands
e-mail: l.j.de.kok@rug.nl

M. Kebert
Institute of Lowland Forestry and Environment (ILFE), Antona Čehova 13, 21000 Novi Sad, Serbia

L.J. De Kok et al. (eds.), *Sulfur Metabolism in Higher Plants - Fundamental, Environmental and Agricultural Aspects*, Proceedings of the International Plant Sulfur Workshop, DOI 10.1007/978-3-319-56526-2_8

other studies DMSP content of *Spartina* species was not affected by salinity exposure (Van Diggelen et al. 1986; Otte and Morris 1994). In the current study a sensitive method was used to assess the occurrence of DMSP within the plant kingdom. Moreover, the impact of the salt marsh factors salinity, atmospheric H_2S and anoxia on the DMSP content of the non-DMSP accumulator maize (*Zea mays*) was investigated.

In order to assess the occurrence of the secondary sulfur compound DMSP within the plant kingdom, plant shoots from a wide variety of plant species were collected in the northern part of the Netherlands. In addition, nine-day old seedlings of maize (*Zea mays* subsp. *mays*, cv. Ricardinio, Van der Wal, Hoogeveen, The Netherlands) were grown on aerated 25% Hoagland nutrient solutions and exposed to 100 mM NaCl and/or 0.25 μl l^{-1} H_2S for 7 days (Ausma et al. 2017). This experiment was later repeated at anoxic conditions (anoxia; solutions were not aerated, but flushed with N_2 for 10 min on each day). For the determination of the DMSP content, fresh plant material was homogenized in a 0.3 M HCl medium using an Ultra Turrax (10 ml g FW^{-1}). The HCl medium contained 1 μM deuterated DMSP as an internal standard. The homogenate was gently and shortly shaken and 0.3 ml was transferred to a small vial containing 8 ml de-ionized water. Subsequently, one pellet of NaOH was added to the vial, after which the vial was quickly closed: alkalization causes DMSP to convert to the gas dimethylsulfide (DMS). The vials were stored at 4 °C for approximately a week until analysis. The amount of DMS gas in each vial was measured using a proton-transfer-reaction mass spectrometer (PTR-MS; Ionicon, GmbH, Innsbruck, Austria; see Stefels et al. 2012). The recovery of DMSP during the treatment was checked with the internal standard and appeared to be higher than 80%. In plants, DMS can also be formed upon degradation of *S*-methylmethionine or other sulfonium compounds. However, the content of these compounds and their conversion efficiency into DMS under the experimental conditions (i.e. the pH) appears to be negligible (Paquet et al. 1995) and when no NaOH was added to maize homogenates, no DMS was detected.

The secondary sulfur compound DMSP appeared to be present in foliar tissue of all investigated plant species (Table 1). DMSP was detectable in shoot tissue from monocots and dicots and in species that occur in freshwater and saline environments. Moreover, DMSP was detectable in the evolutionary ancient species *Dryopteris dilatata*, a fern, and *Equisetum arvense*, a horsetail. However, all shoots only contained trace quantities of DMSP, ranging from a half to several nanomoles per gram fresh weight (Table 1).

The crop plant maize (a glycophyte) also contained low levels of DMSP: the shoot and root content were 0.5 and 1 nmol g FW^{-1}, respectively (Fig. 1). The DMSP content in both shoot and root was not affected by a 7-day exposure to atmospheric H_2S and/or the absence of oxygen in the root environment (anoxia; see Ausma et al. 2017 for more data). Evidently, in maize DMSP did not act as a metabolic sink of excessive foliarly absorbed and metabolized atmospheric H_2S. A 7-day exposure of maize to 100 mM NaCl did not affect DMSP content in the root, whilst that in the shoot was enhanced threefold, also in the presence of H_2S and anoxia (Fig. 1).

Table 1 The occurrence of DMSP in shoots of natural plant species collected in the northern part of the Netherlands

Species	Family	DMSP content
Monocots		
Narcissus pseudonarcissus	Amaryllidaceae	1.0
Convallaria majalis	Asparagaceae	1.4
Carex appropinquata	Cyperaceae	1.8
Carex echinata	Cyperaceae	0.7
Carex nigra	Cyperaceae	0.5
Iris pseudacorus	Iridaceae	4.8
Juncus effusus	Juncaceae	4.2
Juncus gerardii[a]	Juncaceae	0.7
Alopecurus pratensis	Poaceae	2.3
Bromus hordeaceus	Poaceae	2.8
Dactylis glomerata	Poaceae	9.7
Elytrigia atherica[a]	Poaceae	0.4
Festuca rubra[a]	Poaceae	1.2
Holcus lanatus	Poaceae	0.8
Lolium perenne	Poaceae	1.7
Poa annua	Poaceae	2.4
Puccinellia maritima[a]	Poaceae	1.4
Dicots		
Anthriscus sylvestris	Apiaceae	2.1
Artemisia maritima[a]	Asteraceae	0.9
Leucanthemum vulgare	Asteraceae	0.5
Taraxacum officinale	Asteraceae	0.6
Brassica napus	Brassicaceae	1.0
Cannabis sativa	Cannabaceae	1.0
Quercus robur	Fagaceae	3.9
Limonium vulgare[a]	Plumbaginaceae	1.8
Ranunculus repens	Ranunculaceae	0.9
Other		
Dryopteris dilatata	Dryopteridaceae	1.5
Equisetum arvense	Equisetaceae	1.7

DMSP content is expressed as nmol g^{-1} fresh weight
[a]Saline species

The results show that DMSP appears to occur widespread within the plant kingdom, since the shoots of all investigated species contained detectable levels of this secondary sulfur compound. Moreover, it was evident that similar to observations with halophytic DMSP-accumulating *Spartina* species, exposure of the glycophyte and non-DMSP accumulator maize to salinity resulted in an enhanced DMSP content in the shoot (Dacey et al. 1987; Mulholland and Otte 2001). Whether this enhancement in DMSP content upon salinity exposure has any physiological significance needs to be questioned, since its actual content remained

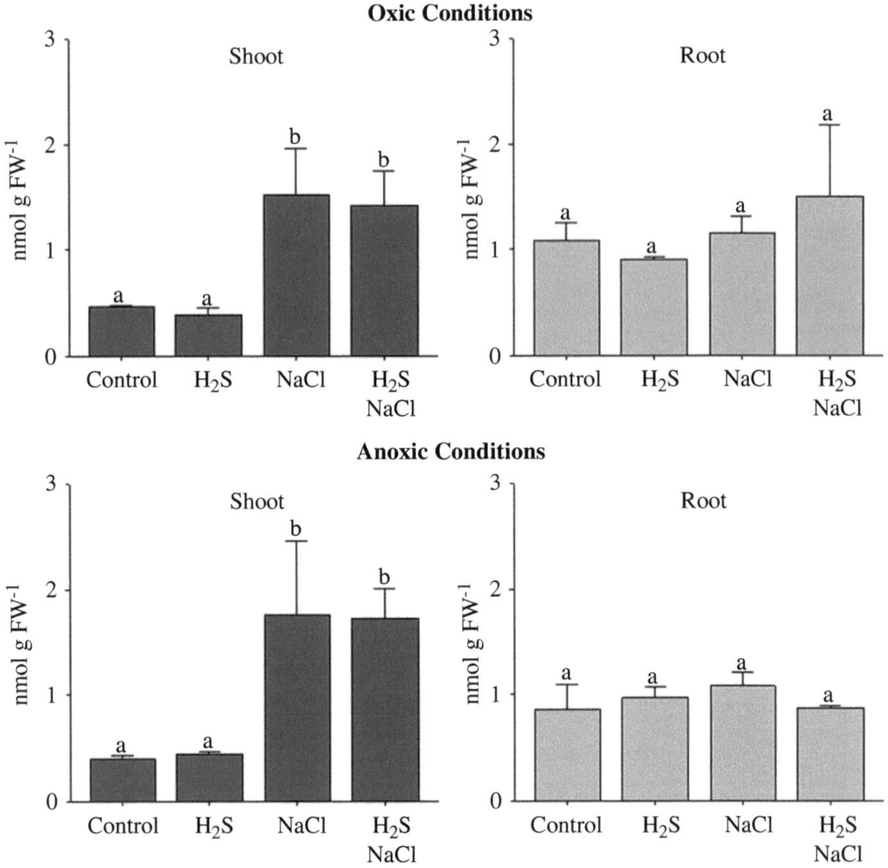

Fig. 1 The impact of atmospheric H_2S, NaCl salinity and anoxia on the DMSP content in the shoot and root of maize (*Zea mays* subsp. *mays*, cv. Ricardinio). Plants were exposed to 0.25 μl l^{-1} H_2S, 100 mM NaCl and anoxia for 7 days (see Ausma et al. 2017 for details). Data represent the mean ± SD of 3 measurements with 3 plants in each. Different letters indicate significant differences ($P \leq 0.01$, Student's t-test)

rather low. Recently, it was reported that in the low-DMSP containing species tomato (*Solanum lycopersicum*) the leaf DMSP content was strongly enhanced upon drought exposure (Catola et al. 2016). Both salinity and drought exposure can cause osmotic stress. The significance of osmotic stress as a trigger of an enhanced DMSP content in foliar tissue deserves to be investigated further.

References

Ausma T, Parmar S, Hawkesford MJ, De Kok LJ (2017) Impact of atmospheric H_2S, salinity and anoxia on sulfur metabolism in Zea mays. In: De Kok LJ, Haneklaus SH, Hawkesford MJ,

Schnug E (eds) Sulfur metabolism in higher plants: fundamental, environmental and agricultural aspects. Springer, Cham, pp 93–101

Catola S, Kaidala Ganesha SD, Calamai L, Loreto F, Ranieri A, Centritto M (2016) Headspace-solid phase microextraction approach for dimethylsulfoniopropionate quantification in *Solanum lycopersicum* plants subjected to water stress. Front Plant Sci 7:1257

Dacey JWH, King GM, Wakeham SG (1987) Factors controlling emission of dimethylsulphide from salt marshes. Nature 330:643–645

Mulholland MM, Otte ML (2001) The effects of nitrogen supply and salinity on [DMSP], [glycine betaine] and [proline] concentrations in leaves of *Spartina anglica*. Aquat Bot 71:63–70

Otte ML, Morris JT (1994) Dimethylsulphoniopropionate (DMSP) in *Spartina alterniflora* Loisel. Aquat Bot 48:239–259

Otte ML, Wilson G, Morris JT, Moran BM (2004) Dimethylsulphoniopropionate (DMSP) and related compounds in higher plants. J Exp Bot 55:1919–1925

Paquet L, Rathinasabapathi B, Saini H, Zamir L, Gage DA, Huang Z, Hanson AD (1994) Accumulation of the compatible solute 3-dimethylsulfoniopropionate in sugarcane and its relatives, but not in other gramineous crops. Aust J Plant Physiol 21:37–48

Paquet L, Lafontaine PJ, Saini HS, James F, Hanson AD (1995) Évidence en faveur de la présence du 3-diméthylsulfoniopropionate chez une large gamme d'Angiospermes. Can J Bot 73:1889–1896

Stefels J (2000) Physiological aspects of the production and conversion of DMSP in marine algae and higher plants. J Sea Res 43:183–197

Stefels J (2007) Sulfur in the marine environment. In: Hawkesford MJ, De Kok LJ (eds) Sulfur in plants, an ecological perspective. Springer, Dordrecht, pp 77–90

Stefels J, Carnat G, Dacey JWH, Goossens T, Elzenga JTM, Tison JL (2012) The analysis of dimethylsulfide and dimethylsulfoniopropionate in sea ice: dry-crushing and melting using stable-isotope additions. Mar Chem 128:34–43

Van Diggelen J, Rozema J, Dickson DMJ, Broekman R (1986) β-3-Dimethylsulphoniopropionate, proline and quaternary ammonium compounds in *Spartina anglica* in relation to sodium chloride, nitrogen and sulphur. New Phytol 103:573–586

Impact of Atmospheric H_2S, Salinity and Anoxia on Sulfur Metabolism in *Zea mays*

Ties Ausma, Saroj Parmar, Malcolm J. Hawkesford, and Luit J. De Kok

Abstract Plants in coastal salt marshes have to deal with salinity, anoxia and excessive reduced sulfur at the same time. Sulfur metabolism is presumed to have significance in plant stress-tolerance. In order to obtain more insight into the physiological significance of sulfur metabolism in plant responses to multiple abiotic stress factors, the glycophyte maize (*Zea mays*) was exposed to atmospheric H_2S, salinity and anoxia. Maize seedlings appeared to be rather unsusceptible for the potentially toxic effects of these stressors. A 7-day exposure to 0.25 μl l^{-1} H_2S and/or anoxia (anoxic root conditions) slightly enhanced plant biomass production, whereas it was not affected upon exposure to 100 mM NaCl. A simultaneous exposure of plants to salinity with H_2S and/or anoxia resulted in a decreased biomass production. The total sulfur content of the shoot and root was hardly affected by H_2S exposure, whereas it was strongly decreased upon anoxia. The total sulfur content of the shoot was decreased upon exposure to salinity. The decreases in total sulfur content could be predominantly ascribed to a decrease in the sulfate content. H_2S exposure only resulted in an enhanced water-soluble non-protein thiol content in shoots, whereas it was not affected by salinity and anoxia. Only a simultaneous exposure of plants to H_2S, salinity and/or anoxia resulted in an enhanced water-soluble non-protein thiol content of the root. Anoxia and salinity exposure induced aerenchyma formation in the root, and the increased root thiol contents might be the result of the direct diffusion of atmospheric H_2S via the stomata through the aerenchyma and subsequent metabolism in the root.

In nature plants are often exposed to multiple abiotic stress factors. For instance, plants in coastal salt marshes not only have to deal with salinity, but also with anoxia and excessive reduced sulfur. Anoxia (*viz.* anoxic root conditions) limits

---

T. Ausma • L.J. De Kok (✉)
Laboratory of Plant Physiology, Groningen Institute for Evolutionary Life Sciences, University of Groningen, P.O. Box 11103, 9700 CC Groningen, The Netherlands
e-mail: l.j.de.kok@rug.nl

S. Parmar • M.J. Hawkesford
Plant Biology and Crop Science Department, Rothamsted Research, Harpenden, Hertfordshire AL5 2JQ, UK

L.J. De Kok et al. (eds.), *Sulfur Metabolism in Higher Plants - Fundamental, Environmental and Agricultural Aspects*, Proceedings of the International Plant Sulfur Workshop, DOI 10.1007/978-3-319-56526-2_9

root respiration and carbohydrates are then broken down via fermentative pathways to yield at least some ATP, which often is insufficient to support optimal plant growth (Yamauchi et al. 2013). An adaptation to anoxia exposure is the formation of aerenchyma: spongy tissue that consists of air spaces and channels in the leaves, stems and roots (Yamauchi et al. 2013). Aerenchyma facilitates an enhanced O_2 diffusion from the shoot to the root and this enables root respiration at anoxic conditions (Yamauchi et al. 2013). Exposure to salinity may negatively affect metabolism and plant growth by affecting the water balance and by causing an accumulation of the toxic cation sodium in the cytosol (Grattan and Grieve 1999; Parida and Das 2005). Exposure to sulfide in soil and atmosphere may also be harmful (Beauchamp et al. 1984; De Kok et al. 2002). Hydrogen sulfide is a potentially phytotoxic gas, since it may react with metalloenzymes (*viz.* cytochrome oxidase; Beauchamp et al. 1984; De Kok et al. 2002). However, at low levels foliarly absorbed H_2S may be metabolized and replace sulfate taken up by the root as sulfur source for growth (De Kok et al. 1997, 1998; Hawkesford and De Kok 2006).

It is presumed that sulfur metabolites may fulfill a role in the tolerance of plants to abiotic and biotic stress (Bloem et al. 2014). A variety of organic sulfur compounds are presumed to have stress-protective functions (Rausch and Wachter 2005). For instance, glutathione and derived metabolites are thought to have diverse functions in stress-protection (Tausz et al. 2004; Noctor et al. 2012). In order to obtain more insight into the physiological significance of sulfur metabolism in plant responses to multiple abiotic stress factors, the glycophyte maize (*Zea mays*) was exposed to atmospheric H_2S, salinity and anoxia.

Maize (*Zea mays* subsp. *mays*, cv. Ricardinio, Van der Wal, Hoogeveen, The Netherlands) was germinated on moistened filter paper at 21 °C in the dark for 2 days. Subsequently, the seedlings were transferred to 15 l containers containing tap water in a climate-controlled room at a day/night temperature of 21 °C/18 °C (±1 °C), a relative humidity of 60–70% and a photoperiod of 14 h at a photon fluence rate of 300 ± 20 μmol m^{-2} s^{-1} (within the 400–700 nm range) at plant height, supplied by Philips GreenPower LED (deep red/white 120) production modules. After 7 days the seedlings were transferred to 13 l containers (ten plant sets per container, six plants per set) containing an aerated (oxic) or non-aerated (anoxic) 25% Hoagland nutrient solution (pH 7.0; for composition see Koralewska et al. 2007; the latter nutrient solution was also daily flushed with N_2 for 10 min). Seedlings were exposed to 0.25 μl l^{-1} H_2S and/or 100 mM NaCl for 7 days. The containers with seedlings were placed in 150 l cylindrical stainless steel cabinets (0.6 m diameter) with a polymethyl methacrylate top. Air exchange inside the cabinets was 40 l min^{-1} and the air inside the cabinets was stirred continuously by a ventilator. Day/night temperatures were 22 °C/19 °C (±1 °C), relative humidity was 20–40% and the photoperiod was 14 h at a photon fluence rate of 340 ± 20 μmol m^{-2} s^{-1} (within the 400–700 nm range) at plant height, supplied by Philips GreenPower LED (deep red/white 120) production modules. Temperature was controlled by adjusting the temperature of the cabinet wall. Temperature, relative humidity and photon fluence rate at plant height were monitored using data loggers (Hobo type U12, Onset Computer Corporation, Bourne, MA, USA). For atmospheric H_2S exposure, pressurized H_2S gas diluted with N_2 gas (1 ml l^{-1}) was injected into the incoming air stream and the concentration in the cabinet was adjusted to the desired concentration of 0.25 μl l^{-1} using mass flow controllers (ASM, Bilthoven, The Netherlands). H_2S concentrations in the cabinets were monitored by an SO_2 analyzer (model 9850) equipped with a H_2S converter (model 8770, Monitor Labs, Measurement Controls Corporation, Englewood, CO, USA). The lids of the containers and plant sets were sealed in order to prevent the absorption of atmospheric H_2S by the solution. On the day before harvest, chlorophyll *a* fluorescence (F_v/F_m ratio) of leaves was measured by using a modulated fluorometer in the morning after a dark-adaptation of at least 1 h (PAM 2000, Walz GmbH, Effeltrich, Germany). Moreover, roots were examined for the presence of aerenchyma under a light microscope. At the day of harvest, shoots and roots of plants were separated and fresh weight was determined. Biomass production was calculated by subtracting final fresh weight from initial fresh weight. For determination of the dry matter content, plants were dried at 80 °C for 24 h. For chlorophyll analysis, pigments were extracted from frozen shoots as described by Shahbaz et al. (2010), and the chlorophyll a and b content were determined according to Lichtenthaler (1987). Water-soluble non-protein thiols were extracted from freshly harvested plants (Shahbaz et al. 2010) and the total water-soluble non-protein thiol content was determined colorimetrically according to De Kok et al. (1988). For determination of total sulfur, sulfate and the mineral nutrient composition, dried shoots and roots were pulverized by using a Retsch Mixer-Mill (type MM2, Haan, Germany). The total sulfur and sulfate content were determined as described by Aghajanzadeh et al. (2016). The mineral nutrient composition was analyzed by inductively coupled plasma mass spectrometry (ICP-MS) and inductively coupled plasma atomic emission spectrometry (ICP-AES) as described by Reich et al. (2016b). Data was statistically analyzed by an unpaired Student's t-test at $P \leq 0.01$.

Maize appeared to be not very susceptible to the toxic effects of H_2S and a 7-day exposure to 0.25 μl l^{-1} H_2S resulted in a slight increase in the plant biomass production, accompanied with a slight decrease in the dry matter content of the shoot and root, whereas the shoot to root ratio was not affected (Table 1). A similar increase in biomass production at low levels of atmospheric H_2S has been observed in some other plant species (Thompson et al. 1979; Durenkamp et al. 2007). The chlorophyll content of the shoot, the chlorophyll a/b ratio and chlorophyll *a* fluorescence, the latter represents the quantum yield of photosystem II, were not affected (Table 1). Moreover, the mineral composition of both shoot and root were hardly affected upon H_2S exposure (Table 3). H_2S exposure did also not affect the total sulfur and sulfate content of both shoot and root (Tables 1 and 3), indicating a down-regulation of the sulfate uptake by the root. The latter was supported by the observation that H_2S exposure resulted in a partial but significant decrease in the sulfate uptake capacity of the root (data not shown). H_2S exposure resulted in an increase in the content of water-soluble non-protein thiols (presumably cysteine and glutathione) in the shoot, whereas that in the root remained unaffected (Table 1). The latter data were similar to observations with other plant species (De Kok et al. 1997, 1998).

Table 1 The impact of atmospheric H_2S and NaCl salinity on biomass production, chlorophyll content, chlorophyll *a* fluorescence and sulfur metabolite content of maize

	Control	H_2S	NaCl	H_2S + NaCl
Plant				
Biomass production	4.06 ± 0.88b	5.80 ± 0.93c	3.72 ± 0.76b	2.92 ± 0.54a
Shoot/root ratio	1.01 ± 0.13a	1.09 ± 0.13a	1.26 ± 0.11b	1.15 ± 0.15ab
Shoot				
DMC	9.6 ± 0.4b	8.8 ± 0.3a	10.9 ± 0.6c	11.2 ± 0.4c
Chl a + b	0.67 ± 0.09a	0.79 ± 0.13a	1.11 ± 0.24b	1.26 ± 0.30b
Chl a/b	2.6 ± 0.4a	2.5 ± 0.5a	2.7 ± 0.2a	2.6 ± 0.1a
F_v/F_m	0.75 ± 0.05a	0.74 ± 0.06a	0.77 ± 0.04a	0.76 ± 0.04a
Thiols	0.45 ± 0.10a	0.59 ± 0.03b	0.48 ± 0.06a	0.75 ± 0.03c
Sulfate	91 ± 6b	61 ± 15ab	52 ± 11a	64 ± 18ab
Total sulfur	168 ± 10b	157 ± 24ab	131 ± 12a	149 ± 7ab
Root				
DMC	7.4 ± 0.3b	6.7 ± 0.3a	7.8 ± 0.4c	8.1 ± 0.6c
Thiols	0.46 ± 0.06ab	0.41 ± 0.04a	0.40 ± 0.05a	0.51 ± 0.03b
Sulfate	81 ± 21ab	89 ± 9ab	108 ± 6b	91 ± 6a
Total sulfur	152 ± 18a	156 ± 22a	166 ± 11a	172 ± 9a
Aerenchyma	Absent	Absent	Present	Present

Plants were exposed to 0.25 $\mu l\,l^{-1}$ H_2S and 100 mM NaCl for 7 days. The initial plant fresh weight was 1.31 ± 0.17 g. Data on biomass production (g fresh weight) and shoot to root ratio (on a fresh weight basis) represent the mean of two experiments with 14 measurements with three plants in each (±SD). Data on dry matter content (DMC; % of fresh weight), chlorophyll content (mg g^{-1} fresh weight) and water-soluble non-protein thiol content (μmol g^{-1} fresh weight) represent the mean of two experiments with 6, 3 and 3 measurements with three plants in each (± SD), respectively. Data on chlorophyll *a* fluorescence (F_v/F_m ratio) represent the mean of two experiments with 12 measurements in each (±SD). Data on the total sulfur and sulfate content (μmol g^{-1} dry weight) represent the mean of 3 measurements with three plants in each (±SD). Different letters indicate significant differences between treatments ($P \leq 0.01$, Student's t-test)

A 7-day exposure of maize to 100 mM NaCl hardly affected plant biomass production and the shoot to root ratio, but resulted in a substantial increase in the dry matter content of both shoot and root, even though aerenchyma had developed in the root (Table 1). Similar to observations with other plant species (Grattan and Grieve 1999; Reich et al. 2016a) exposure to NaCl salinity strongly affected the mineral composition of both shoot and root of maize (Table 3). It resulted in a 70-fold and 28-fold increase in the sodium content in the shoot and root, respectively, accompanied by a strong decrease in the content of calcium, potassium and magnesium in both shoot and root (Table 3). Moreover, salinity exposure resulted in a decrease in the total sulfur content, which could for a greater part be ascribed to a decrease in sulfate content (Tables 1 and 3). Exposure to NaCl salinity resulted in an increase in the chlorophyll content of the shoot, whereas the chlorophyll a/b ratio, chlorophyll *a* fluorescence and the content of the water-soluble non-protein thiols in both shoot and root were unaffected (Tables 1 and 3). Apparently, salinity exposure did not affect the composition and functioning of the photosystems, *viz.* photosynthetic electron transport.

Table 2 The impact of atmospheric H_2S and NaCl salinity on biomass production, chlorophyll content, chlorophyll *a* fluorescence and sulfur metabolite content of maize at anoxic conditions

	Anoxic conditions			
	Control	H_2S	NaCl	H_2S + NaCl
Plant				
Biomass production	5.15 ± 0.72b	7.71 ± 1.12c	2.67 ± 0.47a	2.74 ± 0.33a
Shoot/root ratio	1.42 ± 0.17b	1.77 ± 0.18c	1.23 ± 0.12a	1.23 ± 0.14a
Shoot				
DMC	9.3 ± 0.3a	9.0 ± 0.4a	11.3 ± 0.4b	11.6 ± 0.6b
Chl a + b	0.68 ± 0.10a	0.89 ± 0.08a	0.95 ± 0.10a	1.00 ± 0.03b
Chl a/b	3.1 ± 0.1b	3.0 ± 0.0b	2.7 ± 0.1a	2.8 ± 0.1a
F_v/F_m	0.72 ± 0.07a	0.76 ± 0.09a	0.77 ± 0.04a	0.78 ± 0.06a
Thiols	0.42 ± 0.02a	0.54 ± 0.04b	0.35 ± 0.08a	0.99 ± 0.02c
Sulfate	20 ± 2a	27 ± 5a	36 ± 2b	32 ± 4b
Total sulfur	99 ± 3a	105 ± 3a	108 ± 5a	110 ± 5a
Root				
DMC	6.9 ± 0.5a	6.7 ± 0.5a	7.9 ± 0.6b	8.1 ± 0.7b
Thiols	0.34 ± 0.03a	0.62 ± 0.07b	0.30 ± 0.02a	0.81 ± 0.02c
Sulfate	51 ± 5a	67 ± 12ab	68 ± 3b	57 ± 3a
Total sulfur	111 ± 9a	135 ± 5b	134 ± 5b	127 ± 3ab
Aerenchyma	Present	Present	Present	Present

Plants were exposed to 0.25 μl l^{-1} H_2S and 100 mM NaCl at anoxic root conditions for 7 days. The initial plant fresh weight was 1.23 ± 0.16 g. Data on biomass production (g fresh weight) and shoot to root ratio (on a fresh weight basis) represent the mean of 14 measurements with three plants in each (±SD). Data on dry matter content (DMC; % of fresh weight), chlorophyll content (mg g^{-1} fresh weight) and water-soluble non-protein thiol content (μmol g^{-1} fresh weight) represent the mean of 6, 3 and 3 measurements with three plants in each (±SD), respectively. Data on chlorophyll *a* fluorescence (F_v/F_m ratio) represent the mean of 12 measurements (±SD). Data on the total sulfur and sulfate content (μmol g^{-1} dry weight) represent the mean of 3 measurements with three plants in each (±SD). Different letters indicate significant differences between treatments ($P \leq 0.01$, Student's t-test)

The observed increased plant biomass production upon H_2S exposure did not occur upon a simultaneous exposure to NaCl salinity. Biomass production of these plants was even lower than that of unexposed (control) plants (Table 1). Upon a simultaneous exposure to atmospheric H_2S and NaCl salinity, the chlorophyll content of the shoot, the mineral composition of both shoot and root and the sulfate and total sulfur content of both shoot and root were all quite similar to those observed in the absence of H_2S (Tables 1 and 3). Moreover, the development of aerenchyma in the root was noticeable. Upon a simultaneous exposure to H_2S and NaCl salinity there was not only a substantial increase in the water-soluble non-protein thiol content in the shoot, but also a slight increase in the root (Table 1).

A 7-day exposure of maize to anoxia only slightly affected plant biomass production as compared to oxic conditions and it resulted in an increase in the shoot to root ratio (Tables 1 and 2). Under anoxic conditions, the impact of

Table 3 The impact of atmospheric H_2S and NaCl salinity on the mineral nutrient content of maize

	Control	H_2S	NaCl	H_2S + NaCl
Shoot				
Calcium	198 ± 28b	180 ± 8b	45 ± 4a	46 ± 1a
Copper	0.29 ± 0.03a	0.25 ± 0.03a	0.27 ± 0.01a	0.25 ± 0.02a
Iron	0.74 ± 0.18a	0.96 ± 0.34a	0.87 ± 0.02a	0.86 ± 0.05a
Magnesium	165 ± 19b	152 ± 10b	99 ± 7a	97 ± 1a
Manganese	1.00 ± 0.19a	0.83 ± 0.18a	0.69 ± 0.02a	0.72 ± 0.08a
Molybdenum	0.015 ± 0.001a	0.017 ± 0.003ab	0.026 ± 0.002c	0.020 ± 0.002b
Phosphorus	303 ± 46a	357 ± 37a	309 ± 4a	282 ± 14a
Potassium	1585 ± 38b	1710 ± 26c	1129 ± 157a	907 ± 31a
Sodium	13 ± 3a	19 ± 10a	907 ± 202b	1114 ± 28b
Sulfur	148 ± 20ab	154 ± 7b	123 ± 14a	146 ± 10a
Zinc	0.89 ± 0.13a	0.88 ± 0.24a	0.85 ± 0.13a	0.71 ± 0.07a
Root				
Calcium	257 ± 13b	270 ± 25b	173 ± 11a	140 ± 14a
Copper	0.95 ± 0.16ab	0.67 ± 0.08a	1.02 ± 0.06b	0.92 ± 0.06b
Iron	4.63 ± 0.87ab	2.88 ± 0.37a	6.59 ± 0.41c	4.93 ± 0.58b
Magnesium	217 ± 3b	257 ± 14c	137 ± 13a	112 ± 9a
Manganese	4.09 ± 0.73ab	3.25 ± 0.75a	6.05 ± 0.87b	5.71 ± 0.73b
Molybdenum	0.013 ± 0.001a	0.021 ± 0.003b	0.022 ± 0.001b	0.021 ± 0.002b
Phosphorus	217 ± 2b	239 ± 5c	197 ± 15ab	181 ± 7a
Potassium	1078 ± 38b	1139 ± 14c	378 ± 21a	381 ± 6a
Sodium	72 ± 18a	93 ± 30a	2007 ± 79c	1646 ± 61b
Sulfur	162 ± 9a	174 ± 11a	181 ± 8a	162 ± 7a
Zinc	0.92 ± 0.12b	0.86 ± 0.20ab	0.65 ± 0.12ab	0.53 ± 0.03a

Plants were exposed to 0.25 $\mu l\ l^{-1}$ atmospheric H_2S and 100 mM NaCl for 7 days. Data on the mineral nutrient content ($\mu mol\ g^{-1}$ dry weight) represent the mean of 3 measurements with three plants in each (±SD). Different letters indicate significant differences between treatments ($P \leq 0.01$, Student's t-test)

exposure to atmospheric H_2S, NaCl salinity and their combination on plant biomass production, chlorophyll content, chlorophyll a/b ratio and chlorophyll *a* fluorescence were quite similar to their impact under oxic conditions (Tables 1 and 2). Again, H_2S exposure resulted in an increased plant biomass production (Table 2). However, upon exposure to NaCl salinity and upon a simultaneous exposure to NaCl salinity and H_2S, the plant biomass was reduced and lower than that of unexposed (control) plants (Table 2). Upon exposure to anoxia the development of aerenchyma in the root was observed at all conditions.

Anoxia exposure affected the mineral composition and resulted in a substantial decrease in the total sulfur (Tables 1 and 2), calcium and magnesium content in both shoot and root as compared to oxic conditions (Tables 3 and 4). The decrease in the plant total sulfur content could predominantly be ascribed to a decrease in sulfate

Table 4 The impact of atmospheric H_2S and NaCl salinity on the mineral nutrient content of maize at anoxic conditions

	Anoxic conditions			
	Control	H_2S	NaCl	H_2S + NaCl
Shoot				
Calcium	124 ± 24b	120 ± 2b	47 ± 8a	42 ± 6a
Copper	0.27 ± 0.04a	0.22 ± 0.01a	0.30 ± 0.07a	0.22 ± 0.02a
Iron	0.64 ± 0.04a	0.89 ± 0.06b	0.85 ± 0.19ab	0.93 ± 0.13b
Magnesium	107 ± 11ab	103 ± 3b	101 ± 20ab	87 ± 6a
Manganese	0.77 ± 0.12a	0.62 ± 0.03a	0.87 ± 0.14a	0.62 ± 0.03a
Molybdenum	0.013 ± 0.001a	0.022 ± 0.001c	0.020 ± 0.002bc	0.018 ± 0.001b
Phosphorus	246 ± 24ab	290 ± 14b	217 ± 10ab	208 ± 6a
Potassium	1484 ± 100b	1703 ± 37c	642 ± 76a	705 ± 45a
Sodium	7 ± 1a	7 ± 0a	1097 ± 120b	945 ± 121b
Sulfur	75 ± 5a	90 ± 0b	86 ± 7ab	98 ± 8b
Zinc	0.56 ± 0.14a	0.49 ± 0.06a	0.93 ± 0.26a	0.64 ± 0.12a
Root				
Calcium	131 ± 15a	127 ± 7a	135 ± 39a	102 ± 15a
Copper	1.47 ± 0.17a	1.68 ± 0.23a	1.93 ± 0.52a	1.77 ± 0.25a
Iron	2.25 ± 0.20a	2.51 ± 0.23ab	3.03 ± 0.25bc	3.66 ± 0.37c
Magnesium	135 ± 9b	155 ± 9b	128 ± 22ab	112 ± 5a
Manganese	4.63 ± 0.51a	5.37 ± 0.65a	7.73 ± 1.49a	6.71 ± 1.00a
Molybdenum	0.013 ± 0.002a	0.029 ± 0.005b	0.018 ± 0.005ab	0.018 ± 0.002a
Phosphorus	212 ± 6a	224 ± 13a	198 ± 21a	195 ± 20a
Potassium	1156 ± 67b	1264 ± 25b	411 ± 27a	464 ± 44a
Sodium	49 ± 8a	38 ± 1a	1437 ± 108b	1463 ± 103b
Sulfur	101 ± 6a	132 ± 9b	118 ± 12ab	129 ± 7b
Zinc	0.78 ± 0.13ab	0.82 ± 0.09b	0.63 ± 0.11ab	0.54 ± 0.04a

Plants were exposed to 0.25 μl l^{-1} H_2S and 100 mM NaCl at anoxic root conditions for 7 days. Data on the mineral nutrient content (μmol g^{-1} dry weight) represent the mean of 3 measurements with three plants in each (±SD). Different letters indicate significant differences between treatments ($P \leq 0.01$, Student's t-test)

content (Tables 1 and 2). Again, H_2S exposure hardly further affected the mineral composition of plants under anoxic conditions (Table 4). Similar to observations under oxic conditions, NaCl salinity strongly affected the mineral composition of both shoot and root of maize (Table 4). It resulted in an increase in the sodium content in both shoot and root, accompanied by a strong decrease in the content of calcium in the shoot and potassium in both shoot and root (Table 4). In contrast to oxic conditions, H_2S exposure resulted not only in an increase in the content of water-soluble non-protein thiols in the shoot, but also in the root (Table 2). Again, exposure to NaCl salinity hardly affected the water-soluble non-protein thiol content in both shoot and root (Table 2). However, a simultaneous exposure of maize to H_2S and NaCl salinity resulted in a more strongly enhanced water-soluble non-protein thiol content in both shoot and root than that observed in the absence of NaCl salinity (Table 2).

From the current study it was evident that maize seedlings were rather unsusceptible to the potentially toxic effects of exposure to H_2S, NaCl salinity and anoxia. Only a combination of NaCl salinity with H_2S and/or anoxia negatively affected plant biomass production. This may indicate that under these conditions, the combination of abiotic stress factors negatively affected the balance between carbon use for structural growth and carbon use for the maintenance respiration required to alleviate the negative effects of the stressors. Furthermore, it was evident that not only anoxia but also NaCl salinity induced the formation of aerenchyma in the roots of maize. It was previously observed that several abiotic stress factors might induce aerenchyma formation in roots (Bouranis et al. 2003; Evans 2003). The enhanced water-soluble non-protein thiol content in the root of maize upon the simultaneous exposure to H_2S, salinity and/or anoxia might be the result of the direct diffusion of atmospheric H_2S via the stomata through the aerenchyma and subsequent metabolism in the root.

Acknowledgement Rothamsted Research is supported via the 20:20 Wheat® Programme by the UK Biotechnology and Biological Sciences Research Council.

References

Aghajanzadeh T, Hawkesford MJ, De Kok LJ (2016) Atmospheric H_2S and SO_2 as sulfur sources for *Brassica juncea* and *Brassica rapa*: regulation of sulfur uptake and assimilation. Environ Exp Bot 124:1–10

Beauchamp RO, Bus JS, Popp JA, Boreiko CJ, Andjelkovich DA (1984) A critical review of the literature on hydrogen sulfide toxicity. Crit Rev Toxicol 13:25–97

Bloem E, Haneklaus S, Schnug E (2014) Milestones in plant sulfur research on sulfur-induced-resistance (SIR) in Europe. Front Plant Sci 5:779

Bouranis DL, Chorianopoulou SN, Siyiannis VF, Protonotarios VE, Hawkesford MJ (2003) Aerenchyma formation in roots of maize during sulphate starvation. Planta 217:382–391

De Kok LJ, Buwalda F, Bosma W (1988) Determination of cysteine and its accumulation in spinach leaf tissue upon exposure to excess sulfur. J Plant Physiol 133:502–505

De Kok LJ, Stuiver CEE, Rubinigg M, Westerman S, Grill D (1997) Impact of atmospheric sulfur deposition on sulfur metabolism in plants: H_2S as sulfur source for sulfur deprived *Brassica oleracea* L. Bot Acta 110:411–419

De Kok LJ, Stuiver CEE, Stulen I (1998) Impact of atmospheric H_2S on plants. In: De Kok LJ, Stulen I (eds) Responses of plant metabolism to air pollution and global change. Backhuys Publishers, Leiden, pp 51–63

De Kok LJ, Stuiver CEE, Westerman S, Stulen I (2002) Elevated levels of hydrogen sulfide in the plant environment: nutrient or toxin. In: Omasa K, Saji H, Youssefian S, Kondo N (eds) Air pollution and plant biotechnology: prospects for phytomonitoring and phytoremediation. Springer, Tokyo, pp 3–11

Durenkamp M, De Kok LJ, Kopriva S (2007) Adenosine 5'-phosphosulphate reductase is regulated differently in *Allium cepa* L. and *Brassica oleracea* L. upon exposure to H_2S. J Exp Bot 58:1571–1579

Evans DE (2003) Aerenchyma formation. New Phytol 161:35–49

Grattan SR, Grieve CM (1999) Salinity-mineral nutrient relations in horticultural crops. Sci Hortic 78:127–157

Hawkesford MJ, De Kok LJ (2006) Managing sulphur metabolism in plants. Plant Cell Environ 29:382–395

Koralewska A, Posthumus FS, Stuiver CEE, Buchner P, De Kok LJ (2007) The characteristic high sulfate content in *Brassica oleracea* is controlled by the expression and activity of sulfate transporters. Plant Biol 9:654–661

Lichtenthaler HK (1987) Chlorophylls and carotenoids: pigments of the photosynthetic biomembranes. Methods Enzymol 148:350–382

Noctor G, Mhamdi A, Chaouch S, Han Y, Neukermans J, Marquez-Garcia B, Queval G, Foyer CH (2012) Glutathione in plants: an integrated overview. Plant Cell Environ 35:454–484

Parida AK, Das AB (2005) Salt tolerance and salinity effects on plants: a review. Ecotoxicol Environ Saf 60:324–349

Rausch T, Wachter A (2005) Sulfur metabolism: a versatile platform for launching defence operations. Trends Plant Sci 10:503–509

Reich M, Aghajanzadeh T, Helm J, Parmar S, Hawkesford MJ, De Kok LJ (2016a) Chloride and sulfate salinity differently affect biomass, mineral nutrient composition and expression of sulfate transport and assimilation genes in *Brassica rapa*. Plant Soil 411:1–14

Reich M, Shahbaz M, Prajapati DH, Parmar S, Hawkesford MJ, De Kok LJ (2016b) Interactions of sulfate with other nutrients as revealed by H_2S fumigation of Chinese cabbage. Front Plant Sci 7:541

Shahbaz M, Tseng MH, Stuiver CEE, Koralewska A, Posthumus FS, Venema JH, Parmar S, Schat H, Hawkesford MJ, De Kok LJ (2010) Copper exposure interferes with the regulation of the uptake, distribution and metabolism of sulfate in Chinese cabbage. J Plant Physiol 167:438–446

Tausz M, Sircelj H, Grill D (2004) The glutathione system as a stress marker in plant ecophysiology: is a stress-response concept valid? J Exp Bot 55:1955–1962

Thompson CR, Kats G, Lennox RW (1979) Effects of fumigating crops with hydrogen sulfide and sulfur dioxide. Calif Agric 33:9–10

Yamauchi T, Shimamura S, Nakazono M, Mochizuki T (2013) Aerenchyma formation in crop species: a review. Field Crop Res 152:8–16

Salinity Influences Single Glucosinolate Content in the Halophyte *Lepidium latifolium*

Christian Boestfleisch, Johann Hornbacher, Annekathrin Rumlow, and Jutta Papenbrock

Abstract The influence of salinity on the biosynthesis of secondary metabolites with a focus on single glucosinolates (GSLs) was investigated in *Lepidium latifolium* L., which is a plant species rich in antioxidants. Mature plants were subjected to 0, 15, 22.5, and 35 Practical Salinity Units (PSU) for 1–4 weeks. While phenols, flavonoids, and the oxygen radical absorbance capacity (ORAC) increased with increasing salinity, the ascorbate concentration did not follow a specific pattern. The concentration of single GSLs was influenced by salinity in different ways: While the concentration of aliphatic GSLs like glucoiberin and sinigrin increased, the concentration of aromatic GSLs such as glucobrassicin decreased under salinity stress. Salinity increased the total GSL concentration significantly with sinigrin being the major contributing GSL. The exact molecular role of the different GSLs in abiotic stress defense needs further analysis.

The halophyte *Lepidium latifolium* L. belongs to the Brassicaceae family, known for their high abundance of glucosinolates (GSLs). The role of GSLs and their break-down products under biotic stress, especially their defense function against herbivores, insects and pathogens, is well known: upon tissue damage myrosinase hydrolyzes GSLs, releasing thiocyanates, isothiocyanates and nitriles (Agrawal and Kurashige 2003; Hopkins et al. 2009; Manici et al. 1997; Rask et al. 2000; Tierens et al. 2001). Stress caused by abiotic factors like drought has different effects on GSL composition and content. In several studies (Mewis et al. 2012; Radovich et al. 2005; Schreiner et al. 2009; Tong et al. 2014) either an increase of aliphatic GSLs under drought stress or a decrease, no effect or a less pronounced increase of aromatic including (indolic) GSLs in *Brassica* species and in *Arabidopsis thaliana* L. was shown. Other studies of GSLs in drought-stressed *Brassica* species showed a reduction or insignificant changes in the GSL content (Khan et al. 2010; Robbins et al. 2005). In both studies the content of indolic GSLs was predominant.

Only a few studies exist on the effect of salinity on GSLs. In two *Brassica napus* cultivars the total GSL concentration increased under salinity (Qasim et al. 2003).

C. Boestfleisch • J. Hornbacher • A. Rumlow • J. Papenbrock (✉)
Institut für Botanik, Gottfried Wilhelm Leibniz Universität, Herrenhäuser Straße 2, D-30419 Hannover, Germany
e-mail: jutta.papenbrock@botanik.uni-hannover.de

L.J. De Kok et al. (eds.), *Sulfur Metabolism in Higher Plants - Fundamental, Environmental and Agricultural Aspects*, Proceedings of the International Plant Sulfur Workshop, DOI 10.1007/978-3-319-56526-2_10

Brassica oleracea showed an increased total GSL content under the influence of salt (López-Berenguer et al. 2008, 2009). In these studies measurements of total GSL contents were performed, while other studies also analyzed single GSLs. In radish sprouts, for example, one aliphatic GSL was predominant and its concentration increased by increased salinity (Yuan et al. 2010). Zaghdoud et al. (2012) found that the total GSL content was not altered by salinity in a *B. oleracea* cultivar, while another cultivar showed a decrease in the total GSL content. This cultivar showed a decrease in indolic GSLs but an increase in aliphatic GSLs. Bloem et al. (2014) revealed no significant changes in contents of the aromatic GSL glucotropaeolin in salt stressed *Tropaeolum majus*. All studies previously mentioned applied low salinity to glycophytes, while in this study high salinities to a halophyte species were applied.

Several antioxidants are altered in their accumulation by salinity, often enhanced under stress conditions, like the non-enzymatic metabolites ascorbate, glutathione, carotenoids, tocopherols, and phenolics. These antioxidants serve, besides the enzymatic antioxidants, to scavenge or detoxify reactive oxygen species (ROS) induced by stresses like salinity (Noctor and Foyer 1998; Sharma et al. 2010).

In this study the salt-tolerant species *Lepidium latifolium* was investigated as we were interested in the GSL concentration in relation to salinity and to identify the time point when the GSL content was highest after beginning of the salt treatment. Sometimes classified as a halophyte, *L. latifolium* is native to southern Europe and Asia (Kaur et al. 2013; Zhao et al. 2010). Its tolerance to salt allows it to grow along the coastline but it is also found in the cold Himalayan region (Gupta et al. 2013; Kaur et al. 2013). It has been identified as an invasive species in North America (Francis and Warwick 2007). Fortunately, the GSL spectrum of *L. latifolium* was analyzed previously indicating eight GSLs with sinigrin being the dominant GSL (Kaur et al. 2013). The use as a vegetable and the medicinal utilization were also previously described (Kaur et al. 2013; Navarro et al. 1994). The influence of salinity on biomass, oxygen radical absorbance capacity (ORAC), ascorbate, phenolics and flavonoids, and GSL spectrum and contents of *L. latifolium* influenced by salt were analyzed in this study. The concentration of aromatic GSLs decreased or showed no clear reaction, whereas indolic GSLs decreased and aliphatic GSLs increased or were indifferent towards salinity stress. All other metabolites except ascorbate that showed varying responses were positively affected by salinity. Thus ORAC increased accordingly. The duration of stress application had an influence on the concentration of some of the metabolites analyzed.

Lepidium latifolium seeds (Rühlemann's Kräuter und Duftpflanzen, Horstedt, Germany) were germinated on propagation soil (Einheitserde, Einheitserdewerk Hameln-Tündern, Germany). After a period of 5 weeks, plants were transplanted to sand of 0–2 mm grain size (Hornbach, Hannover, Germany). During the nursing time, the plants were watered with modified Hoagland solution (Epstein 1972). Mature plants were grown under greenhouse conditions at around 22 °C. Sodium vapor lamps (SON-T Agro 400, Philips, Amsterdam, Netherlands) served as an additional light source, providing a photoperiod of 14 h light and a quantum fluence rate of 350 μmol m^{-2} s^{-2}. Finally, the plants were transferred to aerated containers with 13.5 L solution containing 3.57 mM $NaNO_3$, 316 mM $H_2NaPO_4 \times H_2O$ and 23.5 mM Fe-EDDHA (5.7%) (Duchefa, Haarlem, Netherlands). After 1 week of acclimatization to the hydroponic culture, the sea salt mixture (Seequasal GmbH, Münster, Germany) was added stepwise by an increase of 0.75 PSU every day to the desired concentrations of 0, 15, 22.5, and 30 PSU. Four plants of each salinity treatment were harvested at the time the cultivation solutions reached their final concentration (0 weeks) and 1, 2, 3 and 4 weeks after induction. Whole plants (shoot including leaves) were frozen in liquid nitrogen and stored at −80 °C for further analysis. The metabolite extraction and the determination of total phenols, total flavonoids, oxygen radical absorbance capacity (ORAC) and ascorbic acid were performed as described by Boestfleisch et al. (2014). For the determination of GSLs, frozen, ground leaf material was freeze-dried. One milliliter of 80% methanol was added to 10 mg dried plant material. The sample was placed on a shaker until homogenization and then centrifuged for 5 min at 13,000 *g*. The pellet was re-extracted in the same way and the supernatants were combined. The supernatant was loaded onto a column (QIAGEN GmbH, Hilden, Germany) containing 2 ml of a 5% (w/v) suspension of DEAE Sephadex A25 (Sigma-Aldrich, Taufkirchen, Germany) in 0.5 M acetic acid (pH 5). The column was then flushed with 10 ml of HPLC-grade H_2O and 4 ml of 0.02 M acetic acid (pH 5). For desulfating the GSLs overnight at room temperature, 50 μl of sulfatase (Sigma-Aldrich) solution (Thies 1979) was added to 450 μl 0.02 M acetic acid (pH 5) and loaded onto the column as well. Desulfated GSLs were eluted 3 times with 2 ml HPLC-H_2O. Samples were dried in a vacuum centrifuge overnight and resolved in 300 ml HPLC-H_2O. Analysis was performed with HPLC system (Knauer, Berlin, Germany) equipped with an Ultra AQ C-18 column (150 x 4.6 mm, 5 μm particle size) (Restek GmbH, Bad Homburg, Germany). For measuring the samples, a volume of 50 μl was injected. A water (solvent A)-acetonitrile (solvent B) gradient at a flow rate of 1 ml min^{-1} at room temperature was used. Following gradient was applied: 100% A (6 min), 100–70% A (27 min), 70–40% A (0.1 min), 4.9 min 40% A, 40–100% A (0.1 min), and 19.9 min 100% A. Eluents were monitored at a wavelength of 229 nm. Identification of desulfated GSLs was achieved by comparing the retention time with commercially available GSLs (PhytoLab GmbH and Co. KG, Vestenbergsgreuth, Germany) that were treated the same way as the samples. By means of standard curves of these references, desulfated GSLs were quantified. Integration of peaks and elaboration of data were performed using ChromGate Client/Server Version 3.3.1 (Knauer, Berlin, Germany). GSLs were calculated as the mean of four biological replicates with the standard deviation of these four biological replicates. The total amount of GSLs was calculated as the sum of all individual GSLs. To asses precision and reproducibility of GLS analysis, four technical replicates were prepared by measuring GSL contents in each plant sample four times. The standard deviation relative to the individual GSL content was calculated in these replicates. Values were tested for significance ($p = 0.05$) with an analysis of variance (ANOVA) using R (version 3.2.2), displaying significant differences between metabolites corresponding to PSU values at different harvest times.

At the starting point there were no significant differences in biomass production of *L. latifolium* between the different salinities, but after 1 week there was a significant difference between plants grown at 22.5 and 30 PSU and plants grown at 15 PSU, which produced the highest biomass (Fig. 1a). The difference in biomass at the different salinities increased towards week 2 where the highest increase in biomass was observed in plants grown at 0 PSU. This effect was significantly higher than in plants grown at 22.5 and 30 PSU. Plant growth decreased with increasing salinity at this point of experimentation. After 3 and 4 weeks the increase in biomass reached a maximum for plants grown at 15 PSU, followed by 0 PSU. Plants grown at 30 PSU showed the lowest increase in biomass. The ORAC of *L. latifolium* grown at salt stress (30 PSU) was highest at all sampling dates compared to the other levels of salinity (Fig. 1b). The ORAC maximum was determined at week 0 in plants grown at 30 PSU and decreased slowly towards the 4th week. At week 0 and week 1, ORAC values in plants grown at 30 PSU were significantly higher compared to plants grown at 15 and 0 PSU. This difference decreased, but ORAC values in plants grown at 30 PSU were still significantly higher compared to plants grown at 15 PSU. In the 3rd week plants grown at 30 PSU had again a significantly higher concentration compared to lower levels of salinity. In the 4th week there was no significant difference in the ORAC. The total phenol concentration of plants grown at 30 PSU was the highest during the time of the experiment (Fig. 1c). It was significantly higher than the concentration of plants grown at 0 PSU at all sampling dates with the exception of week 3. In most cases plants grown at 0 PSU had the lowest total phenol concentration followed by 15 PSU. Higher salinity concentrations yielded higher total phenol concentrations. This effect was significant over time except for the 3rd week when plants grown at 0 and 30 PSU had higher total phenol concentrations than plants grown at 15 and 22.5 PSU.

Plants grown at higher salinities of 30 and 22.5 PSU produced higher total flavonoid values than plants grown at 0 and 15 PSU (Fig. 1d). There was a significant difference between plants grown at 30 and 0 PSU at all sampling dates. The differences in total flavonoid values between plants grown in high and low salinities were greatest at the start of experimentation and after 1 week, and became smaller after 3 and 4 weeks.

Small differences in the ascorbate concentration were detected at the beginning of the experiment but plants grown at 0 PSU had a lower concentration compared to plants grown at other salinities (Fig. 1e). In the 1st week plants grown at 0 and 15 PSU had insignificantly lower ascorbate values than plants grown at 22.5 and 30 PSU. Plants grown at 15 and 30 PSU had lower ascorbate values compared to plants grown at 0 and 22.5 PSU in the 2nd week. In the 3rd week there was a significant decline in the ascorbate concentration from plants grown at 0 to plants grown at 30 PSU. In the 4th week plants grown at 0 PSU showed significantly lower values than plants grown at all other salinities. The highest ascorbate concentration in this week was detected in plants grown at 22.5 PSU followed by 15 PSU. The ascorbate content in plants grown at these salinities were significantly higher compared to plants grown at 0 PSU at this point of time.

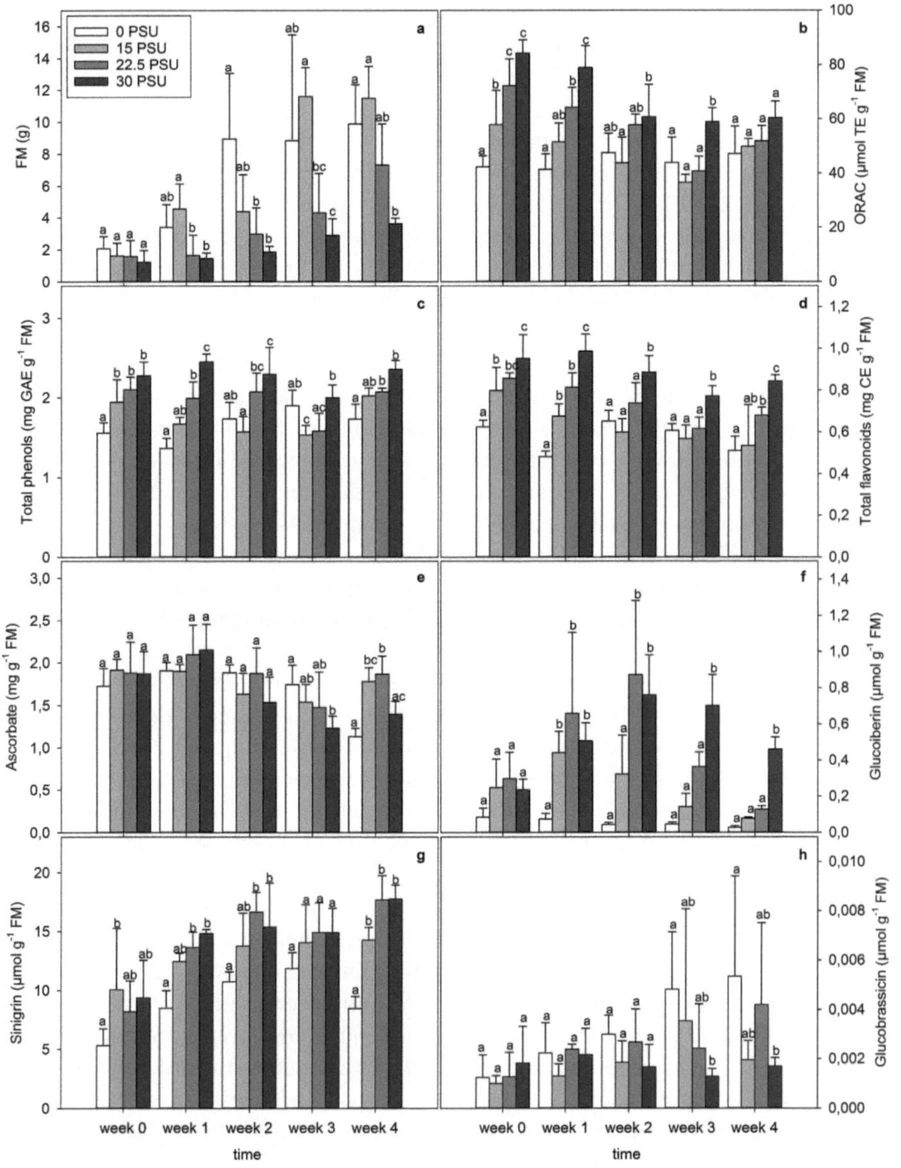

Fig. 1 Biomass production (**a**), oxygen radical absorbance capacity (ORAC, **b**) and contents of total phenols (**c**), total flavonoids (**d**), ascorbate (**e**), glucoiberin (**f**), sinigrin (**g**), glucobrassicin (**h**) of *L. latifolium* plants (n = 4). Six-week-old plants were placed into aerated containers and after an acclimatization time of 1 week the salinity was increased by 0.75 PSU every day to the concentrations of 0 PSU, 15 PSU, 22.5 PSU and 30 PSU. Fresh material was harvested upon reaching the targeted salinities (week 0) followed by a weekly interval. Different letters indicate significant differences (p = 0.05) between different PSU values within one point of time

Salinity caused diverging effects on individual GSLs. There are three ways how the GSL concentration was influenced by salinity. Figure 1f shows the first way, an increase in the concentration of glucoiberin. At the beginning of the experiment there was a trend for a higher glucoiberin concentration in plants grown at 22.5 PSU, but this effect was not significant. This trend continued in the 1st week when plants grown at all salinities had a significantly higher concentration compared to plants grown at 0 PSU. In the 2nd week plants grown at 22.5 and 30 PSU had a significantly higher glucoiberin concentration than plants grown at 0 PSU. One week later the significant maximum in glucoiberin concentration was found in plants grown at 0 PSU. This trend continued in the 4th week while all concentrations decreased.

The temporal pattern of the mean sinigrin concentration (Fig. 1g) was similar to the pattern of glucoiberin. A significant difference in the sinigrin concentration in plants grown at 0 and 15 PSU could be detected in week 0. However, after 1 week a trend was visible in such way that plants grown at higher salinities showed higher sinigrin concentrations. There was a significant difference in the sinigrin concentration between plants grown at 30 PSU and plants grown at 0 PSU in week 1 and 4. In week 2, plants grown at 22.5 PSU reflected a higher and significant (towards 0 PSU) sinigrin concentration. In week 3, plants grown at 22.5 and 30 PSU showed both high sinigrin concentrations, which differed not significantly from that of the 0 PSU treatment.

The second way how GSLs were influenced by salt stress can shown exemplary for glucobrassicin (Fig. 1h). There was no significant difference at the beginning of the experiment until the 2nd week, but a trend was observed after 2 weeks. Plants grown at lower salinities accumulated more glucobrassicin, except for plants grown at 15 PSU. After 3 weeks, there was a significant decrease in the glucobrassicin concentration in plants grown at 0 to 30 PSU, which continued towards the 4th week, but plants grown at 15 PSU had the second lowest concentration.

GSL concentrations of other GSLs are shown in Table 1. For the sake of completeness, the values of Fig. 1f–h are also presented in Table 1. The concentration of gluconapin increased with higher salinity towards the 2nd week, and showed the opposite pattern in the 4th week. Salinity decreased the concentration of glucobrassicin and gluconasturtiin. The decline in the GSL concentration from low to high salinity started around the 1st week with gluconasturtiin, but was more distinctive for glucobrassicin. The concentrations of glucocheirolin, glucoraphanin, and glucotropaeolin did not show a clear pattern for a de- or an increase under salinity stress, which represents the third way GSLs were influenced by salt stress. Within week 1 and 2 the maximum of glucotropaeolin was reached in plants grown at 30 PSU but changed to 0 PSU in week 3 and 4. The highest glucocheirolin concentration accumulated in plants grown at 15 PSU in week 2 and 3 and at 22.5 PSU in week 4. With respect to the relatively high standard deviation it is important to keep in mind that each single determination was done with individual plants.

The measurement of GSL contents in technical replicates resulted in standard deviations not higher than 20% with a mean of 8% relative to the individual GSL content (data not shown). In contrast, relative standard deviations of biological

Table 1 Mean GSL concentration (n = 4) in nmol g^{-1} FM for glucocheirolin, glucoraphanin, gluconapin, glucobrassicin, gluconasturtiin and glucoiberin; in µmol g^{-1} FM for sinigrin, glucotropaeolin and the sum of GSL

Time in weeks	PSU	Gluco-cheirolin	Glucoiberin	Sinigrin	Gluco-raphanin	Gluconapin	Gluco-tropaeolin	Gluco-brassicin	Gluco-nasturtiin	Total GSL amount
0	0	0.65 ± 0.47 a	84.1 ± 49.9 a	5.31 ± 1.45 a	15.9 ± 3.50 a	1.45 ± 1.58 a	0.39 ± 0.17 a	1.25 ± 0.91 a	31.5 ± 19.7 a	5.80 ± 1.68 a
	15	0.93 ± 0.81 a	247 ± 160 a	10.1 ± 5.18 b	34.4 ± 18.7 a	2.23 ± 2.23 a	1.14 ± 0.45 a	1.00 ± 0.31 a	72.5 ± 50.3 a	11.6 ± 5.83 b
	22.5	0.70 ± 0.57 a	297 ± 148 a	8.19 ± 2.61 ab	19.8 ± 9.76 a	3.22 ± 3.03 a	0.99 ± 0.34 a	1.28 ± 0.99 a	69.0 ± 50.2 a	9.60 ± 3.14 ab
	30	0.72 ± 0.36 a	236 ± 58.9 a	9.37 ± 3.18 ab	31.6 ± 9.14 a	6.12 ± 7.46 a	1.18 ± 0.24 a	1.83 ± 1.51 a	69.2 ± 29.9 a	10.9 ± 3.49 b
1	0	1.19 ± 0.73 a	73.9 ± 30.8 a	8.50 ± 1.49 a	21.4 ± 12.5 a	8.87 ± 12.5 a	0.99 ± 0.58 a	2.24 ± 1.23 a	227 ± 183 a	9.80 ± 2.23 a
	15	1.02 ± 0.47 a	442 ± 118 b	12.4 ± 0.70 ab	21.6 ± 3.96 a	3.64 ± 4.79 a	0.89 ± 0.22 a	1.30 ± 0.49 a	48.7 ± 13.0 a	13.8 ± 0.97 ab
	22.5	1.17 ± 0.41 a	658 ± 447 b	13.6 ± 1.31 b	34.7 ± 4.88 ab	9.41 ± 5.91 a	1.39 ± 0.30 ab	2.38 ± 0.20 a	94.8 ± 33.7 a	15.8 ± 1.21 b
	30	1.12 ± 0.48 a	507 ± 100 b	14.8 ± 0.33 b	41.2 ± 3.53 b	16.1 ± 13.0 a	2.48 ± 0.97 b	2.16 ± 1.08 a	83.9 ± 12.1 a	18.0 ± 0.95 b
2	0	1.15 ± 0.22 a	43.5 ± 11.0 a	10.7 ± 0.84 a	37.9 ± 6.13 a	3.01 ± 0.56 a	0.83 ± 0.25 a	3.00 ± 0.77 a	207 ± 41.1 a	11.9 ± 0.89 a
	15	1.23 ± 0.40 a	321 ± 216 a	13.8 ± 2.78 ab	27.1 ± 11.5 a	5.81 ± 3.37 ab	0.80 ± 0.35 a	1.86 ± 0.87 a	133 ± 117 a	15.1 ± 2.51 ab
	22.5	0.87 ± 0.56 a	871 ± 412 b	16.6 ± 1.70 b	29.9 ± 9.12 a	6.21 ± 5.49 ab	1.04 ± 0.38 a	2.67 ± 1.35 a	55.3 ± 17.6 a	18.6 ± 1.64 b
	30	0.51 ± 0.29 a	761 ± 219 b	15.4 ± 3.74 b	31.6 ± 2.51 a	17.2 ± 12.4 b	1.33 ± 0.26 a	1.67 ± 0.91 a	73.2 ± 19.8 a	17.6 ± 4.01 b

(continued)

Table 1 (continued)

Time in weeks	PSU	Gluco-cheirolin	Glucoiberin	Sinigrin	Gluco-raphanin	Gluconapin	Gluco-tropaeolin	Gluco-brassicin	Gluco-nasturtiin	Total GSL amount
3	0	1.01 ± 0.36 a	44.9 ± 12.2 a	11.9 ± 1.34 a	46.8 ± 11.5 a	3.64 ± 1.94 a	1.58 ± 1.51 a	4.82 ± 2.32 a	431 ± 284 a	14.0 ± 1.84 a
	15	1.13 ± 0.70 a	141 ± 73.4 a	14.1 ± 3.21 a	24.9 ± 12.2 b	8.56 ± 11.3 a	0.91 ± 0.84 b	3.54 ± 4.53 ab	279 ± 391 ab	15.4 ± 2.23 a
	22.5	0.85 ± 0.41 a	363 ± 82.4 a	14.9 ± 2.51 a	21.0 ± 8.20 b	6.8 ± 10.4 a	1.08 ± 0.47 b	2.42 ± 1.81 ab	72.3 ± 29.7 b	16.4 ± 2.41 a
	30	0.46 ± 0.24 a	701 ± 170 b	14.9 ± 2.05 a	27.4 ± 8.42 b	3.81 ± 3.61 a	1.01 ± 0.32 b	1.28 ± 0.31 b	53.3 ± 4.73 b	16.7 ± 2.15 a
4	0	0.63 ± 0.29 a	27.1 ± 8.95 a	8.46 ± 1.02 a	23.7 ± 3.98 a	7.63 ± 3.93 a	1.50 ± 0.86 a	5.35 ± 4.07 a	247 ± 116 a	10.3 ± 1.32 a
	15	0.90 ± 0.81 a	78.1 ± 6.46 a	14.3 ± 1.09 b	21.2 ± 9.85 a	4.14 ± 3.52 a	0.32 ± 0.06 a	1.95 ± 0.78 ab	159 ± 5.97 a	14.8 ± 1.14 b
	22.5	1.30 ± 0.89 a	127 ± 19.1 a	17.6 ± 2.10 b	39.8 ± 19.9 a	4.80 ± 5.81 a	0.47 ± 0.16 a	4.20 ± 3.31 ab	176 ± 83.4 a	18.5 ± 2.31 b
	30	0.68 ± 0.60 a	461 ± 67.4 b	17.8 ± 1.18 b	31.0 ± 10.0 a	1.52 ± 0.92 a	0.77 ± 0.17 a	1.70 ± 0.35 b	101 ± 13.5 a	19.1 ± 1.28 b
Time		NS	***	***	NS	.	*	*	*	***
PSU		NS	***	***	NS	NS	*	*	**	***
Time x PSU		NS	***	NS	**	NS	**	NS	NS	NS

Analysis of variance for the influence of time and salinity
NS, *, **, *** non-significant or significant at P $\leq$ 0.01, 0.05, 0.01, or 0.001, respectively

replicates showed values up to 139% with a mean of 38% (data not shown). Because of the low relative standard deviations of technical replicates compared to the ones of biological replicates, the fluctuations in GSL contents emerging from technical procedures were neglected and the standard deviation of GSL contents in biological replicates was used to calculate significant differences between the treatments.

Lepidium latifolium had an optimal growth at 15 PSU in this experiment, as the gain of biomass was the highest at this salinity condition. This was expected as this plant species is a halophyte. Nevertheless, the salt was added in a short time (4 days) and the plant was still able to survive salinity of 22.5 and 30 PSU. The concentrations of the antioxidants were significantly increased by salinity stress. The ORAC increased at the beginning of the experiment and remained on this level for a week. Total phenols and as part of them the flavonoids were affected strongest 1 week after the start of the experiment. The range of concentrations of the mentioned antioxidants was getting smaller towards the end of the experiment closing the difference between high and low salinity. The ascorbate concentration showed a different pattern in comparison to the other antioxidants. There was no significant difference towards the 3rd week of the experiment. Ascorbate concentration quickly changed in *L. latifolium* within 24 h after the beginning of the salinity treatment (Boestfleisch et al. 2014) and returned to a steady state within 4 days.

If the yield of the antioxidants is calculated (multiplying the antioxidant concentration with the biomass produced), there is only one result for all antioxidants: the increase of biomass exceeded the increase in antioxidants (data not shown).

At the beginning of the experiment only changes in the sinigrin content proved to be significant, but 1 week after the induction of salinity stress the content of four out of eight GSLs reacted significantly to salinity. The maximum of glucoiberin shifted along the timescale with increasing salinity from 15 PSU around week 1 to 30 PSU between week 2 and 3 (Fig. 1e and Table 1). Sinigrin contents showed a larger difference at week 4 compared to the beginning, and gluconapin and glucotropaeolin showed an up- and down-regulation, whereas this was more distinctive in the latter one.

GSLs seemed to react partly different than antioxidants. While ORAC, phenols and flavonoids were positively and highly significant intercorrelated ($r > 0.8$; $p < 0.001$), and all of them showed a positive significant though weak correlation with ascorbate ($r = 0.27$–0.38; $p < 0.05$), GSLs showed ten positive and two negative correlations (Pearson correlation, data not shown). The negative ones were between glucoiberin and glucobrassicin and between glucoiberin and gluconasturtiin. There were some low correlations of mainly aliphatic GSLs with ORAC (glucoiberin: $r = 0.36$; $p < 0.001$, glucotropaeolin: $r = 0.32$; $p < 0.005$ and gluconasturtiin: $r = -0.28$; $p < 0.05$), phenol (glucoiberin, sinigrin, glucoraphanin and glucotropaeolin: $r = 0.31$–0.43; $p < 0.05$), flavonoid (glucoiberin, sinigrin, glucoraphanin and glucotropaeolin, $r = 0.23$–0.49; $p < 0.05$) and ascorbate values (glucocheirolin and glucoraphanin $r = 0.28$–0.3; $p < 0.05$). From the temporal patterns of the GSLs and the correlations between them, they can be classified into three different groups: an up regulation (aliphatic GSLs), a down regulation (aromatic including indolic GSLs) and an intermediate reaction (aromatic excluding indolic GSLs) under salinity. This up and down, or the different regulation, was previously shown for salinity (Yuan et al. 2010; Zaghdoud et al. 2012) and drought treatments (Mewis et al. 2012; Radovich et al. 2005; Schreiner et al. 2009; Tong et al. 2014). Jensen et al. (1996) gave an explanation for these observations under salinity and drought stress. They showed that GSL synthesis increased when the leaf water potential was less than −1.4 MPa for extended periods. In our study, the VWC (volumetric water content) decreased from 85% to 77% at 30 PSU and from 92 to 83% at 0 PSU (data not shown). However, only aliphatic GSLs had a significant correlation with the VWC. The decrease of glucobrassicin with increasing salinity might also be the reason for its ease of oxidation, as it has a high antioxidant capacity (Cabello-Hurtado et al. 2012). These authors demonstrated a relatively high ORAC value for glucobrassicin, much higher than ascorbic acid. However, in our study the most abundant GSL was sinigrin, which has according to Cabello-Hurtado et al. (2012) a 6–7 times lower ORAC compared to ascorbic acid. Therefore, the total GSL content did not contribute much to the ORAC resulting in a low correlation ($r = 0.33$; $p < 0.01$). Furthermore many of the studies had only one point of time for the measurement of GSLs, but we could show that GSL contents in salt-stressed plants changed over time.

AS already mentioned sinigrin was the most abundant GSL in *L. latifolium*. Therefore, an increase in salinity elevated the total GSL yield, which would be beneficial for herbivore protection under abiotic stress conditions (Agrawal and Kurashige 2003; Hopkins et al. 2009), in addition to the enhanced growth and antioxidant production at 15 PSU. It was suggested that transient allocation and redistribution of some GSLs indicate a role in signaling mechanisms under abiotic stress conditions to induce fast physiological adaptation to unfavorable conditions (del Carmen Martínez-Ballesta et al. 2013). However, the determination of the exact functions of GSLs in reaction to abiotic stress needs further investigation.

References

Agrawal A, Kurashige NS (2003) A role for isothiocyanates in plant resistance against the specialist herbivore *Pieris rapae*. J Chem Ecol 29:1403–1415

Bloem E, Haneklaus S, Kleinwächter M, Paulsen J, Schnug E, Selmar D (2014) Stress-induced changes of bioactive compounds in *Tropaeolum majus* L. Ind Crop Prod 60:349–359

Boestfleisch C, Wagenseil NB, Buhmann AK, Seal CE, Wade EM, Muscolo A, Papenbrock J (2014) Manipulating the antioxidant capacity of halophytes to increase their cultural and economic value through saline cultivation. AoB Plants 6:1–16

Cabello-Hurtado F, Gicquel M, Esnault M-A (2012) Evaluation of the antioxidant potential of cauliflower (*Brassica oleracea*) from a glucosinolate content perspective. Food Chem 132:1003–1009

del Carmen M-BM, Moreno D, Carvajal M (2013) The physiological importance of glucosinolates on plant response to abiotic stress in *Brassica*. Int J Mol Sci 14:11607–11625

Epstein E (1972) Mineral nutrition of plants: principles and perspectives. Department of Soils and Plant Nutrition, California University, Davis

Francis A, Warwick SI (2007) The biology of invasive alien plants in Canada. 8. *Lepidium latifolium* L. Can J Plant Sci 87:639–658

Gupta SM, Pandey P, Negi PS, Pande V, Grover A, Patade VY, Ahmed Z (2013) DRE-binding transcription factor gene (LlaDREB1b) is regulated by various abiotic stresses in *Lepidium latifolium* L. Mol Biol Rep 40:2573–2580

Hopkins RJ, van Dam NM, van Loon JJ (2009) Role of glucosinolates in insect-plant relationships and multitrophic interactions. Annu Rev Entomol 54:57–83

Jensen CR, Mogensen VO, Mortensen G, Fieldsend JK, Milford GFJ, Andersen MN, Thage JH (1996) Seed glucosinolate, oil and protein contents of field-grown rape (*Brassica napus* L.) affected by soil drying and evaporative demand. Field Crop Res 47:93–105

Kaur T, Hussain K, Koul S, Vishwakarma R, Vyas D (2013) Evaluation of nutritional and antioxidant status of *Lepidium latifolium* Linn.: a novel phytofood from Ladakh. PLoS One 8:e69112

Khan MAM, Ulrichs C, Mewis I (2010) Influence of water stress on the glucosinolate profile of *Brassica oleracea* var. *italica* and the performance of *Brevicoryne brassicae* and *Myzus persicae*. Entomol Exp Appl 137:229–236

López-Berenguer C, Martínez-Ballesta MC, García-Viguera C, Carvajal M (2008) Leaf water balance mediated by aquaporins under salt stress and associated glucosinolate synthesis in broccoli. Plant Sci 174:321–328

López-Berenguer C, Martínez-Ballesta MC, Moreno DA, Carvajal M, García-Viguera C (2009) Growing hardier crops for better health: salinity tolerance and the nutritional value of broccoli. J Agric Food Chem 57:572–578

Manici LM, Lazzeri L, Palmieri S (1997) *In vitro* fungitoxic activity of some glucosinolates and their enzyme-derived products toward plant pathogenic fungi. J Agric Food Chem 45:2768–2773

Mewis I, Khan MAM, Glawischnig E, Schreiner M, Ulrichs C (2012) Water stress and aphid feeding differentially influence metabolite composition in *Arabidopsis thaliana* (L.) PLoS One 7:e48661

Navarro E, Alonso J, Rodriguez R (1994) Diuretic action of an aqueous extract of *Lepidium latifolium* L. J Ethnopharmacol 41:65–69

Noctor G, Foyer CH (1998) Ascorbate and glucosinolate: keeping active oxygen under control. Annu Rev Plant Physiol Plant Mol Biol 49:249–279

Qasim M, Ashraf M, Ashraf MY, Rehman S-U, Rha ES (2003) Salt-induced changes in two canola cultivars differing in salt tolerance. Biol Plant 46:629–632

Radovich TJK, Kleinhenz MD, Streeter JG (2005) Irrigation timing relative to head development influences yield components, sugar levels, and glucosinolate concentrations in cabbage. J Am Soc Hortic Sci 130:943–949

Rask L, Andréasson E, Ekbom B, Eriksson S, Pontoppidan B, Meijer J (2000) Myrosinase: Gene family evolution and herbivore defense in Brassicaceae. Plant Mol Biol 42:93–113

Robbins RJ, Keck A, Banuelos G, Finley JW (2005) Cultivation conditions and selenium fertilization alter the phenolic profile, glucosinolate, and sulforaphane content of broccoli. J Med Food 8:204–214

Schreiner M, Beyene B, Krumbein A, Stützel H (2009) Ontogenetic changes of 2-propenyl and 3-indolylmethyl glucosinolates in *Brassica carinata* leaves as affected by water supply. J Agric Food Chem 57:7259–7263

Sharma P, Jha A, Dubey R (2010) Oxidative stress and antioxidative defense systems in plants growing under abiotic stresses. In: Handbook of plant and crop stress. Taylor & Francis, Boca Raton, pp 89–138

Thies W (1979) Detection and utilization of a glucosinolate sulfohydrolase in the edible snail, *Helix pomatia*. Naturwissenschaften 66:364–365

Tierens KFM-J, Thomma BP, Brouwer M, Schmidt J, Kistner K, Porzel A, Mauch-Mani B, Cammue BP, Broekaert WF (2001) Study of the role of antimicrobial glucosinolate-derived isothiocyanates in resistance of *Arabidopsis* to microbial pathogens. Plant Physiol 125:1688–1699

Tong Y, Gabriel-Neumann E, Ngwene B, Krumbein A, George E, Platz S, Rohn S, Schreiner M (2014) Topsoil drying combined with increased sulfur supply leads to enhanced aliphatic glucosinolates in *Brassica juncea* leaves and roots. Food Chem 152:190–196

Yuan G, Wang X, Guo R, Wang Q (2010) Effect of salt stress on phenolic compounds, glucosinolates, myrosinase and antioxidant activity in radish sprouts. Food Chem 121:1014–1019

Zaghdoud C, Alcaraz-López C, Mota-Cadenas C, Martínez-Ballesta MDC, Moreno D, Ferchichi A, Carvajal M (2012) Differential responses of two broccoli (*Brassica oleracea* L. var *Italica*) cultivars to salinity and nutritional quality improvement. Sci World J 2012:1–12

Zhao K, Song J, Feng G, Zhao M, Liu J (2010) Species, types, distribution, and economic potential of halophytes in China. Plant Soil 342:495–509

The Application of S^0-Coated Fertilizer to Durum Wheat Crop

Styliani N. Chorianopoulou, Georgios I. Saridis, Petros Sigalas, Miltos Margetis, Dimitris Benardos, Haris Mavrogiannis, and Dimitris L. Bouranis

Abstract Elemental sulfur (S^0) is an ideal slow release S fertilizer with a long history as soil amendment. Recently, S^0 has been attached successfully onto the surface of the beads of a commercial fertilizer (F) via a binder (B) by Sulphur Hellas S.A, under the commercial name "Sulfogrow" (FBS^0). F beads act as a core effectively covered by an amount of 2% (w/w) of S^0 yellow dust. To assess and evaluate the effectiveness of FBS^0, we monitored the nutritional dynamics of a durum wheat (*Triticum durum*, Poaceae) commercial crop. The field was divided into two parts; one subject to control F-treatment, and one with the FBS^0-treatment. Rhizosphere pH, organic matter and humic substances contents were monitored, along with the dry mass, and sulfate, total sulfur, organic nitrogen and iron concentrations in the aerial plant part. The FBS^0-treated crop presented denser plantation with more robust plants; the accumulated amounts of iron and organic nitrogen per plant were found to be increased at day 61 after sowing in the aerial part by 120% and by 43% respectively, comprising early effects. After day 100, the accumulated dry mass was twice that of control and all accumulation curves were statistically higher than the control ones. The time-course curve of the relative percentage changes (RC) of iron presented in reverse the pattern of sulfate; in contrast, the time-course curve of organic sulfur RC followed the pattern of organic nitrogen RC precisely.

Elemental sulfur (S^0) is an ideal slow release S fertilizer with a long history as soil amendment (Schnug 1982; Schnug and Finck 1982; Germida and Jansen 1993; Somani and Totawat 1998). The commercially available S^0 is provided in the form

S.N. Chorianopoulou • P. Sigalas • M. Margetis • D.L. Bouranis (✉)
Plant Physiology Laboratory, Crop Science Department, Agricultural University of Athens, 75 Iera Odos, Athens 11855, Greece
e-mail: bouranis@aua.gr

G.I. Saridis
Botanical Institute, Cologne Biocenter, University of Cologne, D-50674 Cologne, Germany

D. Benardos • H. Mavrogiannis
Sulphur Hellas S.A., 142 Leoforos Athinon, 10442 Athens, Greece

L.J. De Kok et al. (eds.), *Sulfur Metabolism in Higher Plants - Fundamental, Environmental and Agricultural Aspects*, Proceedings of the International Plant Sulfur Workshop, DOI 10.1007/978-3-319-56526-2_11

of powder and the minimal guarantee of the product is 95% in total S, which is the highest possible concentration of S in a fertilizer. With regard to S^0 application as fertilizer, two agronomic issues should be taken into account. On the one hand, after application, S^0 needs to be oxidized to sulfate, which is the form of S to be absorbed by the root system; therefore, S^0 is not readily available as nutrient and microbial activity is involved. On the other hand, the application of S^0 in the form of powder is a hindrance due to a number of reasons: it is difficult to apply a fertilizer as powder; in a mix with granular materials segregation is observed; contact with the skin may cause irritation and burning; dust is produced whilst inhalation can cause cough and lung irritation. In order to solve the aforementioned issues, in the last three decades several products have been developed in order to encompass S^0 powder with fertilizers (Vitti and Vale 2014). Taking into account the fact that nowadays fertilizers are considered not only as tools, but also as resources, which require sustainability, stewardship, responsibility and participation (Schnug and Haneklaus 2014), it becomes clear that S^0 is still a challenge for fertilizer production and fertilization, both as a tool and a resource. Recently, S^0 has been attached successfully onto the surface of the beads of a commercial fertilizer (F) via a binder (B) by Sulphur Hellas S.A, under the commercial name "Sulfogrow" (FBS^0). F beads act as a core effectively covered by an amount of 2% (w/w) of S^0 yellow dust. Thus, a new generation of S^0-coated fertilizers is now commercially available. Is "Sulfogrow" more effective in relation to its core fertilizer and in positive case how much more?

To assess and evaluate the effectiveness of FBS^0, we monitored the nutritional dynamics of a durum wheat (*Triticum durum*, cv SIMETO) commercial crop established in Lefktra at Viotia county, Greece, in an area of 2.2 ha with calcareous soil. Sowing day and fertilizer application took place in November 13, 2014 (d0). The field was divided into two parts; one of them was subject of control F-treatment according to the local agricultural practices (control crop), whilst the other one received the corresponding FBS^0-treatment (FBS^0-treated crop). The control crop was fertilized with a commercial 20-10-10 fertilizer (where potassium was provided as K_2SO_4) at a rate of 300 kg ha^{-1}. The FBS^0-treated crop received the equivalent fertilization with the corresponding "Sulfogrow" 20-10-10 commercial fertilizer at the same rate, carrying 2% S^0 (306 kg ha^{-1}). At d146 and d167 after sowing, additional fertilization with commercial ammonium nitrate fertilizer took place at the rates of 270 and 150 kg ha^{-1}, respectively. At d161 and d166 after sowing, herbicide application took place at the rates of 70 g ha^{-1} (Best) and 1.1 L ha^{-1} (Foxtrot 6.9 W), respectively. Both crops received no irrigation. Each crop was divided into five plots and sampling took place from the internal three ones. A number of plants were collected with their root system and the surrounding soil by means of a shovel. The pH (in 10 mM $CaCl_2$), organic matter and humic substances of the rhizosphere were monitored, along with the dry mass (DM), sulfate (SO_4^{2-}), total sulfur (S_{tot}), organic nitrogen (N_{org}) and iron (Fe) concentrations of the aerial plant part. Organic sulfur (S_{org}) was calculated by subtracting SO_4^{2-} content from S_{tot} content.

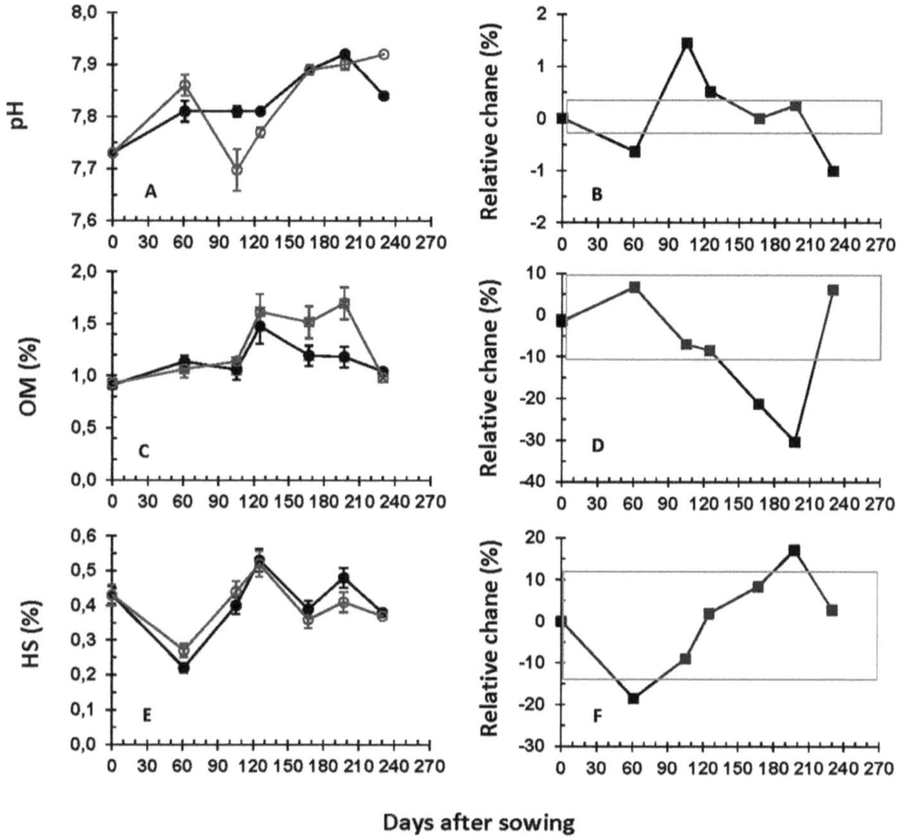

Fig. 1 Time-course of pH (**a**, **b**), organic matter (OM; **c**, **d**) and humic substances (HS; **e**, **f**) in the rhizosphere of the analyzed samples. *Empty circles* and *grey line*: control F-treated crop; *full circles* and *black line*: FBS^0-treated crop. *Bars* represent standard error. Relative changes (i.e. the differences between the corresponding FBS^0-values and control F-values divided by the control F-values) out of the box were statistically significant (95% confidence level)

Before the additional nitrogen fertilization, rhizosphere pH of both control F-treated and FBS^0-treated crops fluctuated around 7.75 and 7.83, respectively. After fertilization, rhizosphere pH of both treatments climbed and stabilized between 7.85 and 7.90. Considering the value of 7.83 as mean level, in overall the fluctuation width was narrower in FBS^0-treated crop (Fig. 1a). Perhaps, this is indicative of an excess uptake of anions over cations, which results to a net removal of protons, thus rising pH (Neumann and Roemheld 2012). Rhizosphere OM content was around 1% up to d105 and then it rose to 1.5%. At d125 the pattern differentiated: in control F-treated crop it was stabilized there until d197 and then it decreased rapidly to the initial value, whilst in FBS^0-treated crop it shifted progressively towards the initial value (Fig. 1c). The FBS^0-treated field was colonized by far less weeds, a fact that could explain the higher OM content of the F-treated

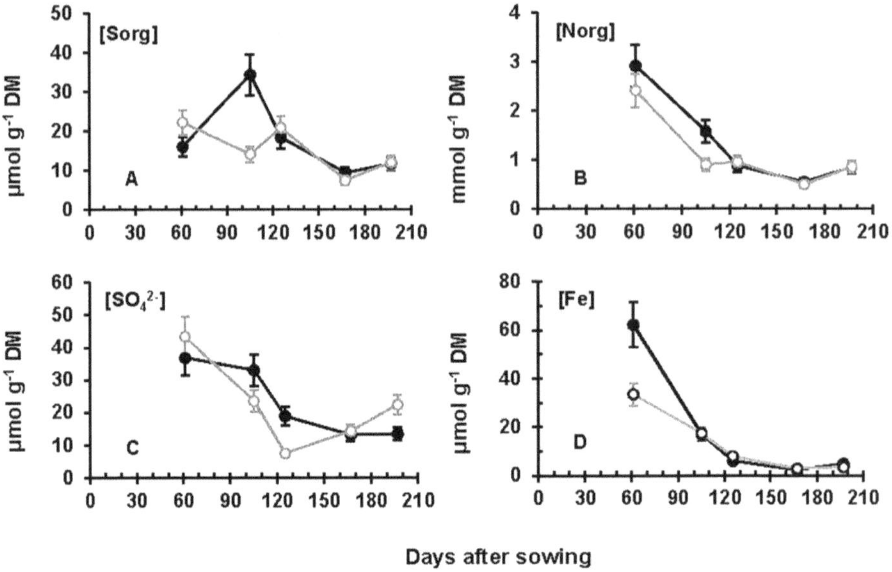

Fig. 2 Time-course of organic sulfur [S_{org}; **a**], organic nitrogen [N_{org}; **b**], sulfate [SO_4^{2-}; **c**], and iron [Fe; **d**] concentrations in the aerial part of wheat plants. *Empty circles* and *grey line*: control F-treated crop, *full circles* and *black line*: FBS^0-treated crop. *Bars* represent standard error

soil after the herbicide treatment, presumably due to the decomposition of the eliminated weeds. Rhizosphere HS content fluctuated around 0.4%, ranging between 0.2 and 0.56% (Fig. 1e). The rhizosphere HS content of the FBS^0-treated crop presented a significant decrease (−18.5%) at d60 and a significant increase (17.1%) at d197 (Fig. 1f) relative to that of F-treated crop. The FBS^0-treated crop presented denser plantation with more robust plants, exhibiting reduced number of tillers. After d100, the accumulated dry mass in the FBS^0-treated crop plants was twice that of control (Figs. 3a and 4a). At d61, iron concentration in FBS^0-treated plants was close to twice than control (Fig. 2d), resulting in a 120% relative change in Fe accumulation in the aerial part, a fact that constitutes an early effect (Fig. 4b). Increased Fe (Fig. 2d) and N_{org} (Fig. 2b) concentrations were early responses of FBS^0-treated crop, whilst the corresponding S_{org} (Fig. 2a) and SO_4^{2-} (Fig. 2c) concentrations were lower that control. Taking into account that Fe availability is highly restricted in calcareous soils due to Fe^{3+} precipitation and immobilization, the fact that the FBS^0-plants accumulated greater amounts of Fe implies that soil reserves of Fe (1) were in adequate amounts to support the enhanced growth of FBS ^{0}-plants and (2) they had been mobilized. The late nitrogen fertilization caused a significant increase in both N_{org} concentration and accumulation in the aerial part. After d100, all accumulation curves were higher than the control ones, with those of N_{org} and Fe being statistically higher from d61 onward (Fig. 3c, e). The time-course curve of the relative percentage changes (RC) of Fe presented in reverse the pattern of that of SO_4^{2-} (Fig. 4b); the maximum RC value of SO_4^{2-} (397%) at d125

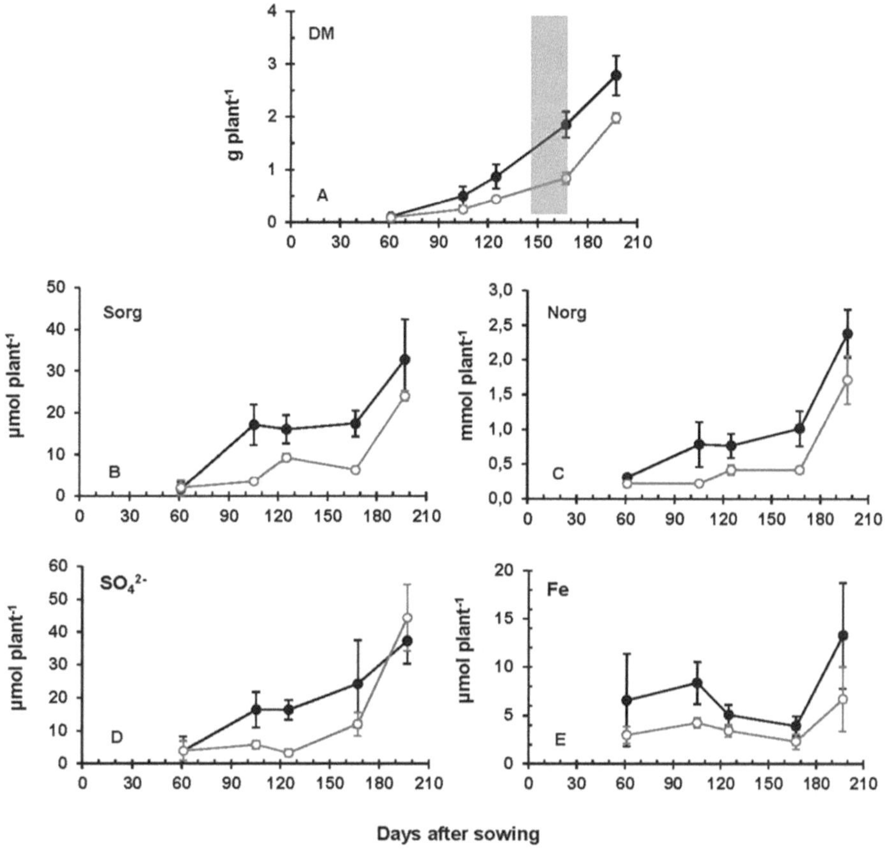

Fig. 3 Time-course of dry mass (DM; **a**), organic sulfur (S_{org}; **b**), organic nitrogen (N_{org}; **c**), sulfate (SO_4^{2-}; **d**), and iron (Fe; **e**) accumulations in the aerial part of wheat plants. *Empty circles* and *grey line*: control F-treated *crop, full circles* and *black line*: FBS^0-treated crop. *Bars* represent standard error. *Gray lane* indicates the timing of herbicide application and additional nitrogen fertilization

coincided with the minimum RC value of iron (47.9%). In contrast, the time-course curve of S_{org} RC followed the pattern of N_{org} RC precisely (Fig. 4c). S_{org} accumulation in the aerial part was in strong linear relationship with N_{org} in both control crop ($R^2 = 0.9667$) and S^0-treated crop ($R^2 = 0.8998$) (Fig. 4d) with the same slope.

The aforementioned data sketch an emerging scenario for the FBS^0-treated crop: Fe was mobilized early and in higher amounts, whilst the N nutrition status of the FBS^0-treated crop at this period was a better one. The metabolic links between S and Fe nutrition are well documented in plants that use the chelation strategy for iron accumulation (Zuchi et al. 2012; Forieri et al. 2013), whilst the link between S and N is well known. It seems that the earlier and higher Fe and N status enhanced the uptake of SO_4^{2-} and its conversion to S_{org}, as it is shown by the higher

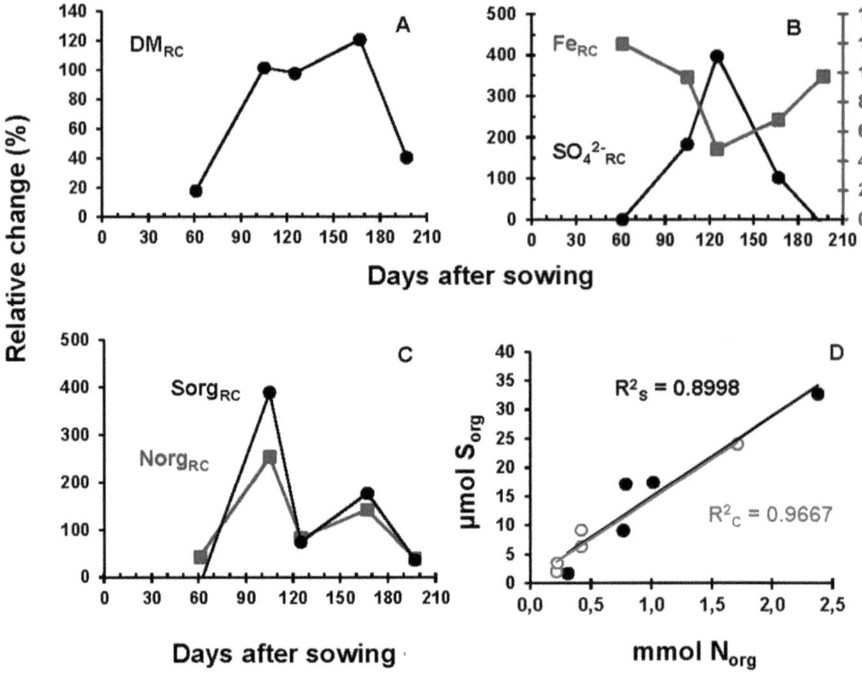

Fig. 4 Time-course of relative percentage changes (RC, %) of DM (**a**), SO_4^{2-} vs. Fe (**b**), and N_{org} vs. S_{org} (**c**) in FBS^0-treated crop plants, along with the correlation between N_{org} and S_{org} accumulations (**d**) in the aerial part of both the control F-treated crop (*empty circles*) and FBS^0-treated crop (*full circles*)

concentrations of SO_4^{2-} and S_{org} at d100. S_{org} seems to have covered effectively the needs for the formation of Fe-S carrying compounds and the stoichiometry with N_{org}. On the other hand it seems that the enhanced accumulation of S_{org} contributed to plant-plant interactions (less weeds) and plant-microbe interactions (robust plants), a fact that seems to be in accordance with the Sulfur-Induced-Resistance concept (Bloem et al. 2014) and the role of S in these interactions.

In conclusion, these data suggest that the "Sulfogrow" product was more effective in comparison to its core fertilizer, as regards the aforementioned parameters. The added 2% S^0 obviously does not act as soil conditioner but rather as enhancer of soil microbial activity, a hypothesis that needs confirmation. Moreover, another point that needs further research is the fact that higher early mobilization of Fe coincides with the statistically significant decrease in rhizosphere HS content. Is this indeed a decrease in HS content or does this reflect a deactivation of the reactive groups of these molecules that are measured by the Mehlich method used? These groups are known to interact with iron in the soil (Li et al. 2012).

References

Bloem E, Haneklaus S, Schnug E (2014) Milestones in plant sulfur research on sulfur-induced-resistane (SIR) in Europe. Front Plant Sci 5:779

Forieri I, Wirtz M, Hell R (2013) Toward new perspectives on the interaction of iron and sulfur metabolism in plants. Front Plant Sci 4:357

Germida JJ, Jansen HH (1993) Factors affecting the oxidation of elemental sulfur in soils. Fert Res 35:101–114

Li Y, Yu S, Strong J, Wang H (2012) Are the biochemical cycles of carbon, nitrogen, sulfur and phosphorus driven by the "Fe^{III}-Fe^{II} redox wheel" in dynamic redox environments? J Soils Sediments 12:683–693

Neumann G, Roemheld R (2012) Rhizosphere chemistry in relation to plant nutrition. In: Marschner P (ed) Marschner's mineral nutrition of higher plants, 3rd edn. Academic, London, pp 347–368

Schnug E (1982) Untersuchungen zum Einfluß bodenversauernder Düngung auf die Spurennährstoff-Versorgung von Kulturpflanzen. Dissertation Agrarwiss Fak, Kiel, Germany

Schnug E, Finck A (1982) Trace element mobilization by acidifying fertilizers. Proc 9th Int Plant Nutr Coll 2:582–587

Schnug E, Haneklaus S (2014) Tomorrow's challenges for fertilizers and fertilization. In: Vinicius de Melo Benites et al (eds) Proceedings of 16th world fertilizer congress of CIEC, pp 17–19

Somani LL, Totawat KL (1998) Mined and industrial waste products capable of generating gypsum in soil. In: Wallace A, Terry RE (eds) Handbook of soil conditioners. Marcel Dekker, Inc, New York/Basel/Hong Kong, pp 257–291

Vitti GC, Vale F (2014). Dynamic of S oxidation from fertilizers in tropical soils. In: de Melo Benites V, de Oliveira A Jr , Pavinato PS, Teixeira PC, Ferreira Moraes M, Villas Bôas de Campos Leite RM, Pereira de Oliveira R (eds) Proceedings of 16th world fertilizer congress of CIEC, pp 33–35

Zuchi S, Cesco S, Astolfi S (2012) High S supply improves Fe accumulation in durum wheat plants grown under Fe limitation. Environ Exp Bot 77:25–32

Re-assessing Systems Biology Approaches on Analyzing Sulfate Metabolism

Rainer Hoefgen and Mutsumi Watanabe

Abstract Mainly driven by the availability of the first Arabidopsis genome sequence in 2000 a rapid development of high throughput analytical techniques were developed and termed transcriptomics. This development was quickly followed by developing metabolite-profiling technologies, metabolomics. The ever-growing data bases made the development of new biostatistical and bioinformatics tools necessary. These 'omics' approaches were also applied to analyze the response of plants towards sulfate deprivation with the aim to gain a more complete, holistic view on plant metabolism and its control in response to varied nutrient supply. Early results though already providing novel results and fostering new routes of investigation were hampered by the incomplete annotation of the genes of the Arabidopsis genome. In recent years this informational gap was largely filled. Thus, we revisit here one old data set obtained at the infancy of 'omics' research and indicate novel conclusions possible when re-assessing these data as well as indicate new possibilities of continued analyses.

Approaches using high throughput methods were first applied on Arabidopsis seedlings at the beginning of this millennium with the arrival of transcriptomics. Arabidopsis seedlings were germinated on or transferred to culture media containing reduced (deprivation) or even zero (starvation) sulfate levels in the culture medium eventually leading to limitation of internal sulfate resources reducing growth and sulfate starvation over time, eventually leading to death of the plants. Plant tissues were analysed by transcriptomics (Nikiforova et al. 2003; Hirai et al. 2003; Maruyama-Nakashita et al. 2003) and metabolomics (Fiehn et al. 2000; Roessner et al. 2001; Nikiforova et al. 2005a). These high through-put experiments and the conclusions drawn from these holistic approaches are commonly summarized under the term systems biology (Fernie et al. 2004). Approaches were then mainly focusing on the model species *Arabidopsis thaliana* as the respective necessary databases and genome wide gene annotations were emerging. The basic principle was, and is, to obtain in an unbiased way as much data as possible

R. Hoefgen (✉) • M. Watanabe
Max Planck Institute of Molecular Plant Physiology, 14476 Potsdam-Golm, Germany
e-mail: hoefgen@mpimp-golm.mpg.de

L.J. De Kok et al. (eds.), *Sulfur Metabolism in Higher Plants - Fundamental, Environmental and Agricultural Aspects*, Proceedings of the International Plant Sulfur Workshop, DOI 10.1007/978-3-319-56526-2_12

from an experimental situation, here usually reduction of available sulfate or resupply of sulfate to deprived plants, or from mutant or transgenic plants. Despite the limited resolution of the first transcriptome studiesas e.g. initially 10 k cDNA arrays were used (Nikiforova et al. 2003; Hirai et al. 2003) before commercial arrays became available these approaches yielded entirely new data, revolutionizing our knowledge base and hence the experimental approach to access the response of plants to sulfate availability. Other than targeted approaches which are necessarily focusing e.g. upon the target pathway or selected compound classes, systems biology approaches usually employ high throughput systems and try to obtain more complete and comprehensive data sets. The intention for capturing a holistic data set is also supported by integrating results of targeted analyses. Additional to obtaining a catalogue of responses the eventual aim is to describe a system and its response to perturbations as complete as possible, and on as many levels, *i.e.* transcripts, metabolites, proteins, enzyme activities, fluxes *etc.*, as possible and to identify yet unprecedented links, usually due to co-behaviour, between metabolic and cellular processes. Characteristically, sometimes seemingly unrelated gene or metabolite alterations are identified additional to the expected or known changes within the pathway. In order to fully comprehend plant metabolism and physiology it is necessary to also integrate and link these processes with the main response route. The sheer size of the data sets produced by such holistic approaches made it necessary to develop new biostatistical approaches first developed for protein analysis then applied to all fields of molecular biology, now usually termed bioinformatics (Hagen 2000). A further potential and power of systems biology approaches lies in its principal possibility to fuse diverse data such as transcriptomics, metabolomics, proteomics and other omics or targeted analysis data as long as ratios to controls are provided (Nikiforova et al. 2005b). This then allowed linking processes but also deducing new gene functions (Hirai et al. 2004, 2005).

Experiments were executed in different ways which makes it necessary to define the terms to be used more carefully. Plants were either exposed to a complete withdrawal of sulfate from the culture medium, usually using seedlings on agar, in submerged cultures, or in hydroponic systems which allow exact control of the sulfate supply or re-supply and of the time points (Nikiforova et al. 2003, 2004; Wulff-Zottele et al. 2010; Hubberten et al. 2012b; Bielecka et al. 2015). Still in all cases internal sulfur pools are available and can be utilized for growth and only when the resources are diluted to an extent which prevents further growth the plant enters full starvation. For big seeds these resources might be substantial as e.g. we were not able to provoke starvation symptoms in pumpkin (*Cucurbita pepo*) seedlings unless we used seeds of mother plants exposed to prolonged sulfate deprivation (unpublished). These resources are either deriving from seeds or from phases where initially sulfate is supplied to foster initial growth to establish uniform plant growth. We would term these conditions constitutive and induced starvation, respectively. Plant responses to induced or constitutive starvation are quite different. When seedlings are exposed to starvation from germination onwards we found that they survive longer and need longer to develop symptoms than seedlings pre-grown on sulfate and then exposed to induced starvation (Nikiforova et al. 2003). Under constitutive starvation plants probably adapt to the adverse condition and constantly economise on their resources while under induced conditions abruptly imposed on the plant seedlings stress responses and accumulation of ROS prevail until adaptation and stress response mechanisms help the plants to achieve a new homeostatic status. Also a controlled minimal supply, which we would term deprivation or depletion of sulfate, is a valid experimental system which can be tuned to prevent e.g. accumulation of excess sulfate in the vacuole and will allow investigating adaptation processes rather than emergency responses induced by starvation of the plants (Forieri et al. 2017). Such a system is likely to be closer to conditions plants might be exposed to in nature but responses might be blurred. Sulfate starvation inevitably leads to senescence and plant death (Watanabe et al. 2010) while plants under sulfate depleted or deprived conditions still manage to survive and enter and finish a reproductive cycle with a major negative effect on yield.

As the response to a condition is dynamic (Whitcomb et al. 2014) a systems approach should at its best aim at describing the entire response space. Usually a set of distinct points along the time line are selected allowing identifying changes in response patterns. When plants are exposed to an either sulfate free (starved) or sulfate reduced (deprived) growth condition on either artificial media or soils this reaches from early responses such as the induction of high affinity sulfate transporters (Maruyama-Nakashita et al. 2004) to adaptation processes and rescue responses for example by downregulating the flux to expendable sulfate pools (Aarabi et al. 2016) to eventually nutrient deprivation induced senescence (Watanabe et al. 2010, 2013) or the response to replenishment of nutrients after insufficient supply (Bielecka et al. 2015). Obviously, initial responses are related to rather targeted changes affecting sulfate assimilation and uptake *per se* while later responses are more pleiotropic and hence overlapping with general stress response patterns (Watanabe et al. 2010). The questions to be still answered in this case are how a specific stress as sulfate depletion is perceived and which elements such as signal molecules, transcription factors or regulation at the protein level through e.g. phosphorylation or protein-protein interaction contribute to the regulation of the various processes, the response network. Some progress has been made with e.g. identifying SLIM1 (sulfur limitation1; Maruyama-Nakashita et al. 2006) in the ethylene-insensitive-like (EIL) transcription factor family and SDI1 and SDI2 (sulfur deficiency induced 1 and 2; Aarabi et al. 2016) containing tetratricopeptide repeat (TPR)-like domain which mediates protein-protein interactions as regulators and *O*-acetylserine as a signal (Hubberten et al. 2012a).

One of the first sulfur networks (Fig. 1) was provided in 2005 (Nikiforova et al. 2005b), which was already exploiting joint transcriptomics and metabolomics data. The network was based on similarity of co-behaviour of expression patterns and metabolite concentrations. Essentially, the connectivity was deduced by proof of association. The transcriptomics analysis provided 6454 transcripts changed by at least 1.5-fold and additionally, 81 metabolites were determined. A correlation analysis aiming at reducing the noise component finally resulted in a correlation matrix containing 541 elements significantly changed and showing stable co-behavior patterns. Within this correlation network genes related to sulfate metabolism formed five distinct groups, three of them related to sulfur (Fig. 1a). When combining this to a hierarchical analysis of elements we find the five groups to span over several layers of the hierarchy as indicated in the table included to Fig. 1b. This enrichment of sulfur related genes and their assorting into connected groups proved the validity of this early network. An important limitation at this stage and in comparable current analyses is the fact that only those elements are identified that show changes between conditions, be it differential transcript or metabolite abundances. Genes such as *SLIM1* (Maruyama-Nakashita et al. 2006) which are hardly regulated but obviously post-translationally modulated upon conditions of reduced sulfate availability will not be picked up and need to be identified by other methods, such as mutagenesis in the case of SLIM1.

A hallmark of such networks is the connectivity, which describes the number of vertices (i.e. genes or metabolites) showing a close co-behavior to a certain gene (Nikiforova et al. 2005b) indicated by a connecting line (edges). Thus, it was possible to derive novel conclusions based on the connectivity of network elements (Nikiforova et al. 2003, 2004, 2005a, b; Sasaki-Sekimoto et al. 2005; Hoefgen and

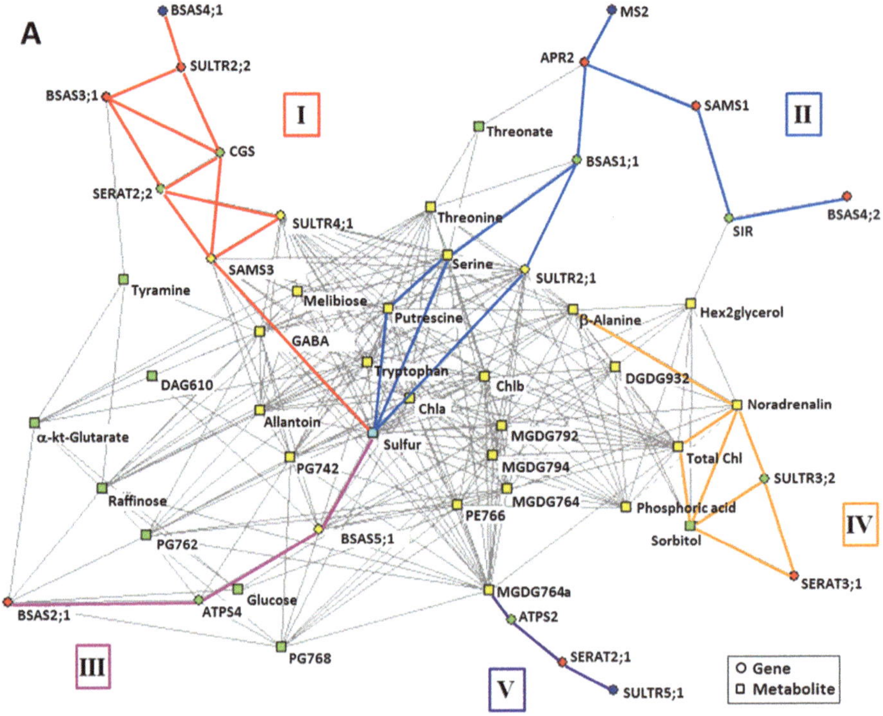

Fig. 1 Combined transcriptomics and metabolomics co-behavior network. Genes and metabolites of sulfur metabolism form five groups within the network, three of them connected to sulfur

B

Layer	Metabolite	Gene	ATG ID	
1	sulfur			
2		SULTR4;1	At5g13550	I
		SAMS3	At3g17390	I
		SULTR2;1	At5g1180	II
		BSAS5;1	At3g03630	III
3		SERAT2;2	At3g13110	I
		CGS	At3g01120	I
		SIR	At5g04590	II
		BSAS1;1	At4g14880	II
		ATPS4	At5g43780	III
		SULTR3;2	At4g02700	IV
		ATPS2	At1g19920	V
4		SULTR2;2	At1g77990	I
		BSAS3;1	At3g61440	I
		APR2	At1g62180	II
		BSAS4;2	At3g04940	II
		SAMS1	At1g02500	II
		BSAS2;1	At2g43750	III
		SERAT3;1	At2g17640	IV
		SERAT2;1	At1g55920	V
5		BSAS4;1	At5g28020	I
		MS2	At3g03780	II
		SULTR5;1	At1g80310	V

Fig. 1 (continued)

Nikiforova 2008; Falkenberg et al. 2008) such as a connectivity of the sulfur depletion response to calmodulin-controlled responses and a link to auxin (Aux) and jasmonic acid (JA). Some genes displaying very high numbers of connections can be assumed to be of central function. Like in a wheel they form a hub or a centre of connectivity. Such a gene is *IAA28* (At5g25890), which shows under conditions of sulfate depletion 28 links. A detailed analysis of IAA28 showed that it controls to a certain extent thiol levels and might be involved in repressing side root development under sulfate deprived conditions (Falkenberg et al. 2008; Hubberten et al. 2012a, b).

The validity and richness of this early dataset can for example further be seen when pulling out metabolite signatures (Nikiforova et al. 2005b). The above mentioned regulator *IAA28* is directly co-behaving with serine, putrescine, GSH, asparagine and g-aminobutyric acid which in turn is linked to the marker molecules, *O*-acetylserine (OAS) and ornithine, which are known to be altered under

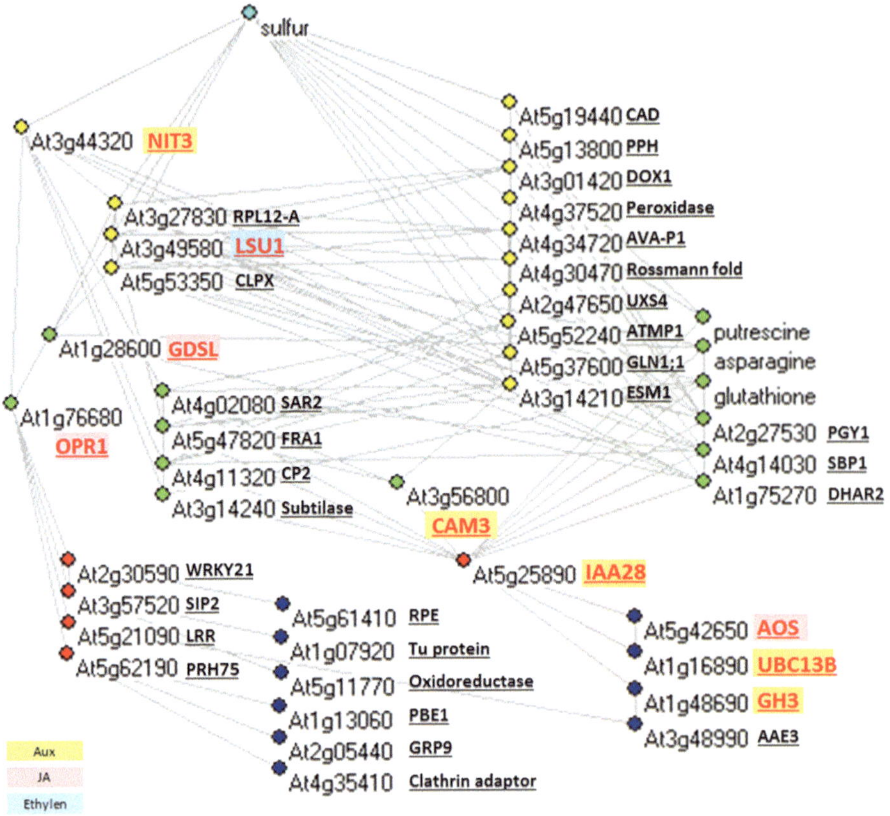

Fig. 2 Reassessing network annotation. Hierarchical clustering within a subfragment of the combined transcriptome and metabolome sulfate response network displaying 39 genes and their connectivities on the basis of co-behaviour. While in the network in 2005 only 7 genes could be annotated, now, all 39 genes can be annotated and functions are at least partially revealed

conditions where sulfate availability is inadequate for sustained plant growth as shown through a detailed metabolomics analysis (Nikiforova et al. 2005a). *IAA28* is co-behaving with GSH, which is in a hierarchical correlation again directly linked to *LSU1* (At3g49580), which is co-behaving directly with sulfur contents under sulfate deprived growth conditions in plants (Sirko et al. 2015). As we could decipher a cause to effect relation in this hierarchy (Nikiforova et al. 2005b) this linear correlation of sulfur-*LSU1*-GSH-*IAA28* would therefore provide a suggestion for an in depth analysis investigating this link. Interestingly LSU has been linked to ethylene metabolism, thus linking auxin and ethylene effects under sulfate depleted conditions (Sirko et al. 2015).

Gene annotation, especially for Arabidopsis has greatly improved over recent years (cf. www.tair.org). When assessing the subnetwork in Fig. 2 we could initially identify only 7 of the 39 displayed gene models (Nikiforova et al. 2005b). However, by now all 39 genes have been annotated and at least putative functions have been assigned which are displayed in the Table 1. This allows re-assessing the data set of which we will display one example.

IAA28 (At5g25890) is linked to the "downstream" genes At5g42650, At1g16890, At1g48690 in this subnetwork (Fig. 2) which are part of the next hierarchical layer. At5g42650 has already been annotated as allene oxide synthase (AOS) being involved in jasmonic acid (JA) biosynthesis providing a link between auxin and JA metabolism when plants are exposed to a reduced availability of sulfate. JAs trigger multiple stress response pathways and influence developmental processes (Malek et al. 2002). At1g16890 has been newly annotated as ubiquitin conjugating enzyme 13B (UBC13B) which functions in root development. An *ubc13* knockout mutant displays shorter primary roots, less lateral roots and fewer root hairs. Aux/IAA (indole-3-acetic acid) proteins, as for example IAA17, accumulate in the mutant leading to a reduced auxin response and, hence, the observed phenotype (Wen et al. 2014). This perfectly relates to the described function of IAA28 under sulfate deprived conditions (Falkenberg et al. 2008) with which UBC13 is linked in the network (Fig. 2). It has still to be elucidated whether IAA28 is acting upstream of UBC13, which we would assume due to its position in the network hierarchy, or whether IAA28 is a target for UBC13 driven poly-ubiquitination. The third gene, At1g48690, has now been annotated as an auxin-responsive GH3 family protein (Gutierrez et al. 2012), again logically linking to the *IAA28* gene. Auxin response factors ARF6, ARF8, and ARF17 form a complex regulatory circuit, which regulates a set of auxin inducible genes, termed Gretchen Hagen3 (*GH3*) genes, *GH3.3*, *GH3.5*, and *GH3.6*, encoding acyl-acid-amido synthetases. These three *GH3* genes have been shown by Gutierrez et al. (2012) to be required for fine-tuning adventitious root initiation in *Arabidopsis thaliana* acting by modulating JA homeostasis. Thus, this subset of genes linked to the auxin responsive IAA28 provides a complex link to root development under sulfate deprivation involving a crosstalk between JA, auxin, and possibly also ethylene. The next steps, thus, will be to analyze the interconnectivity of the mentioned genes to a greater detail to identify their exact position in the regulatory circuit in response to sulfate deprivation.

As such further parts of the network will be worth re-assessing. Likewise, it would be meaningful to perform a network analysis using more complete data sets available now covering more data points. Examples for such kinds of extended investigations are the studies by Watanabe et al. (2010, 2013), Bielecka et al. (2015), Hubberten et al. (2012a) and Aarabi et al. (2016), which are all based on the pioneering work re-assessed in this paper.

In summary, a re-iterative cycle of identification of important elements within a response network, be it genes and metabolites or other parameters such as protein phosphorylation patterns or fluxes, through systems biology approaches is combined to targeted analyses of the identified elements using all available methods, among them again high throughput systems analysis (Fig. 3). It is further important to cross boundaries and determine overlaps between various response modules. One

Table 1 Functional annotation and hierarchical representation of the sub-network presented in Fig. 2

Layer	Metabolite	Gene	Annotation 2005	Annotation 2016 (TAIR)
1	sulfur			
2		At3g44320	+	NIT3 (NITRILASE 3)
		At3g27830		RPL12-A (RIBOSOMAL PROTEIN L12-A)
		At3g49580		LSU1 (RESPONSE TO LOW SULFUR 1)
		At5g53350		CLPX (CLP PROTEASE REGULATORY SUBUNIT X)
		At5g19440		CAD (cinnamyl-alcohol dehydrogenase, putative)
		At5g13800		PPH (PHEOPHYTINASE)
		At3g01420		DOX1 (ALPHA-DIOXYGENASE 1)
		At4g37520		Peroxidase superfamily protein
		At4g34720		AVA-P1 (VACUOLAR H+-PUMPING ATPASE C1)
		At4g30470		NAD(P)-binding Rossmann-fold superfamily protein
		At2g47650		UXS4 (UDP-XYLOSE SYNTHASE 4)
		At5g52240		ATMP1 (MEMBRANE STEROID BINDING PROTEIN 1)
		At5g37600		GLN1;1 (GLUTAMINE SYNTHASE 1;1)
		At3g14210	+	ESM1 (EPITHIOSPECIFIER MODIFIER 1)
3		At1g28600	+	GDSL-motif esterase/acyltransferase/lipase
		At1g76680	+	OPR1 (12-OXOPHYTODIENOATE REDUCTASE 1)
		At4g02080		SAR2 (SECRETION-ASSOCIATED RAS SUPER FAMILY 2)
		At5g47820		FRA1 (FRAGILE FIBER 1)
		At4g11320		CP2 (CYSTEINE PROTEASE 2)
		At3g14240		Subtilase family protein
		At3g56800	+	CAM3 (CALMODULIN 3)
		At2g27530		PGY1 (PIGGYBACK1)
		At4g14030		SBP1 (SELENIUM-BINDING PROTEIN 1)
		At1g75270		DHAR2 (DEHYDROASCORBATE REDUCTASE 2)
	putrescine			
	asparagine			
	glutathione			
4		At2g30590		WRKY21 (WRKY DNA-BINDING PROTEIN 21)
		At3g57520		Raffinose alpha-galactosidase/SIP2 (SEED IMBIBITION 2)
		At5g21090		Leucine-rich repeat (LRR) family protein
		At5g62190		PRH75 (DEAD/DEAH box RNA helicase)
		At5g25890	+	IAA28 (INDOLE-3-ACETIC ACID INDUCIBLE 28)
5		At5g61410		RPE (D-RIBULOSE-5-PHOSPHATE-3-EPIMERASE)
		At1g07920		GTP binding Elongation factor Tu family protein
		At5g11770		NADH-ubiquinone oxidoreductase 20 kDa subunit
		At1g13060		PBE1 (20S PROTEASOME BETA SUBUNIT E1)
		At2g05440		GRP9 (GLYCINE RICH PROTEIN 9)
		At4g35410		Clathrin adaptor complex small chain family protein
		At5g42650	+	AOS (ALLENE OXIDE SYNTHASE)
		At1g16890		UBC13B (UBIQUITIN CONJUGATING ENZYME 13B)
		At1g48690		Auxin-responsive GH3 family protein
		At3g48990		AAE3 (ACYL-ACTIVATING ENZYME 3)

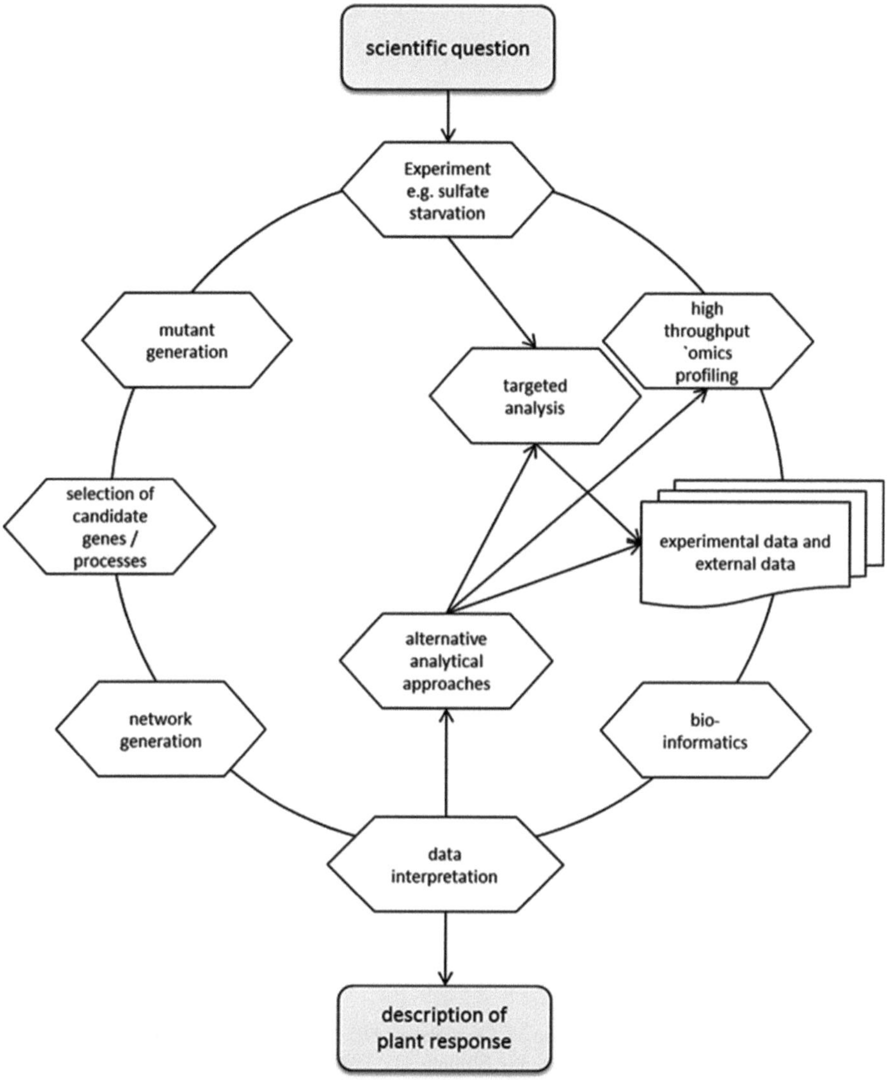

Fig. 3 Process scheme of a systems biology analysis. The process can be repeated (*outer circle*) or alternative analytical approaches can be employed to achieve a consistent answer to the original question

example is that nutrient starvation inevitably culminates into a specific nutrient depletion induced senescence (NuDIS) overlapping, but not identical to developmental senescence. The task will be to determine the regulators controlling how and which parts of certain modules are used to evoke the proper physiological response of a plant to a given environmental or biological challenge.

Acknowledgements We thank the Max Planck Society (MPG) for funding.

References

Aarabi F, Kusajima M, Tohge T, Konishi T, Gigolashvili T, Takamune M, Sasazaki Y, Watanabe M, Nakashita H, Fernie AR, Saito K, Takahashi H, Hubberten HM, Hoefgen R, Maruyama-Nakashita A (2016) Sulfur deficiency-induced repressor proteins optimize glucosinolate biosynthesis in plants. Sci Adv 2:e1601087

Bielecka M, Watanabe M, Morcuende R, Scheible WR, Hawkesford MJ, Hesse H, Hoefgen R (2015) Transcriptome and metabolome analysis of plant sulfate starvation and resupply provides novel information on transcriptional regulation of metabolism associated with sulfur, nitrogen and phosphorus nutritional responses in Arabidopsis. Front Plant Sci 5:805

Falkenberg B, Witt I, Zanor MI, Steinhauser D, Mueller-Roeber B, Hesse H, Hoefgen R (2008) Transcription factors relevant to auxin signalling coordinate broad-spectrum metabolic shifts including sulphur metabolism. J Exp Bot 59:2831–2846

Fernie AR, Trethewey RN, Krotzky AJ, Willmitzer L (2004) Metabolite profiling: from diagnostics to systems biology. Nat Rev Mol Cell Biol 5:763–769

Fiehn O, Kopka J, Dörmann P, Altmann T, Trethewey RN, Willmitzer L (2000) Metabolite profiling for plant functional genomics. Nat Biotechnol 18:1157–1161

Forieri I, Sticht C, Reichelt M, Gretz N, Hawkesford MJ, Malagoli M, Wirtz M, Hell R (2017) System analysis of metabolism and the transcriptome in *Arabidopsis thaliana* roots reveals differential co-regulation upon iron, sulfur and potassium deficiency. Plant Cell Environ 40:95–107

Gutierrez L, Mongelard G, Floková K, Pacurar DI, Novák O, Staswick P, Kowalczyk M, Pacurar M, Demailly H, Geiss G, Bellini C (2012) Auxin controls Arabidopsis adventitious root initiation by regulating jasmonic acid homeostasis. Plant Cell 24:2515–2527

Hagen JB (2000) The origins of bioinformatics. Nat Rev Genet 1:231–236

Hirai MY, Fujiwara T, Awazuhara M, Kimura T, Noji M, Saito K (2003) Global expression profiling of sulfur-starved Arabidopsis by DNA macroarray reveals the role of *O*-acetyl-L-serine as a general regulator of gene expression in response to sulfur nutrition. Plant J 33:651–663

Hirai MY, Yano M, Goodenowe DB, Kanaya S, Kimura T, Awazuhara M, Arita M, Fujiwara T, Saito K (2004) Integration of transcriptomics and metabolomics for understanding of global responses to nutritional stresses in *Arabidopsis thaliana*. Proc Natl Acad Sci USA 101 (27):10205–10210

Hirai MY, Klein M, Fujikawa Y, Yano M, Goodenowe DB, Yamazaki Y, Kanaya S, Nakamura Y, Kitayama M, Suzuki H, Sakurai N, Shibata D, Tokuhisa J, Reichelt M, Gershenzon J, Papenbrock J, Saito K (2005) Elucidation of gene-to-gene and metabolite-to-gene networks in arabidopsis by integration of metabolomics and transcriptomics. J Biol Chem 280:25590–25595

Hoefgen R, Nikiforova VJ (2008) Metabolomics integrated with transcriptomics: assessing systems response to sulfur-deficiency stress. Physiol Plant 132:190–198

Hubberten HM, Klie S, Caldana C, Degenkolbe T, Willmitzer L, Hoefgen R (2012a) Additional role of *O*-acetylserine as a sulfur status-independent regulator during plant growth. Plant J 70:666–677

Hubberten HM, Drozd A, Tran BV, Hesse H, Hoefgen R (2012b) Local and systemic regulation of sulfur homeostasis in roots of *Arabidopsis thaliana*. Plant J 72:625–635

Malek B, van der Graaff E, Schneitz K, Keller B (2002) The Arabidopsis male-sterile mutant *dde2-2* is defective in the ALLENE OXIDE SYNTHASE gene encoding one of the key enzymes of the jasmonic acid biosynthesis pathway. Planta 216:187–192

Maruyama-Nakashita A, Inoue E, Watanabe-Takahashi A, Yamaya T, Takahashi H (2003) Transcriptome profiling of sulfur-responsive genes in Arabidopsis reveals global effects of sulfur nutrition on multiple metabolic pathways. Plant Physiol 132:597–605

Maruyama-Nakashita A, Nakamura Y, Yamaya T, Takahashi H (2004) Regulation of high-affinity sulphate transporters in plants: towards systematic analysis of sulphur signalling and regulation. J Exp Bot 55:1843–1849

Maruyama-Nakashita A, Nakamura Y, Tohge T, Saito K, Takahashi H (2006) Arabidopsis SLIM1 is a central transcriptional regulator of plant sulfur response and metabolism. Plant Cell 18:3235–3251

Nikiforova V, Freitag J, Kempa S, Adamik M, Hesse H, Hoefgen R (2003) Transcriptome analysis of sulfur depletion in *Arabidopsis thaliana*: interlacing of biosynthetic pathways provides response specificity. Plant J 33:633–650

Nikiforova V, Gakière B, Kempa S, Adamik M, Willmitze L, Hesse H, Hoefgen R (2004) Towards dissecting nutrient metabolism in plants: a systems biology case study on sulfur metabolism. J Exp Bot 55:1861–1870

Nikiforova VJ, Kopka J, Tolstikov V, Fiehn O, Hopkins L, Hawkesford MJ, Hesse H, Hoefgen R (2005a) Systems rebalancing of metabolism in response to sulfur deprivation, as revealed by metabolome analysis of Arabidopsis plants. Plant Physiol 138:304–318

Nikiforova VJ, Daub CO, Hesse H, Willmitzer L, Hoefgen R (2005b) Integrative gene-metabolite network with implemented causality deciphers informational fluxes of sulphur stress response. J Exp Bot 56:1887–1896

Roessner U, Luedemann A, Brust D, Fiehn O, Linke T, Willmitzer L, Fernie AR (2001) Metabolic profiling allows comprehensive phenotyping of genetically or environmentally modified plant systems. Plant Cell 13:11–29

Sasaki-Sekimoto Y, Taki N, Obayashi T, Aono M, Matsumoto F, Sakurai N, Suzuki H, Hirai MY, Noji M, Saito K, Masuda T, Takamiya K, Shibata D, Ohta H (2005) Coordinated activation of metabolic pathways for antioxidants and defence compounds by jasmonates and their roles in stress tolerance in Arabidopsis. Plant J 44:653–668

Sirko A, Wawrzyńska A, Rodríguez MC, Sęktas P (2015) The family of LSU-like proteins. Front Plant Sci 5:774

Watanabe M, Hubberten HM, Saito K, Hoefgen R (2010) General regulatory patterns of plant mineral nutrient depletion as revealed by *serat* quadruple mutants disturbed in cysteine synthesis. Mol Plant 3:438–466

Watanabe M, Balazadeh S, Tohge T, Erban A, Giavalisco P, Kopka J, Mueller-Roeber B, Fernie AR, Hoefgen R (2013) Comprehensive dissection of spatiotemporal metabolic shifts in primary, secondary, and lipid metabolism during developmental senescence in Arabidopsis. Plant Physiol 162:1290–1310

Wen R, Wang S, Xiang D, Venglat P, Shi X, Zang Y, Datla R, Xiao W, Wang H (2014) UBC13, an E2 enzyme for Lys63-linked ubiquitination, functions in root development by affecting auxin signaling and Aux/IAA protein stability. Plant J 80:424–436

Whitcomb SJ, Heyneke E, Aarabi F, Watanabe M, Hoefgen R (2014) Mineral nutrient depletion affects plant development and crop yield. In: Hawkesford MJ, Kopriva S, De Kok LJ (eds) Nutrient use efficiency in plants: concepts and approaches. Springer, Cham, pp 205–228

Wulff-Zottele C, Gatzke N, Kopka J, Orellana A, Hoefgen R, Fisahn J, Hesse H (2010) Photosynthesis and metabolism interact during acclimation of *Arabidopsis thaliana* to high irradiance and sulphur depletion. Plant Cell Environ 33:1974–1988

Combining Isotope Labelling with High Resolution Liquid Chromatography-Tandem Mass Spectrometry to Study Sulfur Amino Acid Metabolism in Seeds of Common Bean (*Phaseolus vulgaris*)

Jaya Joshi, Justin B. Renaud, Mark W. Sumarah, and Frédéric Marsolais

Abstract Sulfur plays a crucial role in plant metabolism, protein biosynthesis, homeostasis and defense mechanisms. Like in other grain legumes, the protein quality of common bean is limited by the sub-optimal concentration of methionine and cysteine. *S*-methylcysteine is a non-proteinogenic sulfur amino acid, characteristically found in the *Phaseolus* as well as *Vigna* species. In mature seeds it accumulates in the form of its dipeptide γ-glutamyl-*S*-methylcysteine. In human diet, this sulfur containing amino acid and its dipeptide cannot substitute cysteine or methionine. Despite the inverse relationship between the concentration of cysteine and methionine versus *S*-methylcysteine, to date very little is known about the biosynthesis of the latter in common bean. Here, we developed a method, combining stable isotope label tracking with high resolution liquid chromatography-tandem mass spectrometry to investigate the pathways of *S*-methylcysteine and γ-glutamyl-*S*-methylcysteine biosynthesis in the developing seed of common bean.

Sulfur plays a key role in cell metabolism. Sulfur amino acids are essential for protein synthesis and play a crucial role in cell homeostasis through the biosynthesis of glutathione (GSH; γ-glutamyl-cysteinyl-glycine) (Hernández et al. 2015). In plants, sulfur provided in the form of sulfate is assimilated, reduced and incorporated into amino acids for further downstream protein and metabolite synthesis. Sulfate from soil

J. Joshi • F. Marsolais (✉)
Department of Biology, University of Western Ontario, 1151 Richmond St., London, ON N6A 3K7, Canada

Genomics and Biotechnology, London Research and Development Centre, Agriculture and Agri-Food Canada, 1391 Sandford St., London, ON N5V 4T3, Canada
e-mail: frederic.marsolais@agr.gc.ca

J.B. Renaud • M.W. Sumarah
Genomics and Biotechnology, London Research and Development Centre, Agriculture and Agri-Food Canada, 1391 Sandford St., London, ON N5V 4T3, Canada

L.J. De Kok et al. (eds.), *Sulfur Metabolism in Higher Plants - Fundamental, Environmental and Agricultural Aspects*, Proceedings of the International Plant Sulfur Workshop, DOI 10.1007/978-3-319-56526-2_13

is assimilated into adenosine 5′-phosphosulfate, which is reduced to sulfide for incorporation into cysteine (Mugford et al. 2011). Cysteine acts as the first organic compound containing reduced sulfur synthesized by the plant (Takahashi et al. 2011). During cysteine biosynthesis, the amino acid backbone is derived by serine metabolism via *O*-acetylserine (OAS) while sulfur comes from sulfate uptake and assimilatory sulfate reduction (Leustek et al. 2000). Further, cysteine acts as a precursor of methionine and GSH biosynthesis (Hell and Wirtz 2011). GSH plays a critical role in homeostasis and cellular defense, including redox status, signal transduction and detoxification (Noctor et al. 2011). Some other fates of cysteine include phytochelatins, iron-sulfur clusters, vitamin cofactors, and the biosynthesis of multiple secondary metabolites (Bonner et al. 2005). Apart from these metabolites some plant species also synthesize non-proteinogenic *S*-amino acid derivatives that might act as a storage sink for assimilated sulfur (Table 1). In *Allium* spp. the *S*-alk(en)yl-cysteine sulfoxides give rise to their characteristic aroma and flavour. Common bean (*Phaseolus vulgaris*) and several *Vigna* species accumulate *S*-methylcysteine (*S*-methylCys) and its dipeptide, γ-glutamyl-*S*-methylcysteine (γ-Glu-*S*-methylCys), which are non-proteinogenic in nature (Zacharius et al. 1959; Giada et al. 1998; Taylor et al. 2008).

Protein quality in legumes is compromised due to suboptimal levels of the sulfur amino acids, cysteine and methionine. Among all legumes produced in the world, common bean is considered as one of the best for human consumption (Broughton et al. 2003). However, major seed proteins present in common bean, such as the 7S globulin phaseolin and lectin phytohaemagglutinin have a low concentration of methionine and cysteine (Sathe 2002; Montoya et al. 2010). Despite having a low concentration of methionine and cysteine, common bean seeds accumulate approximately 0.4 nmol/mg of free *S*-methylCys and 9.1 nmol/mg of γ-Glu-*S*-methylCys (Yin et al. 2011), whereas total *S*-methylCys can reach up to 21.8 nmol/mg (Taylor et al. 2008). These *S*-amino derivatives accumulate exclusively in the seeds (Watanabe et al. 1971). They cannot substitute methionine or cysteine in the human diet (Padovese et al. 2001). An inverse relationship was previously observed between total cysteine and methionine vs. *S*-methylCys concentration in common bean. Cysteine was elevated by 70%, and methionine by 10% in SMARC1N-PN1 compared to SARC1, which are genetically related lines that differ in their storage protein composition. Increase in cysteine and methionine was mostly at the expense of the non-protein amino acid *S*-methylCys, suggesting that *S*-methylCys and γ-Glu-*S*-methylCys act as storage forms of excessive sulfur that cannot be accommodated in the protein pool (Taylor et al. 2008).

Although the primary sulfur assimilation pathway in common bean is well studied, the biosynthesis of *S*-methylCys and its dipeptide, γ-Glu-*S*-methylCys is not fully understood. In this study, we developed a method using ^{13}C and ^{15}N labelled serine or cysteine in a feeding experiment to elucidate the biosynthesis of *S*-methylCys and γ-Glu-*S*-methylCys in the developing seed of common bean. The fate of the labelled ($^{13}C_3$, ^{15}N) serine or cysteine was monitored by a combination of targeted and non-targeted high resolution mass spectrometry (HRMS).

For this study, common bean *(Phaseolus vulgaris)* genotype BAT93 plants were grown in growth cabinets (Environmental Growth Chambers, Chagrin Falls, OH,

Table 1 Sulfur amino acid derivatives reported in different plant species

Amino acid derivatives	Plant species	References
S-methylCys	*Phaseolus vulgaris, Vigna unguiculata, and Vigna radiata*	Baldi and Salamini (1973) and Evans and Boulter (1975)
S-methylCys sulfoxides	*Brassica* species	Fales et al. (1987) and Marks et al. (1992)
γ-Glu-*S*-methylCys	*P. vulgaris, V. unguiculata, and V. radiata*	Kasai et al. (1986) and Giada et al. (1998)
γ-Glu-methionine	*Vigna mungo*	Otoul et al. (1975)
γ-Glu-*S*-ethenylcysteine	*Vicia narbonensis*	Arias et al. (2005) and Sanchez-Vioque et al. (2011)
S-Alk(en)yl-cysteine sulfoxides	*Allium* species	Jones et al. (2004), Rose et al. (2005), and Yoshimoto et al. (2015)
S-Methylhomoglutathione	*V. radiata and P. vulgaris*	Kasai et al. (1986) and Liao et al. (2013)

USA) under 16 h light (300–400 μmol photons m^{-2} s^{-1}) at a temperature cycling between 18 and 24 °C as described previously (Pandurangan et al. 2012). Seeds were germinated in vermiculite and 12-day old seedlings were transplanted to pots (17 × 20 cm) containing Pro-Mix BX (Premier Tech, Rivière du Loup, Québec, Canada).

Previously, free amino acids were profiled in BAT93 seeds by HPLC. Free *S*-methylCys biosynthesis takes place during early seed developmental stages, such as stages III – heart stage, and IV and V – cotyledon stages (Walbot et al. 1972). In early developing stages, *S*-methylCys concentration was equal to 0.40 nmol per mg seed weight while accumulation of γ-Glu-*S*-methylCys began later in development, from stage VI – maturation to mature seed (Yin et al. 2011). To track the incorporation of stable isotopes in *S*-methylCys, we chose to feed the labelled precursors, cysteine and methionine, to developing seeds at stage IV – cotyledon. Developing pods were harvested, surface sterilized with 0.5% bleach and soap water for 5 min. This was followed with three washes for 10 min each and dissection for seed collection. Every seed was weighed and immediately the seed coat was removed for improved uptake of the labelled amino acids. Cotyledon weight was taken before transferring them to 25 ml standard line cell culture flasks (VWR, Mississauga, Ontario, Canada). These flasks have a filtered vent to provide oxygen for the developing seeds. Each flask contained six cotyledons from three developing seeds and 2.5 ml of filter-sterilized culture media. Culture media contained all essential components as described previously (Obendorf et al. 1983) with some modification: 0.1 mM Na_2EDTA was replaced with 0.1 mM EDTA ferric sodium salt and 0.1 mM $CoCl_2.6H_2O$ was also added to the media. Seeds were grown with different concentrations of labelled serine or cysteine to determine the optimal concentrations for the feeding experiments. Amino acid extracts from these seeds were analyzed by targeted liquid chromatography-tandem mass spectrometry (LC-MS/

MS) to determine the uptake of cysteine and serine relative to endogenous levels. Supplementation of growth media with 8 mM serine for 24 h resulted in sufficient uptake of ^{13}C and ^{15}N in serine but not in downstream metabolites such as *O*-acetylserine (OAS). This concentration was used in a time course experiment in order to determine the optimal incubation time. Insufficient incubation time would prevent the incorporation into important intermediates and downstream products while a long incubation time could lead to over dilution of the labelled compound. Seeds collected after 24 and 48 h showed efficient incorporation of isotopologues in endogenous serine and cysteine pools as well as in OAS. Therefore, the optimal time of incubation in culture media supplemented by 8 mM amino acid was determined to be 2 days. Using these optimized conditions, 42 seeds in 14 cell culture flasks and 33 seeds in 11 cell culture flasks were incubated with labelled cysteine or serine, respectively. Three groups of treatments were designed based on culture media: serine supplementation, cysteine supplementation and no supplementation. In each treatment group, six cotyledons were grown in a cell culture flask having either labelled or unlabelled serine or cysteine. One group of treatments was not supplemented with serine or cysteine and acted as treatment control. Culture flasks were kept horizontally at room temperature on a shaker. For better seed development, flasks were kept under continuous light and slow shaking (50 rpm). After completion of incubation, seeds were washed three times for 15 min each with sterile water to remove any traces of amino acids on the surface. Seeds were dried and stored at −80 °C following flash freezing in liquid nitrogen. For free amino acid extraction, the frozen seeds were homogenized using steel beads in 1.5 ml Eppendorf tubes. The ground seeds were extracted in ethanol:water (70:30) which is optimal for sulfur containing γ-glutamyl dipeptides (Kasai et al. 1986). Dried amino acids were reconstituted in a 500 µl methanol:water (50:50) solution and filtered through 0.2 µm filters (Pall Life Sciences, Mississauga, Ontario, Canada) into an amber glass HPLC vial. The samples were prepared immediately prior to MS analysis.

In recent years, pairing stable isotope labelling of metabolites with LC-HRMS analysis has provided information on the biosynthesis of targeted compounds and allowed for the relative quantification of cellular compounds (Allen et al. 2009; Creek et al. 2012; Glaser et al. 2014; You et al. 2014; Allen 2016). Here, targeted LC-MS/MS was used to obtain product ion spectra of the unlabelled compounds that could be involved in the biosynthetic pathway of *S*-methylCys. The formula of the major product ions of these compounds were determined by accurate mass and used to map MS/MS dissociation pathways of the precursor ion. These dissociation pathways were then used to predict the number of stable isotopes on the ions in the products that would be synthesized if $^{13}C_3$, ^{15}N cysteine or $^{13}C_3$, ^{15}N serine were incorporated. High resolution LC-MS/MS was then used to monitor both, the unlabelled and predicted isotopically labelled compounds (Table 2).

A challenge in detecting *S*-methylCys in the samples was the presence of homocysteine, which is isobaric to *S*-methylCys. With the use of HRMS/MS we were able to meet this challenge. A secondary product of *S*-methylCys is $C_3H_5O_2$ (73.02900) that arises following the neutral loss of CH_2S. Lost CH_2S carbon is from

Table 2 Product ions tracked for unlabelled and labelled (*) compounds

Name	Neutral formula	*m*/*z*	*m*/*z**	Product ion 1	*m*/*z*	*m*/*z**	Product ion 2	*m*/*z*	*m*/*z**
Serine	$C_3H_7NO_3$	106.0499	110.0570	C_2H_6ON	60.0451	63.0481	$C_3H_6O_2N$	88.0397	92.0464
Homoserine	$C_4H_9NO_3$	120.0655	124.0726	C_3H_8ON	74.0606	77.0638	$C_4H_8O_2N$	102.0552	106.0605
Cysteine	$C_3H_7NO_2S$	122.0270	126.0341	C_2H_6NS	76.0222	79.0253	C_3H_3OS	86.9904	90.0000
S-MethylCys	$C_4H_9NO_2S$	136.0427	140.0498	$C_4H_7O_2S$	119.0161	122.0262	$C_3H_5O_2$	73.0290	76.0385
OAS	C_5H9NO_4	148.0604	152.0675	$C_3H_8O_3N$	106.0499	110.0570	$C_3H_6O_2N$	88.0396	92.0464
Methionine	$C_5H_{11}NO_2S$	150.0583	154.0654	$C_4H_{10}NS$	104.0530	107.0566	$C_5H_9O_2S$	133.0316	136.0418
S-Methylmethionine	$C_6H_{13}NO_2S$	164.0740	168.0811	$C_6H_{11}O_2S$	147.0474	150.0575			
Cystathionine	$C_7H_{14}N_2O_4S$	223.0747	227.0818	$C_4H_8O_2NS$	134.0268	137.0304			
γ-Glu-*S*-methylCys	$C_9H_{16}N_2O_5S$	265.0853	269.0924	$C_4H_7O_2S$	119.0160	122.0262	$C_4H_{10}O_2NS$	136.0423	140.0498
GSH	$C_{10}H_{17}N_3O_6S$	308.0911	312.0982	$C_5O_3N_2SH_{11}$	179.0485	183.0556			
HomoGSH	$C_{11}H_{19}N_3O_6S$	322.1067	326.1138	$C_6H_{13}O_3N_2S$	193.0637	197.0712			
S-Methylhomoglutathione	$C_{12}H_{21}N_3O_6S$	336.1224	340.1295	$C_7H_{15}O_3N_2S$	207.0793	211.0867			
S-Adenosylmethionine	$C_{15}H_{22}N_6O_5S$	399.1445	403.1516	$C_{10}H_{12}O_3N_5$	250.0928	250.0928	$C_{10}H_{20}O_7NS$	298.0962	298.0962

S-methyl and is not labelled as expected. This product ion does not occur with homocysteine and was used to distinguish *S*-methylCys from homocysteine. The secondary product ion of a $^{13}C_3{}^{15}N$ labelled *S*-methylCys would be $^{13}C_3H_5O_2$ (76.03899) after losing CH_2S. The product ions observed strongly agree with the expected positions of labelled atoms in *S*-methylCys and helped to distinguish between two sulfur containing amino acids *S*-methylCys and homocysteine, which share the same molecular weight.

MS data were acquired with a Q-Exactive Orbitrap mass spectrometer (Thermo Fisher Scientific, Mississauga, Ontario, Canada) coupled to an Agilent 1290 HPLC system (Mississauga, Ontario, Canada). Two μL were injected onto an Agilent Zorbax Eclipse Plus RRHD C18 column (2.1 × 50 mm, 1.8 μM) which was maintained at 35 °C. Mobile phase A (0.1% formic acid in LC-MS grade H_2O, Thermo Fisher Scientific) was maintained at 100% for 1.25 min. Mobile phase B (0.1% formic acid in LC-MS grade acetonitrile, Thermo Fisher Scientific) was increased to 50% over 1.75 min, and 100% over 0.5 min. Mobile phase B was held at 100% for 1.5 min and returned to 0% over 0.5 min). The following heated electrospray ionization (HESI) parameters were optimized for the analysis of *S*-methylCys: spray voltage, 3.9 kV; capillary temperature, 250 °C; probe heater temperature, 450 °C; sheath gas, 30 arbitrary units; auxiliary gas, 8 arbitrary units; and S-Lens RF level, 60%. MS/MS was performed at 17,500 resolution, automatic gain control (AGC) target of $1e^6$, maximum injection time (IT) of 60 ms and isolation window of 1.0 *m*/*z*. A top 5 data-dependent acquisition (DDA) method was comprised of a full MS scan at 35,000 resolution, AGC target of $3e^6$, maximum IT of 125 ms, scan range between *m*/*z* 70–450 and intensity threshold of $7.7e^5$. The MS/MS scan conditions were identical to the targeted MS/MS method. Normalized collision energy of 25 was used for both MS/MS and DDA methods. Data were analyzed and all theoretical masses were calculated with Xcalibur™ software. The proportion of labelled compound present was calculated by dividing the intensity of the labelled signal over the sum of labelled and unlabelled intensities.

Following amino acid extraction, 80 ± 4% of the serine content within the $^{13}C_3{}^{15}N$ serine treatment was labelled. Similarly, 68 ± 15% of the cysteine content was labelled in the $^{13}C_3{}^{15}N$ cysteine treatment. These data suggest efficient uptake of serine and cysteine by developing seeds from the growth media. To track cysteine biosynthesis in developing seeds, an initial target was the level of isotope labels in OAS. In the treatment group supplemented with labelled serine, 78 ± 6% of the OAS pool was labelled, while labelled cysteine treated seeds showed no incorporation of isotopes in OAS which is in harmony with prior studies where OAS was shown to be a precursor of cysteine biosynthesis (Hell and Wirtz 2011).

Under the proposed pathways, both a $^{13}C_3{}^{15}N$ serine or $^{13}C_3{}^{15}N$ cysteine precursor would yield a $^{13}C_3{}^{15}N$ *S*-methylCys. The major product ion of *S*-methylCys arising following the neutral loss of NH_3 (17.0265 Da) is $C_4H_7O_2S$ (119.0161). The major product ion of a $^{13}C_3{}^{15}N$ labelled *S*-methylCys would be $^{13}C_3CH_7O_2S$ (122.0262) after losing $^{15}NH_3$ (18.0236 Da; Fig. 1). MS/MS of both the labelled and unlabelled *S*-methylCys showed the carbon on the CH_2S neutral fragment is unlabelled, indicating that the *S*-methyl group is not derived from the labelled

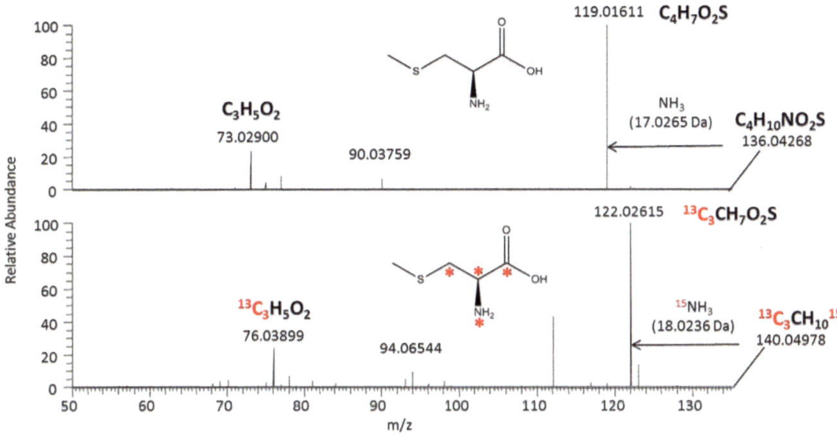

Fig. 1 Representative dissociation of protonated unlabelled *S*-methylCys and labelled *S*-methylCys. *Upper panel*: Product ions of unlabelled *S*-methylCys. *Lower panel*: Product ions of labelled *S*-methylCys

serine. The ability to determine the locations of labelled and unlabelled atoms within a larger compound demonstrates the benefits of isotope tracking by MS/MS over single stage MS.

The dipeptide *γ*-Glu-*S*-methylCys, which accumulates in maturing common bean seed was also monitored by LC-MS/MS in order to identify its biosynthetic precursors. *γ*-Glu-*S*-methylCys ($C_9H_{16}N_2O_5S$) has *m*/*z* of 265.0853 for its protonated product ion. Upon collision induced dissociation, two major product ions were produced including a *S*-methylCys fragment $C_4H_{10}O_2NS$ (*m*/*z* 136.0423) and deaminated *S*-methylCys $C_4H_7O_2S$ (*m*/*z* 119.0160); the incorporation of labelled serine or cysteine into *γ*-Glu-*S*-methylCys would result in product ions $^{13}C_3CH_{10}O_2$ ^{15}NS (*m*/*z* 140.0493) and $^{13}C_3CH_7O_2S$ *m*/*z* 122.0261, respectively (Fig. 2).

Interestingly, results of our feeding experiment with serine or cysteine showed similar percentage of isotope label incorporation in either labelled cysteine or serine treatments in *γ*-Glu-*S*-methylCys. These results point out needs for looking at labelling patterns in other sulfur containing compounds such as homoglutathione and its derivatives to investigate the possibility of more than one pathway for *γ*-Glu-*S*-methylCys biosynthesis.

Single stage HRMS is commonly used in biosynthetic studies to monitor stable isotope incorporation, however, we determined that using high resolution MS/MS greatly reduced the level of background interferences and furthermore, provided evidence for the location of the labelled isotopes within the larger molecule itself. In addition to targeted LC-MS/MS, samples were also analyzed with a non-targeted LC-DDA method to provide a dataset which could be mined retrospectively for other compounds that were not initially predicted to be involved in biosynthetic pathways.

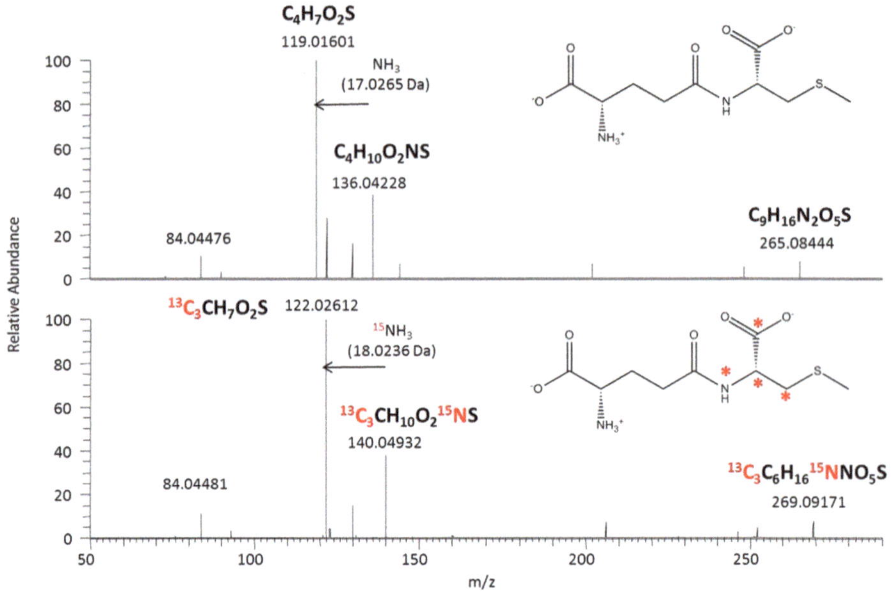

Fig. 2 Product ion spectra of protonated unlabelled γ-Glu-*S*-methylCys and labelled γ-Glu-*S*-methylCys. *Upper panel*: Product ions of unlabelled γ-Glu-*S*-methylCys. *Lower panel*: Product ions of labelled γ-Glu-*S*-methylCys

In summary, in the past, several strategies have been developed to understand the sulfur amino acid metabolism in legumes. Use of high resolution MS and isotope labelling techniques will help to overcome limitation with sulfur containing metabolite detection and will lead to better understanding of cysteine and methionine derivatives which play crucial role in protein quality and protein quantity of legumes.

References

Allen DK (2016) Quantifying plant phenotypes with isotopic labeling and metabolic flux analysis. Curr Opin Biotechnol 37:45–52

Allen DK, Ohlrogge JB, Shachar-Hill Y (2009) The role of light in soybean seed filling metabolism. Plant J 58:220–234

Arias M, Simo C, Ortiz LT, De Ios M-PM, Barbas C, Cifuentes A (2005) Detection and quantitation of a bioactive compound in *Vicia narbonensis* L seeds by capillary electrophoresis-mass spectrometry: a comparative study with UV detection. Electrophoresis 26:2351–2359

Baldi G, Salamini F (1973) Variability of essential amino acid content in seeds of 22 *Phaseolus* species. Theor Appl Genet 43:75–78

Bonner ER, Cahoon RE, Knapke SM, Jez JM (2005) Molecular basis of cysteine biosynthesis in plants: structural and functional analysis of *O*-acetylserine sulfhydrylase from *Arabidopsis thaliana*. J Biol Chem 280:38803–38813

Broughton WJ, Hernández G, Blair M, Beebe S, Gepts P, Vanderleyden J (2003) Beans (*Phaseolus* spp.) – model food legumes. Plant Soil 252:55–128

Creek DJ, Chokkathukalam A, Jankevics A, Burgess KEV, Breitling R, Barrett MP (2012) Stable isotope-assisted metabolomics for network-wide metabolic pathway elucidation. Anal Chem 84:8442–8447

Evans IM, Boulter D (1975) *S*-Methyl-L-cysteine content of various legume meals. Qual Plant 24:257–261

Fales SL, Gustine DL, Bosworth SC, Hoover RJ (1987) Concentrations of glucosinolates and *S*-methylcysteine sulfoxide in ensiled rape (*Brassica napus* L.) J Dairy Sci 70:2402–2405

Giada MDLR, Miranda MTM, Marquez UML (1998) Sulphur γ-glutamyl peptides in mature seeds of common beans (*Phaseolus vulgaris* L.) Food Chem 61:177–184

Glaser K, Kanawati B, Kubo T, Schmitt-Kopplin P, Grill E (2014) Exploring the Arabidopsis sulfur metabolome. Plant J 77:31–45

Hell R, Wirtz M (2011) Molecular biology, biochemistry and cellular physiology of cysteine metabolism in *Arabidopsis thaliana*. Arabidopsis Book 9:e0154

Hernández LE, Sobrino-Plata J, Montero-Palmero MB, Carrasco-Gil S, Flores-Cáceres ML, Ortega-Villasante C, Escobar C (2015) Contribution of glutathione to the control of cellular redox homeostasis under toxic metal and metalloid stress. J Exp Bot 66:2901–2911

Jones MG, Hughes J, Tregova A, Milne J, Tomsett AB, Collin HA (2004) Biosynthesis of the flavour precursors of onion and garlic. J Exp Bot 55:1903–1918

Kasai T, Shiroshita Y, Sakamura S (1986) γ-Glutamyl peptides of *Vigna radiata* seeds. Phytochemistry 25:679–682

Leustek T, Martin MN, Bick JA, Davies JP (2000) Pathways and regulation of sulfur metabolism revealed through molecular and genetic studies. Annu Rev Plant Biol 51:141–165

Liao D, Cram D, Sharpe AG, Marsolais F (2013) Transcriptome profiling identifies candidate genes associated with the accumulation of distinct sulfur γ-glutamyl dipeptides in *Phaseolus vulgaris* and *Vigna mungo* seeds. Front Plant Sci 4:60

Marks HS, Hilson JA, Leichtweis HC, Stoewsand GS (1992) *S*-Methylcysteine sulfoxide in *Brassica* vegetables and formation of methyl methanethiosulfinate from Brussels sprouts. J Agric Food Chem 40:2098–2101

Montoya CA, Lallès J-P, Beebe S, Leterme P (2010) Phaseolin diversity as a possible strategy to improve the nutritional value of common beans (*Phaseolus vulgaris*). Food Res Int 43:443–449

Mugford SG, Lee BR, Koprivova A, Matthewman C, Kopriva S (2011) Control of sulfur partitioning between primary and secondary metabolism. Plant J 65:96–105

Noctor G, Queval G, Mhamdi A, Chaouch S, Foyer CH (2011) Glutathione. Arabidopsis Book 9: e0142

Obendorf RL, Rytko GT, Byrne MC (1983) Soybean seed growth and maturation by *in vitro* pod culture. Ann Bot 51:217–227

Otoul E, Marechal R, Dardenne G, Desmedt F (1975) Des dipeptides soufres differencient nettement *Vigna radiata* de *Vigna mungo*. Phytochemistry 14:173–179

Padovese R, Kina SM, Barros RMC, Borelli P, Lanfer Marquez UM (2001) Biological importance of γ-glutamyl-*S*-methylcysteine of kidney bean (*Phaseolus vulgaris L*). Food Chem 73:291–297

Pandurangan S, Pajak A, Molnar SJ, Cober ER, Dhaubhadel S, Hernandez-Sebastia C, Kaiser WM, Nelson RL, Huber SC, Marsolais F (2012) Relationship between asparagine metabolism and protein concentration in soybean seed. J Exp Bot 63:3173–3184

Rose P, Whiteman M, Moore PK, Zhu YZ (2005) Bioactive *S*-alk(en)yl cysteine sulfoxide metabolites in the genus *Allium*: the chemistry of potential therapeutic agents. Nat Prod Rep 22:351–368

Sanchez-Vioque R, Rodriguez-Conde MF, Vioque J, Giron-Calle J, Santana-Meridas O, De-Los-Mozos-Pascual M, Izquierdo-Melero ME, Alaiz M (2011) A colorimetric method for determination of gamma-glutamyl-*S*-ethenyl-cysteine in narbon vetch (*Vicia narbonensis* L) seeds. Anal Biochem 418:180–183

Sathe SK (2002) Dry bean protein functionality. Crit Rev Biotechnol 22:175–223
Takahashi H, Kopriva S, Giordano M, Saito K, Hell R (2011) Sulfur assimilation in photosynthetic organisms: molecular functions and regulations of transporters and assimilatory enzymes. Annu Rev Plant Biol 62:157–184
Taylor M, Chapman R, Beyaert R, Hernández-Sebastià C, Marsolais F (2008) Seed storage protein deficiency improves sulfur amino acid content in common bean (*Phaseolus vulgaris* L.): redirection of sulfur from γ-glutamyl-*S*-methyl-cysteine. J Agric Food Chem 56:5647–5654
Walbot V, Clutter M, Sussex I (1972) Reproductive development and embryogeny in *Phaseolus*. Phytomorphology 22:59–68
Watanabe T, Tsugawa M, Takayama N, Furukawa Y (1971) Changes of free amino acids of each organ in the development and growth of kidney bean plant. J Hokkaido Univ Ed Sect II A 22:45–54
Yin F, Pajak A, Chapman R, Sharpe A, Huang S, Marsolais F (2011) Analysis of common bean expressed sequence tags identifies sulfur metabolic pathways active in seed and sulfur-rich proteins highly expressed in the absence of phaseolin and major lectins. BMC Genomics 12:268
Yoshimoto N, Yabe A, Sugino Y, Murakami S, Sai-Ngam N, Sumi S, Tsuneyoshi T, Saito K (2015) Garlic γ-glutamyl transpeptidases that catalyze deglutamylation of biosynthetic intermediate of alliin. Front Plant Sci 5:1–11
You L, Zhang B, Tang YJ (2014) Application of stable isotope-assisted metabolomics for cell metabolism studies. Meta 4:142–165
Zacharius RM, Morris CJ, Thompson JF (1959) The isolation and characterization of γ-L-glutamyl-*S*-methyl-L-cysteine from kidney beans (*Phaseolus vulgaris*). Arch Biochem Biophys 80:199–209

Investigations on the Current Sulfur and Sulfate Intake of Cattle in Germany: Are There Any Risks for a Consumption Exceeding Recommended Upper Limits?

Josef Kamphues, Anne Dohm, F. von und zur Mühlen, and Petra Wolf

Abstract The primary goal of the investigation was a deeper knowledge on the current sulfur and sulfate intake of dairy cows and beef cattle fed common rations in Germany based on roughages and supplemented with concentrates (as protein and energy sources). Finally, specific feeding practices/conditions should be identified that could result in adverse effects in cattle because upper levels for sulfur intake ($\geq$4 g/kg dry matter) were exceeded. In grass silages (especially from the 3rd/4th cut, at using S containing fertilizers) values near to 4 g S/kg dry matter were not rare; unexpectedly high contents of sulfur and sulfate were measured in compound feeds (median: 3.70 g S/kg dry matter; >40% derived from sulfates!). These high values are caused by higher proportions of components that contain higher amounts of sulfur and sulfates physiologically (like rapeseed products) or contain high sulfate levels due to diverse processes in the production (DDGS/sugar beet pulp/corn gluten feed) for example like the use of sulfuric acid or sulfur dioxide. Finally, it is recommended to test more frequently and systematically the sulfate content in grass and grass silages (indicating the S supply of plants) but also in concentrates (with an unexpected high variation of sulfur and sulfate) to identify further routes of sulfur entrance.

Sulfur is an essential element for the ruminal microflora to produce S containing amino acids and further nutrients like vitamins (for example biotin and thiamine). Therefore, the sulfur content should reach values of ~2 g S/kg dry matter in the entire ration (McDowell 2003); on the other hand the S content should not exceed values >4 g/kg dm (NRC 2005) in the whole ration due to predisposing effects

J. Kamphues (✉) • A. Dohm • F. von und zur Mühlen
Institute for Animal Nutrition, University of Veterinary Medicine Hannover, Foundation, Bischofsholer Damm 15, 30173 Hanover, Germany
e-mail: josef.kamphues@tiho-hannover.de

P. Wolf
University of Rostock, Faculty for Agricultural and Environmental Sciences, Chair of Nutrition Physiology and Animal Nutrition, Justus-von-Liebig-Weg 6b, 18059 Rostock, Germany

L.J. De Kok et al. (eds.), *Sulfur Metabolism in Higher Plants - Fundamental, Environmental and Agricultural Aspects*, Proceedings of the International Plant Sulfur Workshop, DOI 10.1007/978-3-319-56526-2_14

regarding the polioencephalomalacia (PEM) and further undesired consequences (McAllister 2004). According to Dänicke and Schenkel (2009) the sulfur levels of grass and grass silages (samples from southern parts of Germany) did not exceed upper levels (~4 g S/kg dry matter); in whole plant corn silages the S contents often vary around ~1 g/kg dry matter, not meeting the S requirements of high producing cattle. In America distinct concentrates (e.g. DDGS, corn gluten feed) are on debate for their high and risky sulfur contents (Drewnoski et al. 2014; Amat et al. 2013, 2014). Regarding sulfate contents of feedstuffs very few data exist up to now contrary to the sulfate contents in water for drinking with a high variation due to geological conditions (NRC 2005; Kamphues et al. 2007). Kamphues et al. (1999) reported high sulfate levels in distinct milk replacers for calves resulting in diarrhea due to laxative effects when young calves were fed liquid diets with more than 600–800 mg SO_4^{2-}/l diet. On the other hand Stemme et al. (2003) published a review on condensed molasses (from sugar beet) sometimes containing excessive levels of sulfates (up to 100 g/kg dm) that were tolerated by fattening bulls without any sign of illness. There are two trends that might affect the intake of sulfur and sulfate under current conditions of feeding dairy cows but also of feeding beef cattle: More and more S containing fertilizers are used in green fodder production, including grassland management. Furthermore, increasing amounts of by-products are used in the feed industry producing the compound/supplementary feeds and concentrates for cattle. Thus, there were two main goals of these investigations: The primary aim was to generate basic data on the sulfur and sulfate contents in feedstuffs representing common feeding practices for dairy cows and for beef cattle in Germany. The second goal was a critical look on the relevance of sulfate as an inorganic sulfur source and its contribution to the total S content of feedstuffs and thereby to the total S intake of cattle.

For analyzing the sulfur and sulfate contents in roughages (green fodder, grass, hay, and silages) samples of different origin were available: The majority of samples were grown in northern parts of Germany, most of them submitted by farmers and veterinary practitioners between 2012 and 2015 for quality control (nutritive value, hygienic status, success of preservation). Secondary, there were samples available from further institutions (LUFA North-West, Oldenburg; Lower Saxony Chamber of Agriculture) sent in as dried and ground samples. The different concentrates (especially protein rich components) also were of diverse origin (submitted for analyzing other nutrients), some of them were taken at farms, further ones were conceded by other institutions engaged in feed control and animal nutrition. Finally, various feedstuffs derived from own sampling in the field or at feed manufacturers. The dried and ground samples of feedstuffs were analyzed on the total S content (as well as on nitrogen content) by the Dumas technique [oxidative combustion and determination of both elements (N, S) by thermal conductivity detection]. The sulfate content was measured gravimetrically (boiling up the dried, ground samples with hydrochloric acid (37%) followed by precipitation of sulfates with $BaCl_2$, and weighing as $BaSO_4$). To demonstrate the relevance of sulfate as a source for total sulfur content its relative proportion was calculated in general (SO_4-S in percent of total S). The results are expressed as median value due to the non-normal distribution found in most feedstuffs.

The most important roughages used for cattle feeding in Germany and their contents of sulfur and sulfate are presented in Table 1. At first it has to be underlined that there was a large variation in the sulfur and in the sulfate contents of roughages, especially in grass and grass silages. Furthermore, there was an influence of the date of harvesting (first cut compared to the following ones). In grass 41–51% of the total S derived from sulfates, for grass silages these values were even higher (58–59%). In samples of grass but also of grass silages the total S contents varied on a high level (1st vs. following cuts: median of grass 2.19 vs. 3.78 g/kg dry matter; median of grass silages 2.58 vs. 3.76 g/kg dry matter) when compared to upper recommended levels (max. 4 g/kg dry matter). On the other hand, in corn silage lower values were found in general (median: 1.20 g/kg dry matter). Very low sulfate levels (0.31 up to 1.13 g) were measured in samples of whole plant corn silage, too.

From a field study regarding the impact of fertilization of grassland with sulfur, samples of grass were available for the investigation (Table 2). There was a marked effect of fertilization (N, S) on the total sulfur and sulfate content in grass: depending on the S input (0, 20, 40 kg S/ha) the sulfur and sulfate values increased, but additionally reliant on the intensity of nitrogen fertilization: At moderate N levels (220 kg N/ha) the sulfur content of grass increased by about 45% (0 vs. 20 kg S/ha) and by further 8.3% (20 vs. 40 kg S/ha). At high nitrogen fertilization (300 kg N/ha) there was a similar trend, but on a slightly lower level (+28% vs. +10%). It has to be underlined that S from sulfate related to total sulfur reached values of about 60% (Fig. 1).

As expected, low levels of sulfur without detectable sulfate contents were found in cereals. In dried sugar beet pulp largely moderate levels of sulfur (2.37 g S/kg dry matter) occurred but in some samples also high sulfate levels were proved (up to 12.7 g/kg dry matter). Two thirds of total sulfur stemmed from sulfates. Similar sulfur contents (median 4.4 g/kg dry matter) were measured in soybean meal; in this feed material unexpected sulfate values appeared, too (up to 5.87 g/kg dry matter) and here about 30.9% of total sulfur stemmed from sulfates. In general, high levels of sulfur and of sulfate were measured in rapeseed meal, namely about 7 g sulfur and about 7 g sulfate/kg dry matter. More than 30% of the total sulfur were contributed by sulfates (Table 3).

The highest sulfur and sulfate contents were found in DDGS, but with a huge variation: up to 12 g sulfur and up to 18.9 g sulfates per kg dry matter were determined in this by-product of bioethanol production. Here it is noteworthy that samples of DDGS differed markedly regarding the S content (Table 4).

It has to be emphasized that there was no correlation between the nitrogen and the sulfur content as it might be expected assuming that S containing amino acids (S-AA) are the main source of sulfur. Depending on the available roughages, the entire ration for dairy cows also consists of concentrates up to proportions of 50% of the total dm. Besides the use of cereals, dried sugar beet pulp, and soybean meal it is very common to include supplementary compound feeds in the rations (up to

Table 1 Sulfur/sulfate contents in roughages for cattle (samples from northern parts of Germany; median values and variation, min-max)

Feed material	n	Sulfur (g/kg dry matter)		Sulfate (g/kg dry matter)		Sulfate-S (%)[a]
Grass		Median	(min–max)	Median	(min–max)	
1st cut	25	2.19	(1.15–4.44)	3.07	(0.94–8.68)	41
Following cuts	17	3.78	(2.08–8.97)	6.52	(2.04–18.0)	51
Grass silage[b]						
1st cut	42	2.58	(2.23–4.16)	5.18	(2.37–8.16)	59
3rd cut	12	3.76	(2.35–4.58)	6.08	(4.56–9.36)	58
Corn silage[c] (whole plant)	13	1.20	(0.75–1.49)	0.89	(0.31–1.13)	25

[a]Sulfate-S, in percent of total S
[b]LUFA North-West: 14% of all samples (n = 681) of grass silage >3.5 g S/kg dry matter
[c]LUFA North West: 127 samples of corn silage: all in the range 0.9–1.3 g S/kg dry matter

Table 2 Sulfur and sulfate contents in grass related to the intensity of fertilization with nitrogen and sulfur

Fertilization (kg/ha)		Contents (g/kg dm)			
N	S	Crude protein	Sulfur	Sulfate	
220	0	150	2.90	4.96	(57.0)[a]
300		184	2.67	3.44	(43.0)[a]
220	20	160	4.21	7.55	(59.7)[a]
300		186	3.41	5.29	(51.7)[a]
220	40	156	4.56	8.34	(60.9)[a]
300		194	3.76	6.98	(61.8)[a]

Samples of the 3rd cut from July 2014; Lange (2015)
[a]Sulfate-S, in percent of total S

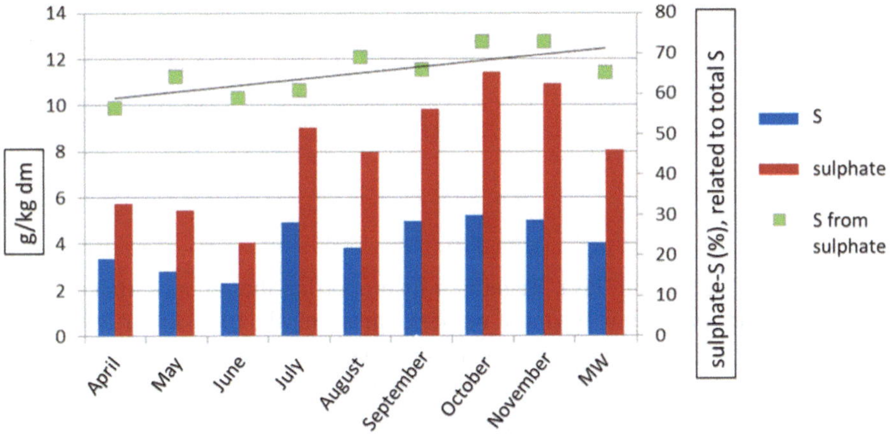

Fig. 1 Sulfur and sulfate content (g/kg dry matter) as well as sulfate-S related to total S (%) in grass from one single location in Lower Saxony in the course of a year

Table 3 Sulfur and sulfate contents (g/kg dry matter) of feed materials used in the production of compound/supplementary feeds for cattle

Feed material	n	Sulfur (g/kg dry matter)			Sulfate (g/kg dry matter)			Sulfate-S[a]
		Median	Min	Max	Median	Min	Max	%
Cereals	11	1.51	1.15	1.92	Not detectable			–
Dried sugar beet pulp	10	2.37	1.58	4.42	4.76	2.34	12.7	66.5
Soybean meal, extr.	9	4.37	3.78	4.84	4.46	1.89	5.87	30.9
Rapeseed meal, extr.	13	7.22	6.63	7.59	6.93	3.97	8.76	31.9
DDGS[b]	22	6.42	3.15	11.8	3.38	0.289	18.9	26.6

[a]Sulfate-S, in percent of total S
[b]Dried distillers grains with solubles

Table 4 Characterization of DDGS based on the differences in their total sulfur content[a]

	<5 g S/kg dry matter	>5 g S/kg dry matter
n	9	13
Sulfur (g/kg dry matter)	3.15–4.05	6.11–11.8
Nitrogen (g/kg dry matter)	43.2–60.5	49.4–91.5
N/S ratio	12.3–15.3	5.05–9.73
Sulfate (g/kg dry matter)	0.29–4.68	0.66–18.9
Sulfate-S (%) of total sulfur	2.78–43.8	3.08–65.0
Sulfur, derived from S-AA, %[b]	60.4–75.4	24.8–47.8

[a]Samples available from the Institute of Animal Production, University of Bonn (Agricultural Faculty)
[b]Calculated, assuming a constant amino acid pattern, independent of protein content

Table 5 Sulfur and sulfate contents of common supplementary compound feeds (n = 20) for dairy cows (Dohm 2015)

	Sulfur (g/kg dry matter)			Sulfate (g/kg dry matter)		
	Median	Min	Max	Median	Min	Max
Supplementary compound feeds for dairy cows	3.70	2.37	6.32	4.84	1.82	8.06

10–12 kg per animal per day). Regarding its proportion these concentrates are the most important ones, thus having a marked impact on the total sulfur and sulfate intake (Table 5). Supplementary compound feeds for beef cattle have a similar composition regarding sulfur and sulfate contents but the amounts fed per day are much lower (up to 3 kg). In all samples of supplementary feeds the nitrogen content (→ crude protein content) was additionally analysed and compared to the sulfur content: there was a significant weak correlation ($r^2 = 0.253$); but it has to be underlined that in the sample with the lowest protein content (126 g/kg dry matter) the highest level of sulfate (8.06 g/kg dry matter) was found.

Contrary to the majority of published data regarding the sulfur contents in feedstuffs it seems that sulfate belongs to the "forgotten" constituents of feedstuffs and ingredients. Although, Scharrer and Jung (1955) observed that the sulfate content of grass (*Lolium perenne*) increased much more than the content of S containing amino acids when the intensity of fertilization with sulfur was elevated. Up to 79% of the total S content in their experiments were contributed by sulfate and only 20% by the S containing amino acids. Furthermore, it is well known that sulfates serve as stored reserves for plants (or parts of plants), and that sulfates accumulate with increasing maturity (Hell 2002). This might explain the trend for higher S contents during the entire vegetation.

Since decades, by-products from the process of bioethanol production are marketed as DDGS and used in the feed industry as protein source for supplementary feeds for dairy cows and beef cattle. Comparing the low sulfur and sulfate levels (near zero) in cereals to the high values in the by-products of cereal fermentation (like DDGS) indicates that further ways of entrance for sulfur and/or sulfates must exist. Here the technology in the production of bioethanol earns special attention: the differences in the composition of DDGS are mainly caused by the use of sulfuric acid in the process of bioethanol production by yeasts on the basis of diverse cereals. Regarding rapeseed meal there is a different reason for the high values of sulfur: The high contents of cysteine and methionine result in corresponding high sulfur contents (methionine: 21.5% S; cysteine: 26.6% S), but only 60% of the sulfur content are accounted to this fraction. Thus, further constituents that result in higher sulfur values must be taken into consideration. Here the glucosinolates are worth to be mentioned: in spite of a high variation in the composition of the diverse glucosinolates the average proportion of sulfur is about 16% (Henkel and Mosenthin 1989). According to Schnug and Haneklaus (1990) there is a strong correlation between the S content and the glucosinolate content in rapeseed products (extracted meal, cakes). Based on the current findings on sulfates in rapeseed meal, about 32% of the total S is accounted for sulfates and about 60% for the S-AA.

As reported in earlier studies, very low levels of sulfur were found in samples of corn silage (whole plant corn silage) with a mean of 1.2 g/kg dry matter, the minimum value was 0.75 g/kg dm and the highest one reached 1.49 g/kg dry matter (Dohm 2015). Also in southern regions of Germany low levels (0.9–1.1 g/kg dry matter, according to Wessels 2002; 0.91–0.93 g S/kg dry matter, Walch 1998) that could result in a marginal S supply of the ruminal flora were found. Regarding the sulfur content of grass silages it seems that there is a trend for some higher values in northern parts of Germany (Müller and Engling 2015), especially with values increasing in the course of the year: in the 3rd and 4th cut the median values reached nearly 4 g/kg dry matter, presumably due to the fact that it is more common to use S containing fertilizers in this region. It has to be specially emphasized that sulfate is a main source for sulfur in grass and grass silages: 40 up to 60% of the total sulfur came from sulfate and not from the protein fraction (S-AA). Such high sulfate levels might indicate a surplus of available sulfur for the plants and maybe an impaired protein synthesis in the plants due to other reasons. Especially in dairy cow rations with high proportions of grass silage the sulfur levels may touch or exceed the recommended safe upper levels (max. 4 g S/kg dry matter).

Regarding the sulfur and sulfate levels in common compound feeds for dairy cows (used as supplementary feeds) it has to be underlined that the values vary near the "upper" level of 4 g/kg dry matter. But these findings are not astonishing because of the use of rapeseed products, DDGS, and by-products of the wet milling industry (like corn gluten feed). In this last group of by-products Myer and Hersom (2008) found a high variation of S contents (3.3–7.3 g S/kg dry matter). Furthermore, dried sugar beet pulp and condensed molasses may contain high levels of sulfate as found in earlier studies (cited by Stemme et al. 2003). Based on current data regarding the sulfur and sulfate contents in diverse feed materials and compound feeds, rations for dairy cows based on grass silage and supplemented with high amounts of compound feeds predispose for an intake of sulfur and sulfate that is near the recommended "upper" level of 4 g S/kg dm in the total ration. On the other hand, whenever higher proportions of corn silage supplemented with cereals and soybean meal are fed, the intake of sulfur and sulfate should be moderate. Only in cases of high sulfate levels in drinking water (in Germany rather rare!) there could be an unintended higher intake of sulfur.

According to Kamphues et al. (2014) high sulfur and/or sulfate intake of ruminants may be associated with different consequences/undesired effects:

- Lower dry matter intake (due to a loss of palatability or due to effects of H_2S in the central nervous system).
- Reduced utilization of distinct trace elements (especially of copper, selenium, and zinc).
- Impaired formation/metabolism of S-containing vitamins (especially of thiamine; fat soluble vitamins?)
- Impaired quality/composition of faeces, watery diarrhoea (primarily calves are affected due to "laxative effects of sulfate", adult ruminants are much more tolerant)
- Clinical symptoms due to polioencephalomalacia (as a consequence of the forced H_2S formation in the rumen and its inhalation from eructated gas → histological alterations in the brain, described as malacia and necrosis)

Summarizing lots of publications related to the sulfur intake Kamphues et al. (2014) stated that S contents per 1 kg dm of the whole ration:

≤1 g:	S requirements of the ruminal flora are not met
2 g:	even at high performance of cattle S-requirements are met in any case
3 g:	increased proneness to secondary effects (trace element utilisation↓) and to PEM when low fibre diets are fed (concentrate rich diets!)
4 g:	higher risk for PEM, even at higher roughage intake
≥5 g:	diverse clinically obvious reactions are to expect (diarrhoea, apathy, . . .), such rations should not be fed to ruminants in higher proportions.

Although it is known for decades that plants can accumulate sulfate as a physiological constituent it did not find the interest in feed science. According to different textbooks on animal nutrition most authors suppose that S containing amino acids are the predominant source of sulfur, but that assessment does not fit, at least not for various kinds of green fodder and some further concentrates. For the feed industry the fact that diverse by-products commonly used for producing compound/ supplementary feeds are characterized by highly variable contents of sulfur and sulfates is of special interest. Therefore, it is recommended to test the sulfate contents of roughages (need for S containing fertilizers?) but also of distinct concentrates (further way of entrance!) more frequently to avoid secondary effects of sulfur/sulfates exceeding the "normal" values and to detect an unintended high intake of sulfur and/or sulfates.

References

Amat S, McKinnon JJ, Olkowski AA, Penner GB, Simko E, Shand PJ, Hendrick S (2013) Understanding the role of sulfur-thiamine interaction in the pathogenesis of sulfur-induced polioencephalomalacia in beef cattle. Res Vet Sci 95:1081–1087

Amat S, McKinnon JJ, Penner GB, Hendrick S (2014) Effects of dietary sulfur concentration and forage-to-concentrate ratio on ruminal fermentation, sulfur metabolism, and short-chain fatty acid absorption in beef heifers. J Anim Sci 92:712–723

Dänicke S, Schenkel H (2009) Zur Bedeutung von Schwefel als unerwünschtem Stoff in der Milchkuhfütterung. Züchtungskunde 81:442–450. ISSN:0044-5401

Dohm A (2015) Der Schwefel- und Sulfat-Gehalt in Grund- und Kraftfuttermitteln – Einflüsse und mögliche Bedeutung (Sulfur and sulfate contents in roughages and concentrates – influences and possible significance). Dissertation, Tierärztliche Hochschule, Hannover

Drewnoski ME, Pogge DJ, Hansen SL (2014) High-sulfur in beef cattle diets: A review. J Anim Sci 92:3763–3780

Hell R (2002) Der assimilatorische Schwefelstoffwechsel in Pflanzen. Biospektrum-Heidelberg 8:248–251

Henkel H, Mosenthin R (1989) Rapssaat und Rapsprodukte in der Tierernährung. Übersichten zur Tierernährung 17:139–190

Kamphues J, Stolte M, Tschentscher A, Rust P (1999) Untersuchungen zu Sulfatgehalten in Milchaustauschern und Molkeprodukten sowie zur Bedeutung der Sulfataufnahme für die Kotbeschaffenheit bei Kälbern (Investigations on sulfate concentration in milk replacers and dried whey products and its effect on feces composition in calves). Dtsch Tierärztl Wschr 106:466–470

Kamphues J, Böhm R, Flachowsky G, Lahrssen-Wiederholt M, Meyer U, Schenkel H (2007) Empfehlungen zur Beurteilung der hygienischen Qualität von Tränkwasser für Lebensmittel liefernde Tiere unter Berücksichtigung der gegebenen rechtlichen Rahmenbedingungen. (Recommendations for evaluating the hygienic quality of drinking water for food producing animals in accordance with current regulatory frame work). Landbauforschung Völkenrode 57:255–272

Kamphues J, Dohm A, Zimmermann J, Wolf P (2014) Schwefel- und Sulfatgehalte in Futtermitteln – Noch oder wieder von Interesse? (Sulphur and sulphate contents in feedstuffs – still or again of interest?). Übersichten zur Tierernährung 42:81–139

Lange G (2015) Examinations and experiments regarding feed production during 2014 at the lower Saxony Chamber of Agriculture; personal communication 2015-03-02

McAllister MM (2004) Sulfur. In: Plumlee KH (ed) Clinical veterinary toxicology. Mosby-Elsevier, St. Louis, pp 133–138

McDowell LR (2003) Sulfur. In: Minerals in animal and human nutrition, 2nd edn. Elsevier Verlag, Amsterdam, pp 179–202

Müller W, Engling F-P (2015) personal communication 2015-05-26

Myer B, Hersom M (2008) Corn gluten feed for beef cattle; AN201, Department of Animal Science, UF/IFAS Extensions; www.edis.ifas.utl.edu. Last downloaded 2016-01-06

NRC (National Research Council) (2005) Sulfur – mineral tolerance of animals, 2nd revised edn. National academic Press, Washington, DC, pp 372–379

Scharrer K, Jung J (1955) Zur Bestimmung des Gesamt-, Protein- und Sulfat-Schwefels in pflanzlichen Futtermitteln. J Anim Physiol Anim Nutr 10:25–31

Schnug E, Haneklaus S (1990) A rapid method for the evaluation of the glucosinolate quality of extracted rapeseed meal by total sulfur determination via X-ray fluorescence spectroscopy. Lipid Sci Technol 92:57–61

Stemme K, Gerdes B, Harms A, Kamphues J (2003) Zum Futterwert von Zuckerrübenvinasse (Nebenprodukt aus der Melasseverarbeitung). (The nutritive value of condensed molasses soubles – by-product of sugar beet molasses manufacutring). Übersichten zur Tierernährung 31:169–201

Walch K (1998) Kationen-Anionen-Bilanz in Rationen trockenstehender Milchkühe und ihre Bedeutung für die Gebärparese. Diplomarbeit, Technische Universität, München-Weihenstephan

Wessels A (2002) Zum Einfluss einer unterschiedlichen Schwefelversorgung auf Leistungsparameter, Verdaulichkeit und Spurenelementgehalt in verschiedenen Geweben bei wachsenden Rindern. Dissertation, Technische Universität, München

Manganese Toxicity Hardly Affects Sulfur Metabolism in *Brassica rapa*

Mariana I. Neves, Dharmendra H. Prajapati, Saroj Parmar, Tahereh A. Aghajanzadeh, Malcolm J. Hawkesford, and Luit J. De Kok

Abstract Manganese (Mn) is an essential plant nutrient, though at elevated levels in plant tissues it may become toxic. The physiological basis for phytotoxicity is largely unclear. Exposure of *Brassica rapa* to elevated levels of Mn^{2+} in the nutrient solution resulted in decreased biomass production at $\geq$20 μM and chlorosis. The Mn content in the shoot increased with the Mn^{2+} concentration in the nutrient solution and became toxic when it exceeded a four-fold concentration of the control. In contrast to observations with Cu and Zn, elevated and toxic Mn^{2+} levels did not affect the water-soluble non-protein thiols in both root and shoot and the expression the sulfate transporters, Sultr1;1 and Sultr1;2, in the root.

Mn is an essential plant nutrient and its availability in soil strongly affects plant growth and development (Kováčik et al. 2014; Sadana et al. 2003). Mn functions in several physiological processes, viz. in photosynthesis, where it is associated with the water-oxidizing complex of photosystem II, which catalyzes the photosynthetic O_2 evolution (Mukhopadhyay and Sharma 1991; Millaleo et al. 2013). Moreover, Mn is an important cofactor of several enzymes, e.g., manganese-dependent superoxide dismutase (MnSOD), catalases, glycotransferases, pyruvate carboxylase, nitrate reductase and is involved in amino acid and lignin synthesis (Marschner 1995; Pedas et al. 2005; Humphries et al. 2006; Pittman 2008). Mn is taken up by the plant root as Mn^{2+}, which availability is strongly affected by the pH of the soil (Humphries et al. 2006; Socha and Guerinot 2014). In alkaline soils (high pH) Mn availability to plants may be low and deficiency may occur, whereas in in acidic soils (low pH) excessive availability may result in toxicity (Humphries et al. 2006;

M.I. Neves • D.H. Prajapati • L.J. De Kok (✉)
Laboratory of Plant Physiology, Groningen Institute for Evolutionary Life Sciences, University of Groningen, P.O. Box 11103, 9700 CC Groningen, The Netherlands
e-mail: l.j.de.kok@rug.nl

S. Parmar • M.J. Hawkesford
Plant Biology and Crop Science Department, Rothamsted Research, Harpenden, Hertfordshire AL5 2JQ, UK

T.A. Aghajanzadeh
Department of Biology, Faculty of Basic Science, University of Mazandaran, Babolsar, Iran

L.J. De Kok et al. (eds.), *Sulfur Metabolism in Higher Plants - Fundamental, Environmental and Agricultural Aspects*, Proceedings of the International Plant Sulfur Workshop, DOI 10.1007/978-3-319-56526-2_15

Mundus et al. 2012). Little is known about manganese transporters in plants, though the iron (Fe^{2+}), zinc (Zn^{2+}) and calcium (Ca^{2+}) family transporters seem to be the most probable candidates (Socha and Guerinot 2014). In the xylem, Mn is transported as Mn^{2+} ion or as complex with citrate or malate; in the shoot high levels of Mn may accumulate in the vacuole (Pittman 2005). The physiological basis for Mn toxicity is largely unclear. Differential tolerance of plants to manganese cannot solely be explained by a restricted Mn^{2+} uptake and transport to the shoot but additionally by intrinsic strategies that enhance cellular accumulation capacity (Foy et al. 1978). Sequestration into the vacuole, activity of antioxidant enzymes and formation of chelation complexes in the cytosol are some of the strategies proposed to promote toxic metal tolerance (Pittman 2005). Sulfur metabolites play a role in the detoxification of potential toxic metals (Yadav 2010). Complexation of sulfur compounds (e.g., cysteine, phytochelatins, metallothioneins) with toxic metal ions as a mechanism to overcome their toxicity is widely described for different elements and plant species (Ernst et al. 2008; Yadav 2010; Leitenmaier and Küpper 2013). Other sulfur metabolites, such as glutathione, are also crucial for antioxidant protection against reactive oxygen species, of which levels might be induced upon toxic metal stress (Na and Salt 2011). Some species capable of high manganese accumulation, such as *Phytolacca americana*, show a positive relationship between sulfur and manganese (Peng et al. 2008; Yadav 2010). Moreover, it has been observed that toxic metals, e.g., Cu, Zn may directly induce changes in sulfur uptake by affecting the activity of the sulfate transporters and affect the regulation of enzymes involved in S assimilation and activity of sulfate transporters (Nocito et al. 2002; Sun et al. 2007; Schiavon et al. 2008; Shahbaz et al. 2010, 2013; Na and Salt 2011; Stuiver et al. 2014). All changes induced in sulfur status of the plant could be linked a toxic metal-induced change in activity of the sulfate transporters (Yoshimoto et al. 2002; Sun et al. 2007; Stuiver et al. 2014). *Brassica* species have high sulfur requirements for growth (Ernst 2000) and are generally considered to be susceptible to Mn toxicity (Foy et al. 1978; Humphries et al. 2006; Lee et al. 2011). In the current paper the interaction between Mn and sulfur metabolism was studied in the *Brassica rapa*.

Brassica. rapa var. perviridis (Komatsuna) seeds were germinated in vermiculite and were subsequently transferred to an aerated 25% Hoagland nutrient solution, containing supplemental concentrations of 0, 10, 20, 50 and 100 μM $MnCl_2$ (pH 5.9) in 30 l plastic containers (20 sets of plants per container, three plants per set) in a climate-controlled room for 10 days. Day and night temperatures were 21 and 18 °C (±1 °C), respectively, relative humidity was 70–80%. The photoperiod was 14 h at a photon fluence rate of 300 ± 20 μmol m^{-2} s^{-1} (400–700 nm) at plant height, supplied by Philips GreenPower LED (deep red/white 120) production modules. After 10 days of Mn^{2+} exposure, plants were harvested 3 h after the start of the light period and shoots and roots separated and weighed. Shoot and root biomass production was calculated by subtracting pre-exposure weight from that after Mn^{2+} exposure. Shoot/root ratio was calculated from the shoot and root fresh weight after the exposure. For the determination of pigments and anions, plant material was frozen in liquid N_2 immediately after harvest and stored at −80 °C. For analysis of water-soluble non-protein thiols freshly harvested material was used. Chlorophyll a + b content, chlorophyll a fluorescence and the content of sulfate, nitrate, water-soluble non-protein thiol content, free amino acid and the Mn and S mineral nutrient content were determined as described by Shahbaz et al. (2010) and Stuiver et al. (2014). Total RNA was isolated from shoots and roots of *B. rapa* plants as described by Aghajanzadeh et al. (2014). The full length sequences of sulfur transporter genes are found under the following accession numbers: Sulfur transporter 1.1 (Sultr1;1 XM009128953), Sulfur transporter 1.2 (Sultr1;1 XM009108197, XM009108195 and XM009108196). Transcription was determined by quantitative real-time polymerase chain reaction (qRT-PCR; see Reich et al. 2017). Statistical analysis of the results was performed using unpaired Student's t-test. Different letters indicate significant differences at $P < 0.01$ between different treatments.

Exposure of *B. rapa* to elevated levels of Mn^{2+} in the nutrient solution resulted in decreased biomass production at $\geq$20 μM (Table 1). Shoot growth was relatively slightly more affected upon Mn^{2+} exposure than root growth resulting in a decrease in shoot to root ratio. Mn^{2+} exposure also resulted in a substantial increase in dry

Table 1 Impact of Mn^{2+} exposure on biomass production, dry matter content, chlorophyll a fluorescence and contents of pigments, nitrate and sulfate, amino acids and water-soluble non-protein thiols of *Brassica rapa*

	Mn^{2+} concentration (μM)				
	0	10	20	50	100
Shoot					
Biomass production (g FW)	0.97 ± 0.18c	0.85 ± 0.11c	0.73 ± 0.19bc	0.59 ± 0.14b	0.32 ± 0.06a
Dry matter content (%)	7.7 ± 0.3a	7.9 ± 0.3a	8.3 ± 0.3ab	8.9 ± 0.7bc	9.4 ± 0.4c
Chl a + b (mg g^{-1} FW)	0.69 ± 0.04b	0.61 ± 0.05b	0.54 ± 0.13ab	0.52 ± 0.09a	0.47 ± 0.05a
Chl a/b	2.4 ± 0.2a	2.6 ± 0.1a	2.5 ± 0.3a	2.2 ± 0.4a	2.7 ± 0.3a
Chl a + b/Car	3.0 ± 0.0a	2.9 ± 0.1ab	2.9 ± 0.2ab	2.9 ± 0.2ab	2.8 ± 0.1b
F_v/F_m	0.81 ± 0.03a	0.81 ± 0.05a	0.82 ± 0.04a	0.80 ± 0.05a	0.82 ± 0.05a
Manganese (μmol g^{-1} DW)	3.3 ± 0.2a	9.8 ± 0.6b	16.4 ± 1.1c	30.5 ± 0.4d	53.6 ± 4.7e
Sulfur (μmol g^{-1} DW)	241 ± 21a	277 ± 7a	265 ± 7a	289 ± 25ab	300 ± 10b
Sulfate (μmol g^{-1} FW)	14 ± 2a	13 ± 1a	15 ± 4a	26 ± 3b	34 ± 4b
Thiols (μmol g^{-1} FW)	0.56 ± 0.01a	0.57 ± 0.09a	0.56 ± 0.05a	0.59 ± 0.04a	0.60 ± 0.06a
Nitrate (μmol g^{-1} FW)	84 ± 7a	105 ± 1b	102 ± 4b	85 ± 3a	85 ± 0a
Amino acids (μmol g^{-1} FW)	14 ± 1a	14 ± 2a	15 ± 1a	15 ± 2a	16 ± 2a

(continued)

Table 1 (continued)

	Mn^{2+} concentration (μM)				
	0	10	20	50	100
Root					
Biomass production (g FW)	0.17 ± 0.03c	0.18 ± 0.03c	0.16 ± 0.05bc	0.12 ± 0.04b	0.07 ± 0.02a
Dry matter content (%)	6.5 ± 0.3a	7.1 ± 0.4a	7.0 ± 0.5a	7.9 ± 1.1ab	8.6 ± 0.5b
Manganese (μmol g^{-1} DW)	49 ± 5a	162 ± 6b	168 ± 75b	180 ± 53b	154 ± 19b
Sulfur (μmol g^{-1} DW)	241 ± 21a	277 ± 7a	265 ± 7a	289 ± 25ab	300 ± 10b
Sulfate (μmol g^{-1} FW)	28 ± 2a	23 ± 2a	27 ± 2a	25 ± 4a	22 ± 2a
Thiols (μmol g^{-1} FW)	0.42 ± 0.03a	0.41 ± 0.08a	0.44 ± 0.02a	0.45 ± 0.06a	0.53 ± 0.06a
Nitrate (μmol g^{-1} FW)	44 ± 5a	39 ± 2a	48 ± 3b	43 ± 4ab	42 ± 3a
Amino acids (μmol g^{-1} FW)	15 ± 1a	15 ± 2ab	18 ± 2ab	18 ± 1b	17 ± 2ab
Plant					
Shoot/root ratio	5.6 ± 0.4a	4.7 ± 0.3b	4.5 ± 0.7b	4.9 ± 0.9b	4.4 ± 0.8b

Ten day-old seedlings were grown on a 25% Hoagland nutrient solution containing supplemental concentrations of 0, 10, 20, 50 and 100 μM $MnCl_2$. The initial shoot and root fresh weights were 0.100 ± 0.01 g and 0.040 ± 0.01 g, respectively. Data on biomass production, dry matter, pigment and amino acid content represent the mean of two independent experiments, with a total of 12, 6, 6, and 6 measurements with 3 plants in each, respectively (± SD). Data on chlorophyll a fluorescence represents the mean of 10 measurements (± SD). Data on nitrate, sulfate and water-soluble non-protein thiol content represent the mean of 3 measurements with 3 plants in each (± SD). Different letters indicate significant differences at $P < 0.01$ between different treatments

matter content of both root and shoot at 100 μM. Mn^{2+} exposure resulted in interveinal chlorosis (especially of the younger leaves) at ≥50 μM, however the chlorophyll a/b and the chlorophyll/carotenoid ratios were hardly affected (Table 1). Chlorophyll a fluorescence (measured as the quantum yield of the photosynthetic system II photochemistry, F_v/F_m ratio) was not affected upon Mn^{2+} exposure, even at toxic levels (Table 1). The Mn content in the shoot increased with the Mn^{2+} concentration in the nutrient solution (Table 1). However, in the root there was a strong increase in the Mn content at 10 μM Mn^{2+}, which hardly increased further at higher Mn^{2+} concentrations. This increase was markedly higher in the shoot than in the root (Table 1). Evidently, Mn became toxic and reduced the biomass production when the content in the shoot was ≥16 μmol g^{-1} dry weight and exceeded four-fold of that of the control. Exposure of plants to toxic Mn^{2+} levels hardly affected the content of other essential mineral nutrients in both root and shoot (data not presented); there was only a 30% and 19% decrease at 100 μM in the K content of the root and shoot, respectively. Moreover, the Zn content increased

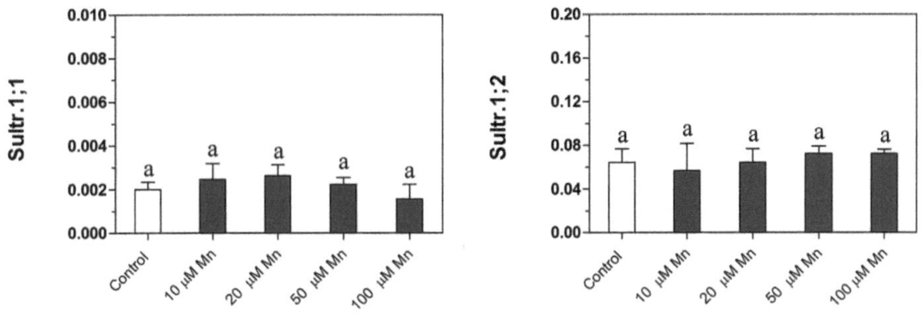

Fig. 1 Impact of Mn^{2+} exposure on the transcript levels of Sultr1;1 and Sultr1;2 in the root of *B. rapa*. For experimental details, see legends of Table 1. Relative gene expression of these genes was determined by qRT-PCR and the mRNA levels were compared to actin. Data on relative expression represent the mean of 3 measurements with 3 plants in each (±SD). *Different letters* indicate significant differences at $P < 0.01\%$ between different treatments

two-fold in both root and shoot; a similar increase in the Cu content was observed in the root. Mn^{2+} hardly affected the nitrate and free amino acid content of the plants (Table 1). The total sulfur and sulfur metabolite contents of *B. rapa* were only slightly affected at toxic Mn^{2+} levels. There was only a 1.25-fold increase in sulfur content in the shoot at 100 µM Mn^{2+}, which could be attributed to an increase of the sulfate content. Mn^{2+} exposure did not affect the total sulfur and sulfate content of the root and the water-soluble non-protein thiols in both root and shoot (Table 1). The sulfate transporters Sultr1;1 and Sultr1;2 are involved in the primary uptake of sulfate by the roots, though the transcript level of Sultr1;2 in the roots of *B. rapa* was 30-fold higher than that of Sultr1;1 (Fig. 1). The expression of these sulfate transporters were not affected upon Mn^{2+} exposure.

Similar to other essential potentially toxic metals, viz. Cu and Zn, exposure of *B. rapa* to elevated Mn^{2+} levels in the nutrient solution resulted in a strong accumulation of the metal in both root and shoot, resulting in decreased plant biomass production and chlororis of the shoot. *B. rapa* was much less susceptible to Mn than *B. pekinenis* to Cu and Zn toxicity: Mn^{2+} became toxic at ≥20 µM, whereas Cu^{2+} and Zn^{2+} already affected plant biomass production at ≥2 µM (Shahbaz et al. 2010, 2013, 2014; Stuiver et al. 2014). The decrease in biomass production due to toxic metal exposure was accompanied or even preceded by a decrease in pigment content (Foy et al. 1978, Shahbaz et al. 2010, 2013; Stuiver et al. 2014), although chlorophyll a fluorescence upon Cu^{2+} (Shahbaz et al. 2010) and Mn^{2+} exposure remained unaffected, which indicated that development rather than chloroplast functioning was negatively affected. High Mn levels also reduced the pigment content in tobacco (Clairmont et al. 1986), mungbean (Sinha et al. 2002), Chinese cabbage (Lee et al. 2011), spearmint (Asrar et al. 2005), tomato (Shenker et al. 2004) by affecting the chlorophyll, carotenoid and flavonoid biosynthesis (Clairmont et al. 1986; González and López 2013).

Exposure of *B. pekinensis* to elevated levels of Cu^{2+} and Zn^{2+} in the nutrient solution substantially affected the uptake, distribution and metabolism of sulfur (Shahbaz et al. 2010, 2013, 2014; Stuiver et al. 2014). Cu^{2+} and Zn^{2+} exposure resulted in an up-regulation of the activity of sulfate transporters and expression of the Group 1 sulfate transporters, viz. Sultr1;2, which are involved in the uptake of sulfate by the root in *Brassica* species (Shahbaz et al. 2010, 2013, 2014; Stuiver et al. 2014). The up-regulation of the sulfate transporters was most likely not due to a higher plant sulfur requirement upon Cu^{2+} and Zn^{2+} exposure, since it was accompanied by a substantial increase in the sulfate content of the shoot (Shahbaz et al. 2010, 2013, 2014; Stuiver et al. 2014). It was presumed that the up-regulation of the sulfate transporters was the consequence of a direct interference of these metal ions with the signal transduction pathway resulting in a disturbed regulation of the transporters (Shabaz et al. 2014; Stuiver et al. 2014). However, exposure of *B. rapa* to elevated Mn^{2+} levels did not affect the transcript levels of the sulfate transporters Sultr1;1 and Sultr1;2 in the root, despite a slight increase in the sulfate content in the shoot. The impact of elevated Mn^{2+} levels was also measured in *B. juncea* and all results on growth, pigment content and metabolite content were quite similar to that in *B. rapa*, with the exception that Mn toxicity was not accompanied with higher sulfur and sulfate contents in the shoot, and that it was even slightly decreased (data not presented). Apparently, in contrast to Cu^{2+} and Zn^{2+}, Mn^{2+} exposure did not interfere with the signaling of the regulation of the sulfate transporters. In general under normal conditions glutathione is the major water-soluble non-protein thiol compound present in plant tissues. Exposure of *B. pekinensis* to Cu^{+} (Shahbaz et al. 2010, 2013, 2014) and Zn^{2+} (Stuiver et al. 2014) resulted a strong increase in the water-soluble non-protein thiol content of the root and to a lesser extent in the shoot. This increase could partially be ascribed to an increase in phytochelatins (Shahbaz et al. 2010) and cysteine (Stuiver et al. 2014). An increase in water-soluble non-protein thiols (e.g., cysteine and glutathione) is expected as a defense mechanism against heavy metal toxicity (Leitenmaier and Küpper 2013). Moreover, cysteine and glutathione are the precursors for the synthesis of phytochelatins, which may complex with metals and increase toxic metal tolerance (Ernst et al. 2008). However, Mn^{2+} exposure did not affect the water-soluble non-protein content of both root and shoot of *B. rapa* (and *B. juncea*, data not presented). Apparently, an exposure of *B. rapa* to elevated and toxic Mn^{2+} levels did not trigger the synthesis of thiols (e.g. cysteine, glutathione and/or phytochelatins).

In conclusion, in contrast to Cu^{2+} and Zn^{2+}, elevated and toxic Mn^{2+} levels in the root environment hardly affected the uptake and metabolism of sulfate in *Brassica*.

Acknowledgement Rothamsted Research is supported via the 20:20 Wheat® Programme by the UK Biotechnology and Biological Sciences Research Council.

References

Aghajanzadeh T, Hawkesford MJ, De Kok LJ (2014) The significance of glucosinolates for sulfur storage in Brassicaceae seedlings. Front Plant Sci 5:704

Asrar Z, Khavari-Nejad RA, Heidari H (2005) Excess manganese effects on pigments of *Mentha spicata* at flowering stage. Arch Agron Soil Sci 51:101–107

Clairmont KB, Hagar WG, Davis EA (1986) Manganese toxicity to chlorophyll synthesis in tobacco callus. Plant Physiol 80:291–293

Ernst WHO (2000) Agriculture aspects of sulfur. In: Lens P, Pol LH (eds) Environmental technologies to treat sulfur pollution. IWA Publishing, London, pp 355–376

Ernst WHO, Krauss G-J, Verkleij JAC, Wesenberg D (2008) Interaction of heavy metals with the sulphur metabolism in angiosperms from an ecological point of view. Plant Cell Environ 31:123–143

Foy CD, Chaney RL, White MC (1978) The physiology of metal toxicity in plants. Annu Rev Plant Physiol 29:511–566

González MB, López JG (2013) Beneficial plant-microbial interactions: ecology and applications. CRC Press, London

Humphries JM, Stangoulis CR, Graham Robin D (2006) Manganese. In: Allen V, Barker DJP (eds) Handbook of plant nutrition. CRC Press, Boca Raton, pp 351–374

Kováčik J, Štêrbova D, Babula P, Švec P, Hedbavny J (2014) Toxicity of naturally-contaminated manganese soil to selected crops. J Agric Food Chem 62:7287–7296

Lee TJ, Luitel BP, Kang WH (2011) Growth and physiological response to manganese toxicity in Chinese cabbage (*Brassica rapa* L. ssp campestris). Hortic Environ Biotechnol 52:252–258

Leitenmaier B, Küpper H (2013) Compartmentation and complexation of metals in hyperaccumulator plants. Front Plant Sci 4:374

Marschner H (1995) Mineral nutrition of higher plants. Academic, London

Millaleo R, Reyes-Diaz M, Alberdi M, Ivanov AG, Krol M, Huner NP (2013) Excess manganese differentially inhibits photosystem I versus II in *Arabidopsis thaliana*. J Exp Bot 64:343–354

Mukhopadhyay MJ, Sharma A (1991) Manganese in cell metabolism of higher plants. Bot Rev 57:117–149

Mundus S, Lombi E, Holm PE, Zhang H, Husted S (2012) Assessing the plant availability of manganese in soils using Diffusive Gradients in Thin films (DGT). Geoderma 183-184:92–99

Na G, Salt DE (2011) The role of sulfur assimilation and sulfur-containing compounds in trace element homeostasis in plants. Environ Exp Bot 72:18–25

Nocito FF, Pirovano L, Cocucci M, Sacchi GA (2002) Cadmium-induced sulfate uptake in maize roots. Plant Physiol 129:1872–1879

Pedas P, Hebbern CA, Schjoerring JK, Holm PE, Husted S (2005) Differential capacity for high-affinity manganese uptake contributes to differences between barley genotypes in tolerance to low manganese availability. Plant Physiol 139:1411–1420

Peng K, Luo C, You W, Lian C, Li X, Shen Z (2008) Manganese uptake and interactions with cadmium in the hyperaccumulator *Phytolacca americana* L. J Hazard Mater 154:674–681

Pittman JK (2005) Managing the manganese: molecular mechanisms of manganese transport and homeostasis. New Phytol 167:733–742

Pittman JK (2008) Mechanisms of manganese accumulation and transport. In: Jaiwal PK, Singh RP, Dhankher OP (eds) Plant membrane and vacuolar transporters, 1st edn. CABI, Wallingford/Cambridge, pp 173–204

Reich M, Aghajanzadeh T, Helm J, Parmar S, Hawkesford MJ, De Kok LJ (2017) Chloride and sulfate salinity differently affect biomass, mineral nutrient composition and expression of sulfate transport and assimilation genes in *Brassica rapa*. Plant Soil 411:319–332

Sadana US, Samal D, Claassen N (2003) Differences in manganese efficiency of wheat (*Triticum aestivum* L) and raya (*Brassica juncea* L) as related to root-shoot relations and manganese influx. J Plant Nutr Soil Sci 166:385–389

Schiavon M, Pilon-Smits EA, Wirtz M, Hell R, Malagoli M (2008) Interactions between chromium and sulfur metabolism in *Brassica juncea*. J Environ Qual 37:1536–1545

Shahbaz M, Tseng MH, Stuiver CEE, Koralewska A, Posthumus FS, Venema JH, Parmar S, Schat H, Hawkesford MJ, De Kok LJ (2010) Copper exposure interferes with the regulation of the uptake, distribution and metabolism of sulfate in Chinese cabbage. J Plant Physiol 167:438–446

Shahbaz M, Parmar S, Stuiver CEE, Hawkesford MJ, De Kok LJ (2013) Copper toxicity and sulfur metabolism in Chinese cabbage are affected by UV radiation. Environ Exp Bot 88:60–70

Shahbaz M, Stuiver CEE, Posthumus FS, Parmar S, Hawkesford MJ, De Kok LJ (2014) Copper toxicity in Chinese cabbage is not influenced by plant sulphur status, but affects sulphur metabolism-related gene expression and the suggested regulatory metabolites. Plant Biol 16:68–78

Shenker M, Plessner OE, Tel-Or E (2004) Manganese nutrition effects on tomato growth, chlorophyll concentration, and superoxide dismutase activity. J. Plant Physiol 161:197–202

Sinha S, Mukherji S, Dutta J (2002) Effect of manganese toxicity on pigment content, Hill activity and photosynthetic rate of *Vigna radiata* L. Wilczek seedlings. J Environ Biol 23:253–257

Socha AL, Guerinot ML (2014) Mn-euvering manganese: the role of transporter gene family members in manganese uptake and mobilization in plants. Front Plant Sci 5:106

Stuiver CEE, Posthumus FS, Parmar S, Shahbaz M, Hawkesford MJ, De Kok LJ (2014) Zinc exposure has differential effects on uptake and metabolism of sulfur and nitrogen in Chinese cabbage. J Plant Nutr Soil Sci 177:748–757

Sun XM, Lu B, Huang SQ, Mehta SK, Xu LL, Yang ZM (2007) Coordinated expression of sulfate transporters and its relation with sulfur metabolites in *Brassica napus* exposed to cadmium. Bot Stud 48:43–54

Yadav SK (2010) Heavy metals toxicity in plants: an overview on the role of glutathione and phytochelatins in heavy metal stress tolerance of plants. S Afr J Bot 76:167–179

Yoshimoto N, Takahashi H, Smith FW, Yamaya T, Saito K (2002) Two distinct high-affinity sulfate transporters with different inducibilities mediate uptake of sulfate in Arabidopsis roots. Plant J 29:465–473

Localization of Sulfate Uptake and pH Changes at Sulfur-Deprived Roots of Intact *Brassica pekinensis* Seedlings by Using H^+-Selective Microelectrodes

Martin Reich, Marten Staal, Luit J. De Kok, and J. Theo M. Elzenga

Abstract Proton-selective microelectrodes were used to determine sulfate uptake by roots of intact plant seedlings. The response of H^+ fluxes to sulfate addition showed to be a good proxy for sulfate uptake by the sulfate/H^+ co-transport system. H^+ influx and increase in root surface pH was much higher in sulfate-deprived seedlings than in seedlings grown with sufficient sulfate. The opposite was true for the response of H^+ fluxes to nitrate addition. By using this method sulfate uptake could be mapped along the root axis, which revealed higher uptake rates in mature regions. Sulfate deprived roots showed a lower root surface pH, which correlated strongly with the response to sulfate addition. A possible contribution of this component to a higher sulfate uptake capacity under sulfur deficiency was further tested by using the fungal toxin fusicoccin, which permanently activates the plasma membrane H^+-pumping ATPase. Application of fusicoccin lowered the pH of sufficient roots to the level of deficient roots, indicating a more activated state of the ATPase under sulfur deficiency rather than a higher abundance.

Sulfate serves as the main source of sulfur to plants and is taken up by plant roots via a H^+-coupled symporter in the plasma membrane (Hawkesford et al. 1993; Smith et al. 1995; Hawkesford et al. 2003) and. The regulation of sulfate transport is well studied on the molecular, biochemical and whole plant level (Saito 2000; Buchner et al. 2004; Hopkins et al. 2005; Koralewska et al. 2008; Rouached et al. 2008; Reich et al. 2015) but knowledge on the distribution of uptake along the root axis is still scarce. The few studies on localization of sulfate fluxes along roots are rather indirect via localization of the sulfate transporter genes using *in situ* hybridization or GFP fusion (Takahashi et al. 1997; Yoshimoto et al. 2002). Electrophysiological studies applying ion selective microelectrodes offer the possibility to measure ion uptake in real-time and with high temporal and spatial resolution

M. Reich • M. Staal • L.J. De Kok • J.T.M. Elzenga (✉)
Laboratory of Plant Physiology, Groningen Institute for Evolutionary Life Sciences, University of Groningen, P.O. Box 11103, 9700 CC Groningen, The Netherlands
e-mail: j.t.m.elzenga@rug.nl

L.J. De Kok et al. (eds.), *Sulfur Metabolism in Higher Plants - Fundamental, Environmental and Agricultural Aspects*, Proceedings of the International Plant Sulfur Workshop, DOI 10.1007/978-3-319-56526-2_16

(Plassard et al. 1999; Newman 2001). Localization along the root axis of different plant species have been conducted for fluxes of e.g. nitrate, ammonium, potassium and H^+ (Henriksen et al. 1990; Taylor and Bloom 1998; Garnett et al. 2001; Rubinigg et al. 2002; Chen et al. 2005; Staal et al. 2011) and gave important insights into the acquisition of these nutrients in relation to nutrient supply and developmental processes. Application of ion-selective microelectrodes relies on highly sensitive and selective ion exchangers. To date, no ion exchanger with the required selectivity is available for sulfate. Instead H^+-selective micro-electrodes were used in the present study to determine and localize sulfate uptake activity along roots of intact seedlings of Chinese cabbage, a plant that is known for its high sulfur need and sulfate content (Koralewska et al. 2008). Because sulfate transporters are H^+/sulfate co-transporters (Hawkesford et al. 1993) sulfate addition should cause an immediate response of H^+ fluxes proportional to the coupled influx of sulfate. To test this assumption we compared fluxes at roots of plants that had been grown with sufficient sulfate in the medium to fluxes of sulfate-deprived plants. As in sulfate-deprived plants the sulfate uptake system is strongly upregulated (Buchner et al. 2004), larger H^+ fluxes in response to sulfate addition were expected. As an additional verification of the method $Mg(NO_3)_2$ was added to the roots instead of $MgSO_4$. Nitrate uptake is known to be depressed under sulfate deficiency and the response of H^+ fluxes in response to $Mg(NO_3)_2$ is expected to be higher in sulfate sufficient than in sulfate deficient roots. After these verifications the method was used to localize sulfate uptake along the root axis of sulfate-sufficient and sulfate-deprived roots. At the same time, differences in root surface pH were assessed.

Seeds of Chinese cabbage (*Brassica pekinensis* Rupr., cv. Michico) were germinated on filter paper, moistened with tap water and placed in the dark at 21 °C. After 4 days, seedlings were placed in a climate controlled room with a day/night temperature 21 °C/18 °C (±1 °C), a photoperiod of 14 h and a photon flux of 340 μmol m^{-2} s^{-1}. Seedlings were positioned such that the roots could grow in a layer of water. The seedlings were kept under these conditions for another 3 days, before being transferred from the filter paper to 13 l containers with aerated 25% Hoagland nutrient solution (pH 5.9; for composition see Reich et al. 2016) containing either 0.5 mM (control) or 0 mM $MgSO_4.7H_2O$ (sulfate-deprived, −S). The pH was controlled and adapted every 2–3 days to pH 5.9 with diluted HCl. As root excision will affect nutrient absorption (Bloom and Caldwell 1988; Shabala et al. 2009) intact seedlings were used in this study. Prior to performing an ion flux experiment, a plant was taken from the climate room and placed in a 25 ml petri dish and the primary root was mounted carefully in such a way that it was easily accessible with the ion selective electrode. The shoot was stabilized at the edge of the petri dish. Thereafter the root was covered with a low salt measuring solution (MS; containing 200 μM $MgCl_2$, 100 μM KCl and 100 μM $CaCl_2$). During the entire measurement the shoot of the plant was enclosed in a small cabinet and supplied with LED light with a photon flux of 250 μmol m^{-2} s^{-1}, while the roots were not illuminated (Fig. 1). The MS was continuously exchanged by a perfusion system with a perfusion rate of ca. 3 ml min^{-1}. The tip of the electrode was then moved to the root surface by using a threedimensional micromanipulator mounted

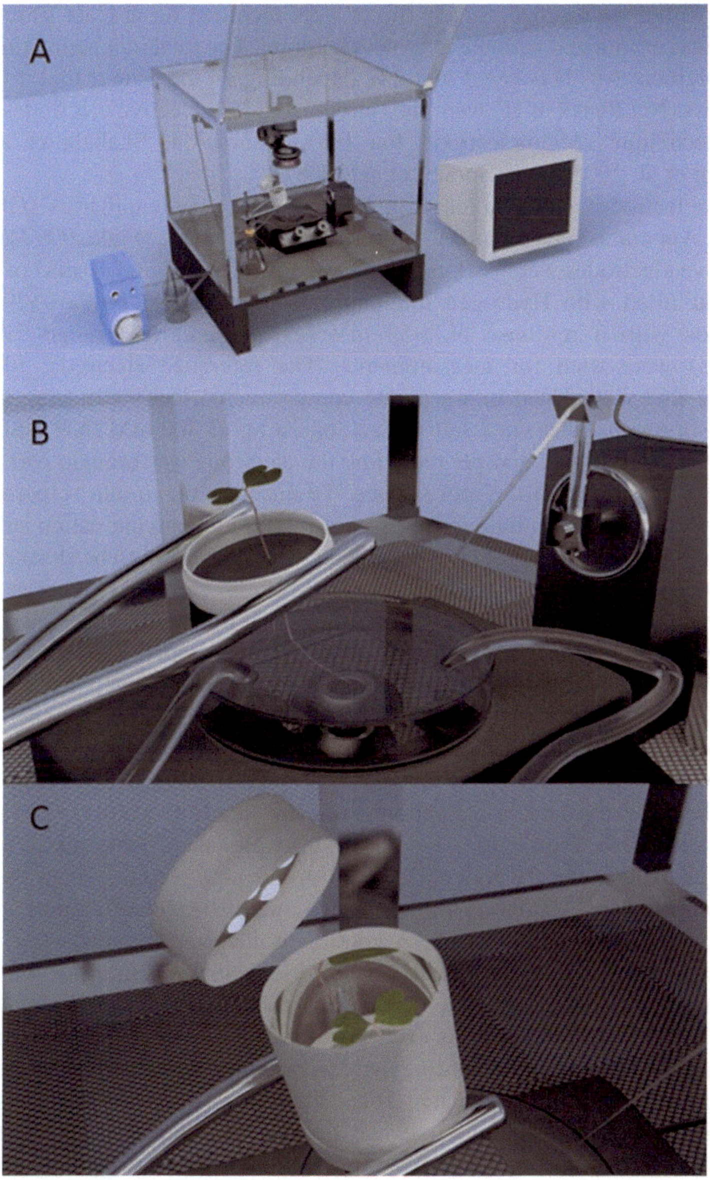

Fig. 1 (**a**) The measurement took place in a Faraday's cage in which an inverted microscope was installed to position the electrode at the root surface. The amplifiers, the pump for the perfusion system (*left*) and the PC for data read-out (*right*) were located outside of the cage. (**b**) The plant shoot was enclosed in the lower part of a small chamber and the root was immobilized in the petri dish which was filled with the measuring solution. The electrode was mounted on a 3D-micromanipulator. Inlet and outlet of the perfusion system were placed at the edges of the dish. (**c**) The plant shoot was fully enclosed and illuminated during the measurement (Visualization by Markus Reich)

on an inverted microscope. Steady fluxes were recorded for at least 10 min before the MS was exchanged with the treatment solution. The total incubation time before a measurement was between 1 and 2 h, depending on the time it took to record a stable flux. Net fluxes of H^+ were measured using H^+-selective electrodes with the MIFE technique (Microelectrode Ion Flux Estimation; Shabala et al. 1997; Vreeburg et al. 2005; Lanfermeijer et al. 2008).

Microelectrodes were pulled from borosilicate glass capillaries (GC150-10; Harvard Apparatus) and silanized with tributylchlorosilane (Fluka 90974). The H^+-selective electrodes were back filled with 15 mM of NaCl and 40 mM of KH_2PO_4 and front filled with Hydrogen Ionophore II (Cocktail A; Fluka 95297). Only electrodes with a response of >53 mV per pH unit (pH range 5.1–7.8, $r^2 > 0.998$) were used for measurements. The reference electrode, filled with 300 mM KCl was placed in a separate compartment electrically connected with the measuring chamber via a salt bridge consisting of 300 mM $(NH_4)_2SO_4$ in 2% (w/v) agar. Prior to flux and pH recording the electrode was brought carefully to a distance of 10 μm from the root surface. To ensure a maximum response of H^+-fluxes to sulfate and a minimum response to its accompanying cation sulfate was supplied in the form of $MgSO_4$, as the uptake Mg^{2+} is known to be slower than any of the other cations, especially in the presence of Ca^{2+} and K^+ (Moore et al. 1961; Schimansky, 1981). For time series recordings, the microelectrode was positioned at the beginning of the differentiated zone, where the first root hairs started to appear (usually 3–5 mm from the root tip). Care was taken not to place the tip of the electrode above an initiating root hair. During the incubation phase before time-series measurement, the position of the electrode and its distance to the root surface was surveyed and corrected if necessary.

For the time series recording sulfate-sufficient (control) seedlings and seedlings which were sulfate deprived for 1–5 days (−S) were taken. To localize the H^+-flux and pH changes along the root axis, a root profile was recorded from the root tip until 7 mm towards the base in 0.5 mm increments between 0–5 mm and 1 mm increments between 5 and 7 mm. A control profile was recorded after 1–2 h of incubation in the sulfate-free measuring solution. Subsequently the solution was exchanged for a solution to which 100 μM $MgSO_4$ was added and after 30 min a second profile was recorded. At each position fluxes and pH were recorded for 3–4 min and averaged. The profiling included the meristematic, the elongation and the beginning of the differentiated zone. For the root profiles sulfate-sufficient seedlings as well as seedlings that had been sulfate deprived for 3 days were taken. For the fusicoccin (FC) experiment excised roots were used instead of seedlings, first because FC is too costly to be applied in the large volume of the perfusion system and second because the response of the H^+-ATPase to FC is expected to be immediate. Roots were placed in measuring chambers and covered with 1 ml MS. After an incubation time of ca. 1 h 1 μl fusicoccin (Sigma, F0537) was added from a 10 mM stock solution (10 μM final concentration) and the solution was vigorously mixed.

A clear response of surface pH, i.e. the concentration of H^+s at the root surface, could be observed if 100 μM $MgSO_4$ was added to the roots via the perfusion

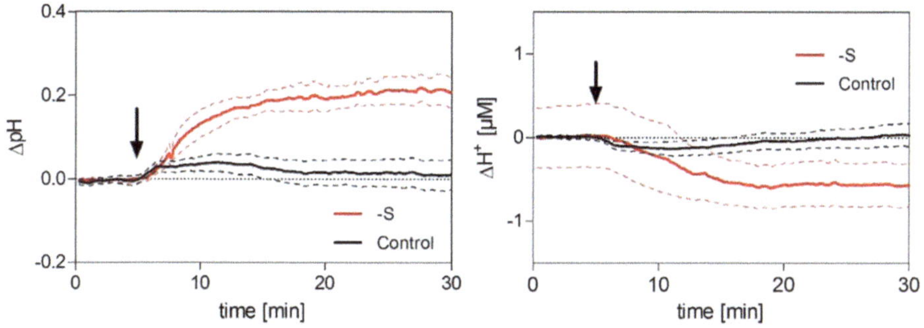

Fig. 2 The change in response of H^+ fluxes and surface pH at roots of intact seedlings to 100 μM sulfate addition (indicated by the *arrow*) over time. Plants were either grown at sulfate-sufficient (control) or sulfate-deprived conditions (−S). Positive values refer to an influx. Mean values ± SE are shown; control, n = 12; −S, n = 15

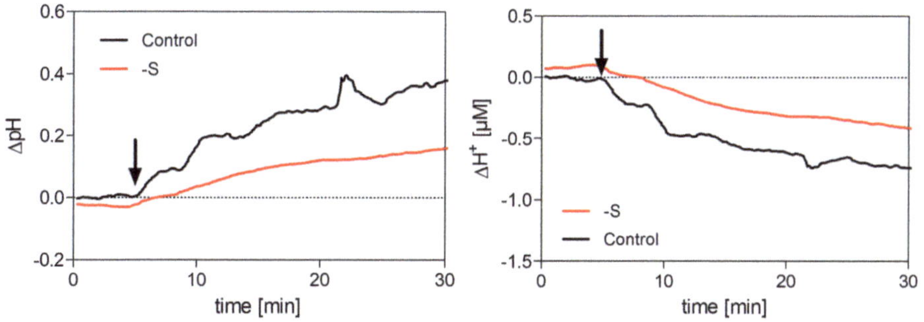

Fig. 3 Tentative results of the change in response to H^+ fluxes and surface pH to 50 μM nitrate addition (indicated by the *arrow*) over time. Plants were either grown at sulfate-sufficient (control) or sulfate-deprived conditions (−S). Positive values refer to an influx (exemplary measurement)

system (Fig. 2). The response was much more pronounced with sulfate-deprived roots with pH increasing about 0.2 units, which corresponds to a decrease in $[H^+]$ of approximately 0.5 μM. Alkalization was also observed if 50 μM $Mg(NO_3)$ was added to the measuring solution, but the sulfur status of the plants had an opposite effect on the NO_3^--induced pH response: sulfate-sufficient plants showed a greater response (ca. 0.3 pH units) than sulfate-deprived plants (ca. 0.1 pH units; Fig. 3). While H^+ fluxes were highly variable, the surface pH, as it is the result of net H^+ fluxes over a somewhat longer time period, was more stable. A response to $MgSO_4$ after 30 min of addition was present at roots of sulfate-deprived seedlings between 3.5 and 7 mm away from the root tip but absent in the first 3 mm which covered the meristematic and the beginning of the elongation zone (Fig. 4). No $MgSO_4$-induced changes of surface pH were observed in sulfate-sufficient plants. A lower surface pH at roots of sulfate-deprived plants compared to sufficient plants was observed during the measurements prior to any compound addition. There was a linear relationship between the concentration of H^+ before and the change in

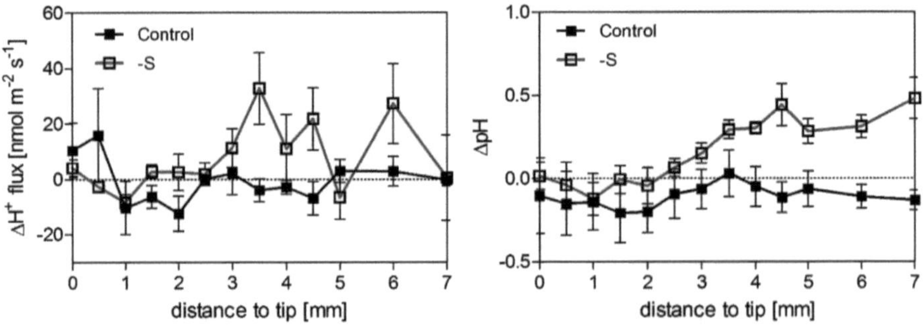

Fig. 4 Root profiles of H^+ fluxes and surface pH in response to sulfate addition at intact roots of sulfate-sufficient (control) and 3 days sulfate-deprived (−S) seedlings. The changes of fluxes and pH in response to sulfate addition at the same position are shown. Positive values refer to an influx. Data represent the mean (± SE, n = 3)

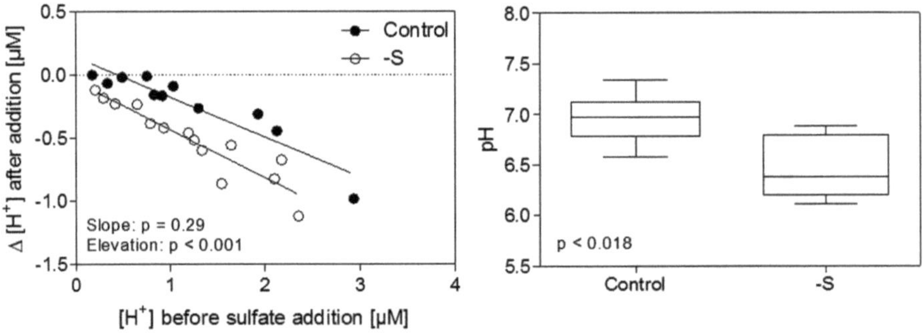

Fig. 5 Roots of sulfate-deprived seedlings (*open symbols*) showed a lower surface pH than roots of sulfate-sufficient seedlings (*closed symbols*) and the concentration of H^+ correlates with the peak response of $[H^+]$ to addition of 100 μM $MgSO_4$ The elevation of the two linear regressions significantly differs (p-value shown in graph). The differences in root surface pH were also measured in the growth medium (25% Hoagland). Data presented as boxes with a 5–95 percentile and whiskers (unpaired Student t-test; control, n = 6; −S, n = 5)

concentration of H^+ in response to sulfate addition (Fig. 5). Additional experiments revealed that the difference in surface pH was also apparent if measurements were performed in the growth medium (25% Hoagland) and therefore can not be considered to be an artifact of the incubation in the measuring solution (Fig. 5). Adding fusicoccin (FC) to the roots resulted in an immediate H^+ efflux and a consecutive decrease in pH. Thirty minutes after addition, the pH at roots of sulfate-sufficient seedlings was approximately at the same level as that of the sulfate-deprived ones (Fig. 6). The response of H^+-flux and surface pH to FC in a typical experiment is presented in Fig. 7.

The results in this study are in agreement with the accepted model of sulfate uptake being depressed under sulfate-sufficient conditions and de-repressed if plants get sulfate-deprived. H^+-selective microelectrodes appear to be a convenient

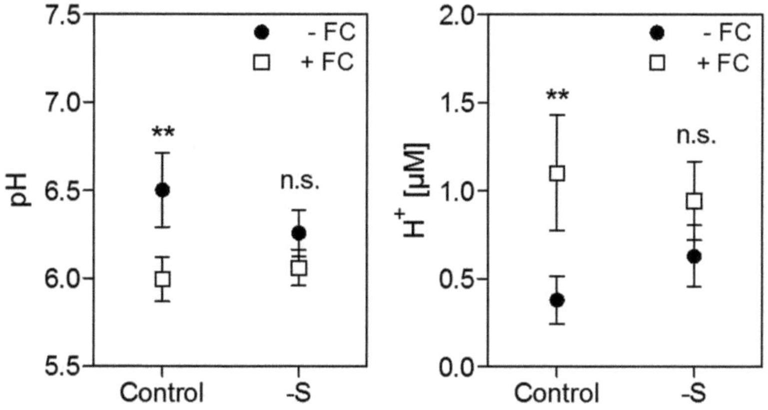

Fig. 6 Average of root surface pH and [H$^+$] of sulfate-sufficient (control) and sulfate-deprived (−S) seedlings 10 min before (*closed symbols*) and peak values after (*open symbols*) the addition of 10 µM FC. Data represent the mean (± SE; ** = $p < 0.01$; One-way-ANOVA; control, n = 3; −S, n = 4)

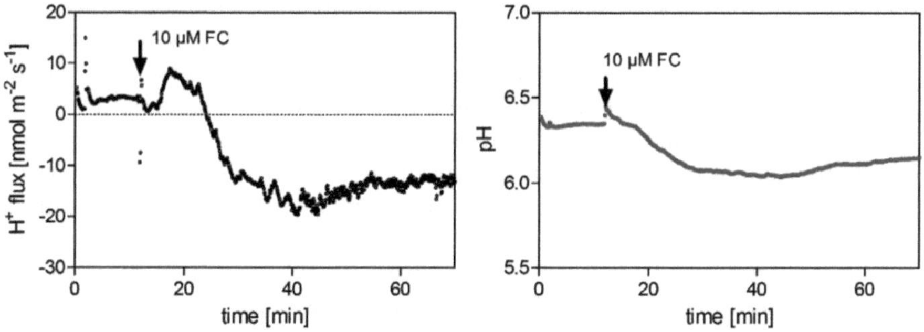

Fig. 7 Over-time response of H$^+$ flux and root surface pH following FC (concentration) addition (indicated by the *arrow*). Negative values correspond to an efflux

tool to determine and localize the uptake of sulfate at the root surface. Two main lines of experimental evidence support this. First, the de-repression of the sulfate uptake system under sulfate deprivation is clearly reflected by the pH changes at the root surface in response to sulfate addition (Fig. 2). The relatively weak alkalization at sulfate-sufficient roots disappeared after ca. 10 min while the strong alkalization at sulfate-deprived roots reached a stable plateau. With sulfate being the only ion newly added to the solution during the measurement, the response in H$^+$-flux seems to be a good proxy for sulfate uptake. Second, addition of nitrate in preliminary experiments resulted in a converse response of sufficient and deprived roots, if compared to sulfate addition. This was expected, as sulfate-deprived plants usually down-regulate nitrate uptake (Clarkson et al. 1989; Prosser et al. 2001). The use of H$^+$-selective microelectrodes appears therefore a reliable method to probe plant roots for differences in sulfate uptake.

In a case study, we applied this method to map sulfate uptake along the first millimeters of the roots of seedlings. We found that 30 min after sulfate addition no active sulfate uptake was apparent in both the meristematic zone and the beginning of the elongation zone (0–3 mm) (Fig. 4). Up to date, the only localization studies concerning sulfate transport at roots of intact plants aimed on localizing transcripts of the genes encoding for the sulfate transporters (e.g. Takahashi et al. 1997; Yoshimoto et al. 2002). In these studies, *in situ* hybridization analyses showed gene expression in the root cap, the epidermal layer, in root hairs and along the root cylinder. Abundance of mRNA transcripts is, however, not necessarily reflecting the final localization and activity of transport proteins. Many post-translational modifications determine the activity of membrane transport-proteins. The results of our analysis of the actual fluxes at intact roots suggest that active sulfate uptake takes place in the more differentiated root zones rather than in the meristematic and elongation zone.

Figure 5 shows that sulfate-deprived plants responded stronger to sulfate addition than sulfate-sufficient plants, indicated by the significant difference in elevation ($p < 0.001$). This is most likely representing the higher abundance of sulfate transporters. Additionally, a strong correlation was found between the H^+ concentration at the root surface prior and after the addition of sulfate (Fig. 5). This dependency of sulfate uptake on the H^+ gradient at the root plasma membrane provides support for the suggested SO_4^{2-}/H^+-symport stoichiometry of sulfate transporters (Hawkesford et al. 1993). Additionally, it raises the question whether a lower external pH and a consequently steeper H^+-gradient could have a physiological function to increase the uptake capacity under sulfate deprivation (Fig. 8). Indeed, a lower root surface pH was also found at seedlings grown in Hoagland solution without sulfate (Fig. 5). While the role of sulfate transporters in sulfate uptake has been studied intensively, the involvement and potential co-regulation of the plasma membrane ATPase has not been investigated. Active rhizosphere acidification has been well described and characterized in plants deprived of phosphate and iron (De Vos et al. 1986; Neumann and Römheld 1999). The uptake of these nutrients is mostly limited by their mobility in the soil solution, which is increased at low pH. Sulfate however, if present, is highly mobile over a wide pH range. A lower root surface pH under sulfate deprivation could also be of passive nature, caused by an altered cation-anion uptake balance (Haynes, 1990): no sulfate and probably less nitrate is taken up and, consequently, also less protons. Our experiments rather suggest an involvement of the activity of the plasma membrane H^+-ATPase: incubation of roots with fusicoccin, a permanent activator of the plasma membrane H^+-ATPase (Johansson et al. 1993), led to a decrease of the surface pH at sulfate-sufficient roots to the same level of sulfate-deprived roots (Fig. 6). Thus it appears that the root plasma membrane H^+-ATPase is not more abundant under sulfate deprivation but in a more active state. Sulfate uptake has been shown to be in a repressed state if sulfate or other sulfur sources are sufficiently abundant, which enables a quick de-repression if sulfate gets limiting (Clarkson and Saker 1989; Herschbach and Rennenberg 1994). The exact cellular and molecular signals involved, still need to be elucidated and some authors

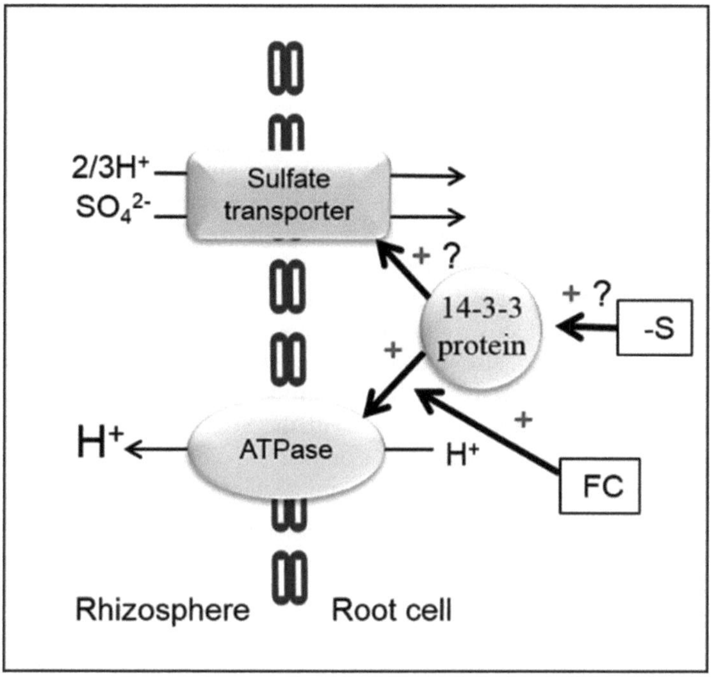

Fig. 8 Lower rhizosphere pH as an adaptive response to sulfate deprivation (−S)? Isoforms of the 14-3-3 protein were shown to stimulate the H^+-ATPase and to interact with the sulfate transporter (in Brassica Sultr1;2) in a yet unknown way (Shin et al. 2011) and might act as a coordinator to increase sulfate uptake under sulfur deprivation. Fusicoccin (FC) permanently stabilizes the binding of 14-3-3 to the H^+-ATPase and can therefore be used to achieve a maximum activation

actually found discrepancies between the sulfate uptake capacity and the expression level of the genes encoding for sulfate transporters (Koralewska et al. 2009). Deeper understanding of the regulation on the protein and membrane level is likely to explain such observed discrepancies. The plasma membrane H^+-ATPase as driver of H^+-coupled sulfate uptake (Fig. 8) might be involved in this regulation and increases sulfate uptake in addition to a higher transcript level of the sulfate transporter genes. Because the plasma membrane H^+-ATPase is building up the H^+-gradient that is utilized also by other transport systems, coordination with the uptake of other nutrients would be needed. A candidate player for such a general regulatory function is the 14-3-3 protein. Isoforms of this protein were shown to interact not only with the plasma membrane H^+-ATPase (Jahn et al. 1997; Bunney et al. 2002), but also with numerous proteins of the nitrogen, phosphorous and sulfur metabolic pathways, including the sulfate transporter Sultr1;2, which is responsible for primary sulfate uptake at the root plasma membrane (Shin et al. 2011). Being "spiders in a web of phosphorylation" (De Boer et al. 2013), 14-3-3 proteins could be a central coordinator of the complex network of different proteins responsible for ion homeostasis and the uptake of nutrients.

Further clarification is needed whether a lower rhizosphere pH under sulfate deprivation is an active response or rather a passive consequence of reduced anion uptake. However, active or passive, a lower pH has the potential to increase sulfate uptake (Fig. 8). Future studies should aim to further integrate cellular and electrophysiological techniques to elucidate the coordination of H^+-ATPase activity and nutrient status. Additionally, methods to assess the nutrient status or gene expression level of homogenized plant tissue reached their limit in explaining the processes under nutrient deficiency. Roots are composed of highly diversified tissues, quickly changing their morphology to adapt to changes in nutrient availability with newly formed roots having a different physiology than older ones. Ion-selective microelectrodes are perfectly suited to localize ion uptake on small scales and should be used more often, also in combination with molecular techniques. As long as there are no highly selective sulfate electrodes available, H^+-electrodes are a reliable alternative to study sulfate uptake with the MIFE technique. Localization of sulfate uptake at different kinds of tissues or the characterization of specific mutants are possible applications. Sulfate-selective microelectrodes will make it possible to investigate the post-translational regulation of sulfate uptake. The putative control by 14-3-3 proteins would be one example.

References

Bloom AJ, Caldwell RM (1988) Root excision decreases nutrient absorption and gas fluxes. Plant Physiol 87:794–796

Buchner P, Stuiver CEE, Westerman S, Wirtz M, Hell R, Hawkesford MJ, De Kok LJ (2004) Regulation of sulfate uptake and expression of sulfate transporter genes in *Brassica oleracea* as affected by atmospheric H_2S and pedospheric sulfate nutrition. Plant Physiol 136:3396–3408

Bunney TD, van den Wijngaard PW, De Boer AH (2002) 14-3-3 protein regulation of proton pumps and ion channels. Plant Mol Biol 50:1041–1051

Chen Z, Newman I, Zhou M, Mendham N, Zhang G, Shabala S (2005) Screening plants for salt tolerance by measuring K^+ flux: a case study for barley. Plant Cell Environ 28:1230–1246

Clarkson DT, Saker LR (1989) Sulphate influx in wheat and barley roots becomes more sensitive to specific protein-binding reagents when plants are sulphate-deficient. Planta 178:249–257

Clarkson DT, Saker LR, Purves JV (1989) Depression of nitrate and ammonium transport in barley plants with diminished sulphate status. Evidence of co-regulation of nitrogen and sulphate intake. J Exp Bot 40:953–963

De Boer AH, van Kleeff PJ, Gao J (2013) Plant 14-3-3 proteins as spiders in a web of phosphorylation. Protoplasma 250:425–440

De Vos CR, Lubberding HJ, Bienfait HF (1986) Rhizosphere acidification as a response to iron deficiency in bean plants. Plant Physiol 81:842–846

Garnett TP, Shabala SN, Smethurst PJ, Newman IA (2001) Simultaneous measurement of ammonium, nitrate and H^+ fluxes along the length of eucalypt roots. Plant Soil 236:55–62

Hawkesford MJ, Davidian J-C, Grignon C (1993) Sulphate/H^+ cotransport in plasma-membrane vesicles isolated from roots of *Brassica napus* L: increased transport in membranes isolated from sulphur-starved plants. Planta 190:297–304

Hawkesford MJ, Buchner P, Hopkins L, Howarth JR (2003) Sulphate uptake and transport. In: Abrol YP, Ahmad A (eds) Sulphur in plants. Kluwer, Dordrecht, pp 71–86

Haynes RJ (1990) Active ion uptake and maintenance of cation-anion balance: a critical examination of their role in regulating rhizosphere pH. Plant Soil 126:247–264

Henriksen GH, Bloom AJ, Spanswick RM (1990) Measurement of net fluxes of ammonium and nitrate at the surface of barley roots using ion-selective microelectrodes. Plant Physiol 93:271–280

Herschbach C, Rennenberg H (1994) Influence of glutathione (GSH) on net uptake of sulphate and sulphate transport in tobacco plants. J Exp Bot 45:1069–1076

Hopkins L, Parmar S, Błaszczyk A, Hesse H, Hoefgen R, Hawkesford MJ (2005) *O*-Acetylserine and the regulation of expression of genes encoding components for sulfate uptake and assimilation in potato. Plant Physiol 138:433–440

Jahn T, Fuglsang AT, Olsson A, Brüntrup IM, Collinge DB, Volkmann D, Larsson C (1997) The 14-3-3 protein interacts directly with the C-terminal region of the plant plasma membrane H (+)-ATPase. Plant Cell 9:1805–1814

Johansson F, Sommarin M, Larsson C (1993) Fusicoccin activates the plasma membrane H^+-ATPase by a mechanism involving the C-terminal inhibitory domain. Plant Cell 5:321–327

Koralewska A, Stuiver CEE, Posthumus FS, Kopriva S, Hawkesford MJ, De Kok LJ (2008) Regulation of sulfate uptake, expression of the sulfate transporters Sultr1;1 and Sultr1;2, and APS reductase in Chinese cabbage (*Brassica pekinensis*) as affected by atmospheric H_2S nutrition and sulfate deprivation. Funct Plant Biol 35:318–327

Koralewska A, Buchner P, Stuiver CEE, Posthumus FS, Kopriva S, Hawkesford MJ, De Kok LJ (2009) Expression and activity of sulfate transporters and APS reductase in curly kale in response to sulfate deprivation and re-supply. J Plant Physiol 166:168–179

Lanfermeijer FC, Staal M, Malinowski R, Stratmann JW, Elzenga JTM (2008) Micro-electrode flux estimation confirms that the *Solanum pimpinellifolium* cu3 mutant still responds to systemin. Plant Physiol 146:129–139

Moore DM, Overstreet R, Jacobson L (1961) Uptake of magnesium and its interaction with calcium in excised barley roots. Plant Physiol 36:290–295

Neumann G, Römheld V (1999) Root excretion of carboxylic acids and protons in phosphorus-deficient plants. Plant Soil 211:121–130

Newman IA (2001) Ion transport in roots: measurement of fluxes using ion-selective microelectrodes to characterize transporter function. Plant Cell Environ 24:1–14

Plassard C, Meslem M, Souche G (1999) Localization and quantification of net fluxes of H^+ along maize roots by combined use of pH-indicator dye videodensitometry and H^+-selective microelectrodes. Plant Soil 211:29–39

Prosser IM, Purves JV, Saker LR, Clarkson DT (2001) Rapid disruption of nitrogen metabolism and nitrate transport in spinach plants deprived of sulphate. J Exp Bot 52:113–121

Reich M, Aghajanzadeh T, Stuiver CEE, Koralewska A, De Kok LJ (2015) Impact of sulfate salinity on the uptake and metabolism of sulfur in Chinese cabbage. In: De Kok LJ, Hawkesford MJ, Rennenberg H, Saito K, Schnug E (eds) Molecular physiology and ecophysiology of sulfur. Springer, Cham, pp 227–238

Reich M, Shahbaz M, Prajapati DH, Parmar S, Hawkesford MJ, De Kok LJ (2016) Interactions of sulfate with other nutrients as revealed by H_2S fumigation of Chinese cabbage. Front Plant Sci 7:541

Rouached H, Wirtz M, Alary R, Hell R, Arpat AB, Davidian J-C, Fourcroy P, Berthomieu P (2008) Differential regulation of the expression of two high-affinity sulfate transporters, SULTR1;1 and SULTR1;2, in Arabidopsis. Plant Physiol 147:897–911

Rubinigg M, Stulen I, Elzenga JTM, Colmer TD (2002) Spatial patterns of radial oxygen loss and nitrate net flux along adventitious roots of rice raised in aerated or stagnant solution. Funct Plant Biol 29:1475–1481

Saito K (2000) Regulation of sulfate transport and synthesis of sulfur-containing amino acids. Curr Opin Plant Biol 3:188–195

Schimansky C (1981) The influence of certain experimental parameters on the flux characteristics of Mg-28 in the case of barley seedlings in hydroculture experiments. Landwirtsch Forsch 34:154–163

Shabala SN, Newman IA, Morris J (1997) Oscillations in H^+ and Ca^{2+} ion fluxes around the elongation region of corn roots and effects of external pH. Plant Physiol 113:111–118

Shabala SN, Pang J, Zhou M, Shabala L, Cuin TA, Nick P, Wegner LH (2009) Electrical signalling and cytokinins mediate effects of light and root cutting on ion uptake in intact plants. Plant Cell Environ 32:194–207

Shin R, Jez JM, Basra A, Zhang B, Schachtman DP (2011) 14-3-3 Proteins fine-tune plant nutrient metabolism. FEBS Lett 585:143–147

Smith FW, Ealing PM, Hawkesford MJ, Clarkson DT (1995) Plant members of a family of sulfate transporters reveal functional subtypes. Proc Natl Acad Sci U S A 92:9373–9377

Staal M, De Cnodder T, Simon D, Vandenbussche F, Van Der Straeten D, Verbelen JP, Elzenga JTM, Vissenberg K (2011) Apoplastic alkalinization is instrumental for the inhibition of cell elongation in the Arabidopsis root by the ethylene precursor 1-aminocyclopropane-1-carboxylic acid. Plant Physiol 155:2049–2055

Takahashi H, Yamazaki M, Sasakura N, Watanabe A, Leustek T, de Almeida EJ, Saito K (1997) Regulation of sulfur assimilation in higher plants: a sulfate transporter induced in sulfate-starved roots plays a central role in *Arabidopsis thaliana*. Proc Natl Acad Sci U S A 94:11102–11107

Taylor AR, Bloom AJ (1998) Ammonium, nitrate, and H^+ fluxes along the maize root. Plant Cell Environ 21:1255–1263

Vreeburg RA, Benschop JJ, Peeters AJ, Colmer TD, Ammerlaan AH, Staal M, Elzenga JTM, Staals RHJ, Darley CP, McQueen-Mason SJ, Voesenek LA (2005) Ethylene regulates fast apoplastic acidification and expansin. A transcription during submergence-induced petiole elongation in *Rumex palustris*. Plant J 43:597–610

Yoshimoto N, Takahashi H, Smith FW, Yamaya T, Saito K (2002) Two distinct high-affinity sulfate transporters with different inducibilities mediate uptake of sulfate in Arabidopsis roots. Plant J 29:465–473

Evidence for Regulation of the Iron Uptake Pathway by Sulfate Supply in S-Deprived Maize Plants

Georgios I. Saridis, Styliani N. Chorianopoulou, Panagiotis Katinakis, and Dimitris L. Bouranis

Abstract Maize plants follow Strategy II to take up Fe from the rhizosphere. In roots, nicotianamine is used as precursor for the production of deoxymugineic acid (DMA). DMA is secreted to the rhizosphere where it chelates Fe(III) and the Fe(III)-DMA complex is then taken up by the root. In this study, non-mycorrhizal and mycorrhizal maize plants were grown in pots with sterile river sand containing $FePO_4$. After a 60-day period of sulfur deficiency, sulfur was provided to the plants in the form of sulfate. The expression profiles of *ZmNAS1*, *ZmDMAS1*, *ZmTOM1* and *ZmYS1*, key genes of the Fe uptake pathway in maize, were monitored 24 and 48 h after sulfate addition in both mycorrhizal and non-mycorrhizal roots in order to estimate the impact of sulfate availability on the Fe uptake pathway of two distinct plant systems. Significant differential responses have been recorded between mycorrhizal and non-mycorrhizal plants, both before as well as 48 h after sulfate addition. However, sulfur repletion resulted in a significant downregulation of all studied genes in all plants 24 h after sulfate addition. This finding suggests a strong correlation between the transcriptional regulation of the Fe uptake pathway genes and sulfate availability no matter if the plants are in mycorrhizal association or not. Sulfate is probably a key component of the signal transduction pathway that regulates the expression of the Fe uptake pathway genes in maize plants.

Iron (Fe) is an essential nutrient for all living organisms. In plants, Fe plays a key role in electron transfer in both photosynthetic and respiratory reactions in chloroplasts and mitochondria. In soils, Fe is sparingly soluble under aerobic conditions at

G.I. Saridis
Botanical Institute, Cologne Biocenter, University of Cologne, D–50674 Cologne, Germany

S.N. Chorianopoulou (✉) • D.L. Bouranis
Plant Physiology Laboratory, Crop Science Department, Agricultural University of Athens, 75 Iera Odos, Athens 11855, Greece
e-mail: s.chorianopoulou@aua.gr

P. Katinakis
Laboratory of General and Agricultural Microbiology, Crop Science Department, Agricultural University of Athens, 75 Iera Odos, Athens 11855, Greece

L.J. De Kok et al. (eds.), *Sulfur Metabolism in Higher Plants - Fundamental, Environmental and Agricultural Aspects*, Proceedings of the International Plant Sulfur Workshop, DOI 10.1007/978-3-319-56526-2_17

high soil pH and exists mainly in its oxidized form, as part of insoluble ferric compounds. In addition to the solubility challenge, the chemical properties of iron require cells to place limitations on its accumulation. Fe(II) and Fe(III) act catalytically to generate hydroxyl radicals that can damage cellular constituents such as DNA and lipids (Halliwell and Gutteridge 1992). Higher plants employ two strategies for Fe uptake from their rhizosphere. Dicots and non-grass monocots take up Fe using ferric-chelate reductases to reduce Fe(III) to Fe(II), which is subsequently taken up by Fe transporters (Connolly et al. 2003). This type of Fe uptake is known as Strategy I pathway for Fe acquisition. Alternatively, graminaceous plants take up iron from their rhizospheres using the Strategy II pathway for Fe acquisition. In this pathway, *S*-adenosyl-methionine (SAM), a sulfur-containing compound, is used for the production of nicotianamine (NA) by the enzyme family of nicotianamine synthases (NAS). NA is then used as precursor for the biosynthesis of deoxymugineic acid (DMA), the first phytosiderophore produced in graminaceous plants and the only one produced in maize. The biosynthesis is conducted in a two-step process by the enzymes of nicotianamine aminotransferase (NAAT) and DMA synthase (DMAS). In maize, the efflux transporter TOM is used for the excretion of the produced DMA to the rhizosphere, where it chelates Fe(III). The complex DMA-Fe(III) is, finally, taken up by the root via a yellow stripe transporter (YS; Nozoye et al. 2013).

Over recent years, sulfur deficiency has become widespread in the world, mainly because of the strong decrease in the inputs of S from atmospheric deposition and the use of S-free fertilizer products such as triple phosphate and urea. Although it is often reported that cereals have a relatively low requirement for S, they can be adversely affected by S deficiency in the field as a result of its effects on growth, grain yield, and quality (Zhao et al. 1999). In plants, cysteine represents the main source of reduced S and is fundamental for the biosynthesis of methionine, Fe–S clusters and glutathione. Thus, there is a strong relationship between S and Fe homeostasis and the role of S in the Fe uptake pathway of graminaceous plants is crucial. It has been demonstrated that sulfur deficiency causes Fe deprivation responses to graminaceous plants which can be reversed, however, by sulfur repletion (Astolfi et al. 2003, 2006, 2010; Bouranis et al. 2003).

In this study, non-mycorrhizal and mycorrhizal maize plants were grown in pots with sterile river sand, in a long-term experiment (Fig. 1). For the mycorrhizal treatment, inoculum of the arbuscular mycorrhizal fungus *Rhizophagus irregularis* was used. Plants were watered with a nutrient solution deprived of Fe and S and containing a minimum P concentration so as to enhance the establishment of the arbuscular mycorrhiza (AM) symbiosis. Iron was provided throughout the experiment in the sparingly soluble form of $FePO_4$. After a 60-day period of sulfur deficiency, sulfur was provided to the plants in the form of sulfate, as described previously (Chorianopoulou et al. 2015). In order to estimate the impact of sulfate availability on the Fe uptake pathway, the expression profiles of *ZmNAS1*, *ZmDMAS1*, *ZmTOM1* and *ZmYS1*, key genes of the Fe acquisition pathway in maize (Fig. 2a), were monitored 24 and 48 h after sulfate addition in mycorrhizal and non-mycorrhizal roots. The gene expression analysis was conducted by means

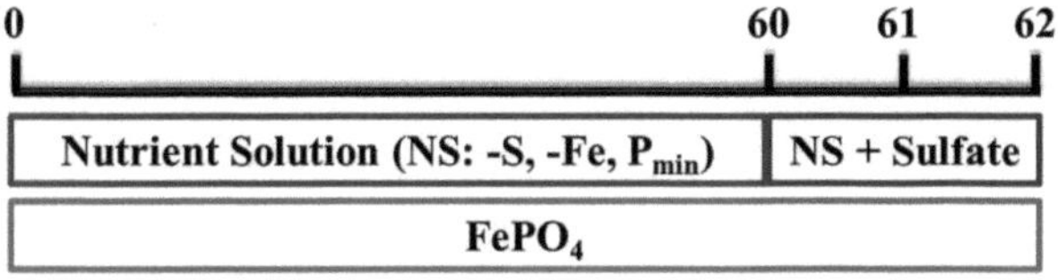

Fig. 1 Depiction of the experimental design of this study. Mycorrhizal and non-mycorrhizal plants were grown under S deficient conditions until day 60 from sowing; all plants were watered with a nutrient solution deprived of Fe and S and containing 10 μM of P (P_{min}) to trigger the establishment of the mycorrhizal symbiosis. On day 60, sulfate was provided to the plants. Fe was available throughout the experiment in the form of sparingly soluble $FePO_4$. Samplings took place on day 60 (before S repletion) as well as on days 61 and 62, 24 and 48 h after sulfate addition, respectively

of Real-Time RT-PCR. The oligonucleotide primers used for RT-qPCR are listed in Table 1. The efficiency of each Real-Time RT-PCR reaction was calculated using the LinRegPCR software (Ruijter et al. 2009). The mathematical formula of Pfaffl (2001) was used for the calculations of the relative expression ratios and ubiquitin was used as reference gene. The results of each sampling were expressed with the use of the previous sampling as control in order to determine the shift in the expression after sulfate supply.

It has been recently shown that S deprived mycorrhizal maize roots suppressed the expression of two key genes of the Fe uptake pathway, ZmNAS1 and ZmYS1, in contrast to non-mycorrhizal ones which enhanced it. In addition, AM symbiosis prevented Fe deprivation responses in the S deprived maize plants suggesting that iron was possibly provided directly to the mycorrhizal plants through the fungal network (Chorianopoulou et al. 2015). Differential response of the four key genes of the Fe uptake pathway in maize was observed between mycorrhizal and non-mycorrhizal roots 48 h after sulfate addition (Fig. 2b, Table 2). Non-mycorrhizal roots upregulated all the respective genes while mycorrhizal roots downregulated only the transporters *ZmTOM1* and *ZmYS1* whilst the relative expression ratios of *ZmNAS1* and *ZmDMAS1* did not alter significantly. These findings revealed that the two systems, mycorrhizal and non-mycorrhizal, respond differently as far as the Fe acquisition pathway is concerned.

Interestingly, despite the observed differential responses between mycorrhizal and non-mycorrhizal plants, a common response of all plants was revealed 24 h after sulfate addition, i.e. an identical decreasing pattern in the expressions of all genes of the Fe uptake pathway. This expression pattern was irrelevant of the symbiotic phenomenon and the short-term common response was related to the sulfate supply itself (Fig. 2b, Table 2).

Thus, our observations suggest a strong relationship between resupply of sulfate and the transcriptional regulation of the Fe uptake pathway genes in S-deprived maize plants, regardless of whether these plants are in mycorrhizal association, or not. This evokes the assumption that sulfate itself is a negative regulator of the Strategy II Fe acquisition pathway and the reduced S compounds (i.e. cysteine,

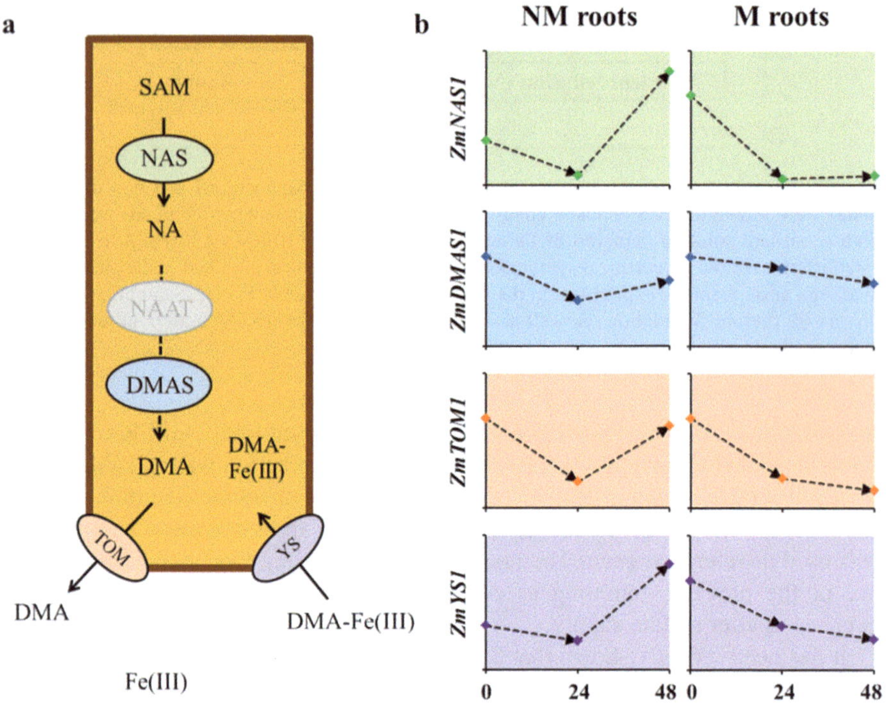

Fig. 2 (**a**) Schematic illustration of the Strategy II Fe uptake pathway in maize roots. (**b**) Expression shift patterns of *ZmNAS1*, *ZmDMAS1*, *ZmTOM1* and *ZmYS1*, 24 and 48 h after sulfate addition in mycorrhizal (*M*) and non-mycorrhizal (*NM*) maize roots. The expression shift values were calculated, each time, relatively to the previous sampling, revealing the pattern of the expression change the first 24 and 48 h after S supply. *Arrows pointing up* or *down* represent increase or decrease, respectively, in the expressions of the genes

methionine) are catalytically important to trigger the production and secretion of DMA.

Given the fact that most of the metabolically active Fe is bound to S in Fe–S clusters, the interaction between Fe and S homeostasis is of particular importance (Forieri et al. 2013). The biosynthesis of Fe–S clusters requires the supply of reduced S (in form of cysteine) and chelated Fe in a defined stoichiometric ratio, strongly suggesting the development of cross-regulatory mechanisms between the metabolisms of the two nutrients. In this line, the provision of the substrates must be tightly regulated in order to meet the plant's changing demands for Fe–S clusters and to avoid potentially toxic free Fe and sulfide.

In conclusion, our work underlines the importance of sulfate as signaling molecule that regulates the Strategy II Fe acquisition pathway; such a role of sulfate has been previously proposed by Forieri et al. (2013) for Fe homeostasis. In addition, we highlight the significance of sulfate reduction for the enhancement of Fe uptake and provide indications that when sulfate is abundant, the Fe uptake pathway is suppressed.

Table 1 The primers used in the real-time RT-PCR

Gene	Forward primer	Reverse primer
ZmNAS1	GGAACTTTTGAGCACCTATGCG	CACTTCACAATGCATAGCATCGAAT
ZmDMAS1	AAGTCCAAGGGCAAGACCG	AGTCCACGATGTCCAGGTTC
ZmTOM1	TGCAGAACTATGCTGTGCCA	GCATCTTGGCGTTTTTGGGT
ZmYS1	GTCTTCCATTCTCGCTCTGG	CAACCAACCACAGTTGATGC

Table 2 Percentage alterations in the relative expression ratios of *ZmNAS1*, *ZmDMAS1*, *ZmTOM1* and *ZmYS1*, 24 and 48 h after sulfate addition in mycorrhizal (M) and non-mycorrhizal (NM) maize roots

	NM roots		M roots	
	24 h	48 h	24 h	48 h
ZmNAS1	-77^{**}	$+1005^{*}$	-93^{**}	+50
ZmDMAS1	-49^{**}	$+44^{*}$	-13^{*}	−19
ZmTOM1	-70^{*}	$+206^{*}$	-66^{*}	-41^{*}
ZmYS1	-35^{**}	$+263^{*}$	-51^{*}	-30^{*}

Values represent the percentage of increase or decrease (indicated with + or –, respectively) of the relative expression ratios of the genes

*/** indicated when the difference between the samplings is statistically significant at $p < 0.05/0.005$, respectively

References

Astolfi S, Zuchi S, Passera C, Cesco S (2003) Does the sulphur assimilation pathway play a role in the response to Fe deficiency in maize (*Zea mays* L.) plants? J Plant Nutr 26:2111–2121

Astolfi S, Cesco S, Zuchi S, Neumann G, Roemheld V (2006) Sulphur starvation reduces phytosiderophores release by Fe-deficient barley plants. Soil Sci Plant Nutr 52:80–85

Astolfi S, Zuchi S, Hubberten H-M, Pinton R, Hoefgen R (2010) Supply of sulphur to S deficient young barley seedlings restores their capability to cope with iron shortage. J Exp Bot 61:99–806

Bouranis DL, Chorianopoulou SN, Protonotarios VE, Siyannis VF, Hopkins L, Hawkesford MJ (2003) Leaf response of young iron-inefficient maize plants to sulfur deprivation. J Plant Nutr 26:1189–1202

Chorianopoulou SN, Saridis YI, Dimou M, Katinakis P, Bouranis DL (2015) Arbuscular mycorrhizal symbiosis alters the expression patterns of three key iron homeostasis genes, ZmNAS1, ZmNAS3, and ZmYS1, in S-deprived maize plants. Front Plant Sci 6:257

Connolly EL, Campbell NH, Grotz N, Prichard CL, Guerino ML (2003) Overexpression of the FRO2 ferric chelate reductase confers tolerance to growth on low iron and uncovers posttranscriptional control. Plant Physiol 133:1102–1110

Forieri I, Wirtz M, Hell R (2013) Toward new perspectives on the interaction of iron and sulfur metabolism in plants. Front Plant Sci 4:357

Halliwell B, Gutteridge JMC (1992) Biologically relevant metal ion dependent hydroxyl radical generation. FEBS Lett 307:108–112

Nozoye T, Nakanishi H, Nishizawa NK (2013) Characterizing crucial components of iron homeostasis in the maize mutants *ys1* and *ys3*. PLoS One 8:e62567

Pfaffl MW (2001) A new mathematical model for relative quantification in real-time RT-PCR. Nucleic Acids Res 29:2002–2007

Ruijter JM, Ramakers C, Hoogaars WM, Karlen Y, Bakker O, van den Hoff MJ, Moorman AF (2009) Amplification efficiency: linking baseline and bias in the analysis of quantitative PCR data. Nucleic Acids Res 37:45

Zhao FJ, Hawkesford MJ, McGrath SP (1999) Sulphur assimilation and effects on yield and quality of wheat. J Cereal Sci 30:1–17

The Sulfur Pathway and Diagnosis of Sulfate Depletion in Grapevine

Sílvia Tavares and Sara Amâncio

Abstract Sulfur is an essential nutrient to all plant species. Plants assimilate sulfur in a well-described pathway, which has been taken up by roots. Regulatory mechanism has been the subject of many research papers. However, recent studies highlighted differences between crop plants and the model plant *Arabidopsis thaliana*. Our work focuses on the identification of genes involved in the sulfur metabolism in the *Vitis vinifera* genome, and their response to sulfur deficiency and other abiotic stress endured by grapevine in the field, namely water stress. Here, we describe the identification and brief characterization of the first assimilation enzymes involved in the sulfur pathway, the enzyme responsible for sulfur activation, ATP sulfurylase (ATPS), and the two enzymes that reduce sulfate to sulfide, Adenosine 5′-phosphosulate reductase (APR) and Sulfite reductase (SiR). A reduction was observed in the number of ATPS and APR isoforms identified in *V. vinifera* genome when compared to *A. thaliana* or *Glycine max* genomes. Two ATPS isoforms were present in the *Vitis* genome, of which only *ATPS1* transcript was detected in the tested tissues, and one APR isoform, suggesting an absence of redundancy in the role of both enzymes. *ATPS1*, *APR* and *SiR* transcript level was up-regulated in response to 2 days exposure to sulfur deficiency in *V. vinifera* cell cultures, which was completely reversed by the addition of GSH to the culture medium. Apparently, oxidative stress triggered GSH has a pivotal role in the regulation of *ATPS1*, *APR* and *SiR* transcription level, since their up-regulation was observed in mRNA from field grapevine berries under water stress, which is known to induce oxidative stress.

Grapevine (*Vitis vinifera* L.) is one of the most important crops worldwide for winemaking and also for table grapes. Grapevines can be successfully grown in a range of different climates and management conditions. Global climate changes are associated with water deficit and high evapotranspiration rates that can affect berry development, yield, and wine quality (Hannah et al. 2013).

S. Tavares • S. Amâncio (✉)
LEAF, Instituto Superior de Agronomia, Universidade de Lisboa, Lisboa 1349-017, Portugal
e-mail: samport@isa.ulisboa.pt

L.J. De Kok et al. (eds.), *Sulfur Metabolism in Higher Plants - Fundamental, Environmental and Agricultural Aspects*, Proceedings of the International Plant Sulfur Workshop, DOI 10.1007/978-3-319-56526-2_18

In plants sulfur (S) is a major essential plant nutrient required for plant growth and development. Unlike animals, vascular plants use sulfate (SO_4^{2-}) taken by the root system as the primary S source for plant growth (Clarkson et al. 1993) and reduce SO_4^{2-} to sulfide (S^{2-}) in plastids of phototrophic organisms, including vascular plants (Shibagaki and Grossman 2008). Sulfide is further assimilated into the amino acids cysteine and methionine. Then most sulfate taken up by plants is incorporated into proteins (Leustek and Saito 1999; Leustek et al. 2000). Organic sulfur can be found in glutathione (GSH), the thiol tripeptide that mediates redox reactions by the interchange of dithiol-disulfide. In addition, several secondary S-metabolites have been suggested to play key roles in defense against pathogens (Hell and Kruse 2007) such as glucosinolates, which are produced mostly by members of the Brassicaceae.

Sulfate assimilation in vascular plants is accomplished in three steps catalyzed by enzymes whose codifying genes were confirmed in *Vitis vinifera* genome: *VvATPS1* and *2* (ATP sulfurylase, EC: 2.7.7.4); *VvAPSR* (APS reductase, EC: 1.8.99.2); VvSiR (sulfite reductase, EC: 1.8.7.1); *VvSERAT1-3* (Serine acetyltransferase, EC 2.2.1.30) and *VvOASTL1-13* (*O*-acetylserine (thiol) lyase, EC 2.5.1.47) (Amâncio et al. 2009).

The Bordeaux mixture, a high-S-content fungicide, has been used since the nineteenth century for the control of downy and powdery mildew in grapevine (Williams and Cooper 2004). Besides its major effect as a fungicide, it was an important source of S. Since sulfur fungicides have been substituted for organic compounds, the S supply to vineyards was reduced. Recent studies of SO_4^{2-} uptake, assimilation, and symptoms of sulfur deficiency in *Vitis* species were reported by our group (Tavares et al. 2008, 2013, 2015; Amâncio et al. 2009). In the first study sulfate uptake was correlated with the expression of sulfate transporter genes in cell systems of two *Vitis* species (*V. vinifera* and *V. rupestris*) (Tavares et al. 2008); the second investigation established a link between sulfur deficiency and phenolic compounds (Tavares et al. 2013) and the third work characterized the serine acetyl-transferase protein family revealing major differences to the best described *A. thaliana* family (Tavares et al. 2015). Altogether the results previously obtained directed our attention to the first steps of sulfate assimilation, namely sulfate activation and reduction, which were explored in different grapevine experimental systems and environmental conditions.

It is largely known that the experimental set up can influence the results. In a study performed to verify differences of plant sensitivity to chemical treatments the sensitivity varied whether species were treated under greenhouse or field conditions (Fletcher et al. 1990). Under water stress different responses were obtained at physiological and transcriptional levels in grapevine plants cultivated in greenhouse and field (Luisa Carvalho, personal communication). Different systems were used to tackle S metabolism in grapevine. Cell suspensions were selected as biological material to obtain a homogeneous experimental system. Studies with maize cells had reported a response to S deficiency following the same trend as intact plants, a de-repression of sulfate uptake (Clarkson et al. 1999). Also cell cultures allow S manipulation in short periods. Cell suspensions of *V. vinifera* var.

Touriga Nacional and *V. rupestris* were obtained as described in Tavares et al. 2008. Cell suspensions were sourced from liquid culture callus material and were grown in 250 ml flasks on a rotary shaker at 100 rpm, in the dark at 25 °C. After at least two weekly cycles in full sulfate (+S, 1.5 mM) sub-cultures were prepared for sulfate treatments: +S conditions (control) and sulfate depletion (−S, sulfate substituted by chloride). Leaves from *V. vinifera* Touriga Nacional subjected to abiotic stress were obtained as described in Rocheta et al. (2014) and Coito et al. (2012). Rooted cuttings were transferred to 3 l pots filled with sterilized soil and placed in the growth chamber. The following growth conditions were adjusted: light intensity 200 μmol m^{-2} s^{-1}, 16 h light and 8 h darkness, 25 °C at day/23 °C at night, and watering with nutrient solution when necessary. The potted plants were subjected to drought and sampled when the pre-dawn leaf water potential was −0.9 MPa. *V. vinifera* cv. Touriga Nacional berries were collected from plants subjected to two irrigation regimes: rainfed (non-irrigation) and full irrigation (water supplied according to evapotranspiration rates) as described in Lopes et al. (2011). Berries were collected at veraison (50% colored berries) and full maturation.

The analysis of the *Vitis vinifera* genome indicates the presence of genes of two ATPS isoforms, one APR isoform with two splicing variants and one SiR isoform. ATPS and APR from plant and algae are encoded by small multigene families. A low number of isoforms is usually found in basal land plants and green algae, the studied species disclosing one or two distinct ATPS isoforms, and a APR unique isoform (Kopriva et al. 2007). In Arabidopsis and soybean genomes, genes encoding four ATPS isoforms and three APR isoforms were identified (Anjum et al. 2015; Hatzfeld et al. 2000; Yi et al. 2010). However, similar to *Vitis vinifera*, in the *Oryza sativa* genome only two ATPS isoforms were identified (Kopriva et al. 2007) and *Selaginella moellendorffii* seems to be the only vascular plant that has a unique ATPS (Kopriva et al. 2009). A single gene encodes SiR in *Vitis vinifera* in accordance with Arabidopsis (Takahashi et al. 2011) while in tobacco and soybean two isoforms were detected (Yi et al. 2010).

In *V. vinifera* the identified genes are located on distinct chromosomes; only ATPS2 shares the same chromosome with two sulfate transporters genes. Both ATPS genes depict five exons and four introns, however ATPS2 is organized in small exons and very long introns, which increased the genomic size from 5.3 to 11 Kbp. SiR is organized in eight exons and seven introns. In contrast the APR gene has only four or five exons. Two variants were identified for the APR gene that differ in a small sequence, variant one interpreted the sequence as an intron, in comparison variant 2 incorporated the sequence in the mRNA (Table 1).

The *Vitis vinifera* sulfate activation and reduction protein sequences present similar traits to other known plant proteins such as protein length and conserved domains (Table 1). *V. vinifera* ATPS coding regions have the N-terminal leader sequences characteristic for plastid-targeting transit peptides and a conserved ATPS catalytic domain (CD00517, Marchler-Bauer et al. 2011), typical of all described plant ATPS. All four ATPS from *Arabidopsis thaliana* (Rotte and Leustek 2000) and *Glycine max* (Yi et al. 2010) have been predicted as chloroplast isoforms

Table 1 Identification of ATP sulfurylase (ATPS), APS reductase (APR) and sulfide reductase (SiR) genes in *Vitis vinifera* genome

Gene	Chr	Genomic Region[a]	mRNA[a]	Protein[a]	Locali-zation
ATPS1	5	NW_003724020.1 5.3 Kbp	XM_002283536.3 1960 bp	XP_002283572.1 467 aa	C
ATPS2	18	NW_003724132.1 11Kbp	XM_002276957.2 2253 bp	XP_002276993.1 483 aa	C
APSr	12	NW_003724079. 3.9 Kbp	XM_002269703.3 (2) 1928 bp XM_010658907.1 (1) 1885 bp	XP_002269739.2 (2) 467 aa XP_010657209.1 (1) 498 aa	C
SiR	6	NW_003724030.1 6.4 Kbp	XM_002285362.2 2690 bp	XP_002285398.1 687 aa	C

Chr chromosome number, [a]NCBI reference numbers, *C* putative chloroplastic localization predicted in TargetP site (http://www.cbs.dtu.dk/services/TargetP/)

although ATPS activity is detected in chloroplasts and cytosol in Arabidopsis (Rotte and Leustek 2000) and spinach leaves (Lunn et al. 1990). Recently, Bohrer et al. (2015) showed that in Arabidopsis leaves, ATPS2 has a dual localization, namely in cytosol and chloroplasts, suggesting that a downstream methionine in the transit peptide sequence could act as an additional initiation translation site. In addition, such methionine is not present on the transit peptide sequence of ATPS1, 3 and 4. ATPS2 seems to be a distinct ATPS isoform, which probably confers different physiological roles; this isoform is also the only ATPS from Arabidopsis that is not a target to miRNA395 post-transcriptional control (Kawashima et al. 2009). Likewise, only the ATPS2 sequence from *Vitis vinifera* has an additional methionine on the transit peptide sequence and, contrary to VvATPS1, it is not a target of Vitis miRNA395 as predicted by the psRNATARGET tool (http://plantgrn.noble.org/psRNATarget/) (Dai and Zhao 2011). In fact, the unrooted tree constructed with *Vitis*, Arabidopsis, soybean and *Populus* ATPS protein sequences (Fig. 1a) also shows a group including the ATPS2 sequences and a second group where the other sequences are not clearly separated. If a certain degree of redundancy in ATPS isoforms is observed among other plant species (Kopriva et al. 2009), apparently in the *Vitis vinifera* genome such redundancy is reduced or absent. A unique APR isoform was identified in the *Vitis* genome, with two variants that share 94% sequence homology. Sulfate reduction occurs exclusively in chloroplasts and, similarly to all plant species *Vitis* APR has a N-terminal transit peptide for plastid-targeting and a conserved multidomain consisting of an reductase domain and a C-terminal thioredoxin-like domain (CD02993, Marchler-Bauer et al. 2011).

Arabidopsis (Kopriva et al. 2009) and soybean (Yi et al. 2010) have three APR isoforms, apparently some level of redundancy exists, since Arabidopsis plants without functional APR1 or APR2 do not show obvious differences when compared with the wild type, however the disruption of *APR2* leads to an 80% decrease in APR activity and an accumulation of sulfate indicating that APR2 is a major

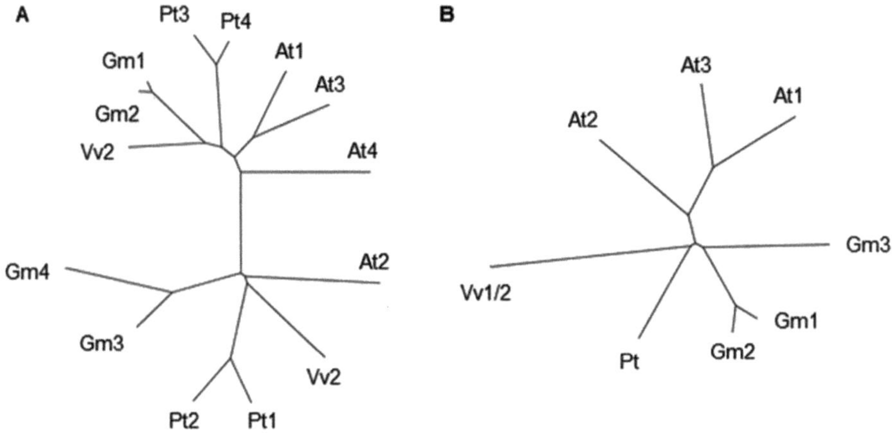

Fig. 1 Phylogenetic analysis of ATPS sulfurylase (ATPS, **a**) and APS reductase (APR; **b**) protein sequences. Unrooted tree constructed using the T-Cofee and PHYLIP programs, PRODIST and NEIGHBOR (http://tcoffee.crg.cat/ and http://mobyle.pasteur.fr/cgi-bin/portal.py#welcome, respectively). At *Arabidopsis thaliana*, Gm *Glycine max*,, Pt *Populus trichocarpa*, Vv *Vitis vinifera*

isoform in Arabidopsis (Loudet et al. 2007). The identification of APR in the *Vitis* genome shows a lack of redundancy and the phylogenetic analysis of the unique *Vitis* APR together with APR protein sequences from Arabidopsis, soybean and *Populus* shows that the *Vitis* protein is closer to Arabidopsis and soybean APR2. The APR protein sequences from the selected species seem to group better inside each species, which could be determined by species-specific evolution of APR (Fig. 1b). Together with APR, SiR is strictly plastidic. *Vitis* SiR contains the siroheme and a [4Fe-4S] cluster typical of plant SiR proteins and the transit peptide for chloroplast targeting (Table 1).

The presence of transcripts encoding enzymes involved in sulfate activation and reduction, ATPS1 and 2, APR and SiR, were examined in different grapevine tissues by RT-PCR (Fig. 2). *ATPS2* was the only gene not detected in RNA isolated from the sampled tissues, namely leaves (young and mature) and roots of potted plants, berries collected in the field and cells from cell culture (Fig. 2b, in berries). *V. vinifera* ATPS2 protein sequence was most similar to the protein sequences of *Populus*, ATPS1 and ATPS2, and AtAPS2 (Fig. 1a) recently proven to have dual subcellular chloroplastic/cytosolic localization, although a physiological role of the cytosolic isoform remains unknown (Bohrer et al. 2015). It has been speculated that cytosolic ATPS may be linked to cytosolic APS kinase in providing PAPS for the secondary metabolism (Rotte and Leustek 2000), namely in the production of glucosinolates. *V. vinifera* invests a great deal of resources in secondary metabolites, such as phenolic compounds, and its genome is enriched in genes devoted to secondary metabolism (Velasco et al. 2007). *V. vinifera* synthesizes no glucosinolates, which might be the reason why *ATPS2* transcripts were not detected in different plant tissues. RNA isolated from berries at *veraison* showed a very

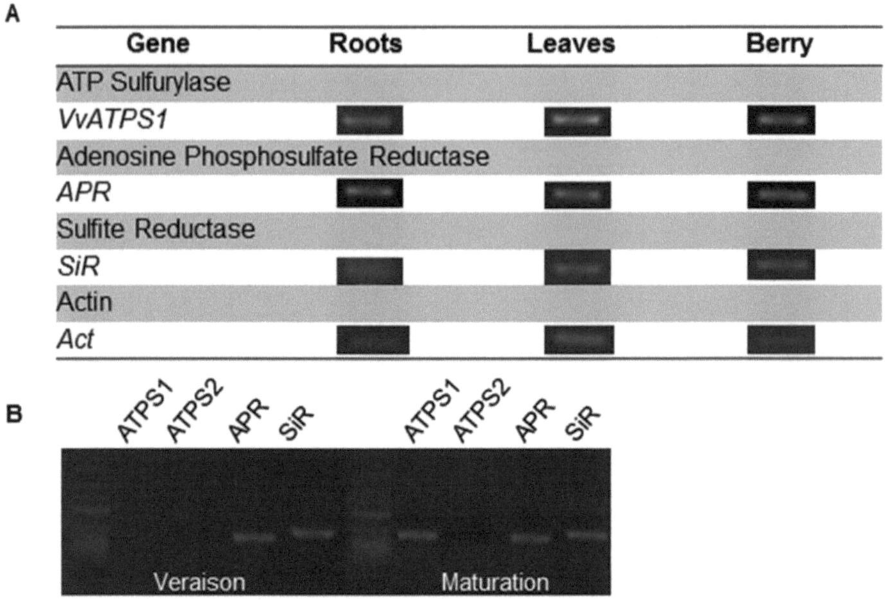

Fig. 2 (**a**) mRNA detection by RT-PCR in different *Vitis vinifera* tissues with specific primers design to ATP sulfurylase1 (ATPS1), APS reductase (APR) and sulfide reductase (SiR) sequences identified in Table 1. (**b**) detection of *ATPS1* and *2*, *APR* and *SiR* transcripts in grapevine berries collected at veraison and full maturation. RNA extraction and RT-PCR were performed according with standard molecular biology techniques

weak signal in RT-PCR using *ATPS1* primers, contrary to mature berries RNA which produced a strong signal indicating presence of the gene (Fig. 2b). Interestingly, a group three sulfate transporter was up-regulated at berry maturation (Guillaumie et al. 2011) and in seeds, transcripts for sulfur metabolism genes, namely group three sulfate transporters, were over-represented when compared with pulp transcripts (Grimplet et al. 2007). Together these results suggest a stimulus of sulfur metabolism at berry maturation, in particular in seeds.

The transcript levels of *VvATPS1*, *VvAPR* and *VvSiR* analyzed by qPCR responded equally to S depletion in cell cultures; all genes were up-regulated in cells after 2 days in an S deficient medium (Fig. 3a, b and c), and an increase in mRNA level was observed in *VvAPR* (Fig. 3b). S deficiency is known to be responsible for increasing the transcript levels of group one sulfate transporters and *APR* in plant species, including *Vitis vinifera* (Tavares et al. 2008), the reason why APR activity and sulfate transport are considered to exert the highest control over the S metabolic pathway (Vauclare et al. 2002). Nevertheless ATPS up-regulation was also observed in several plant species, namely *Arabidopsis thaliana* and *Zea mays* (reviewed by Anjum et al. 2015). The addition of sulfate and GSH to the S deficient medium completely reverses the up-regulation of *VvATPS1*, *VvAPR* and *VvSiR* transcripts. Interestingly, in *Vitis* cells the level of GSH as measured by HPLC (Tavares et al. 2015) was the first S compound to

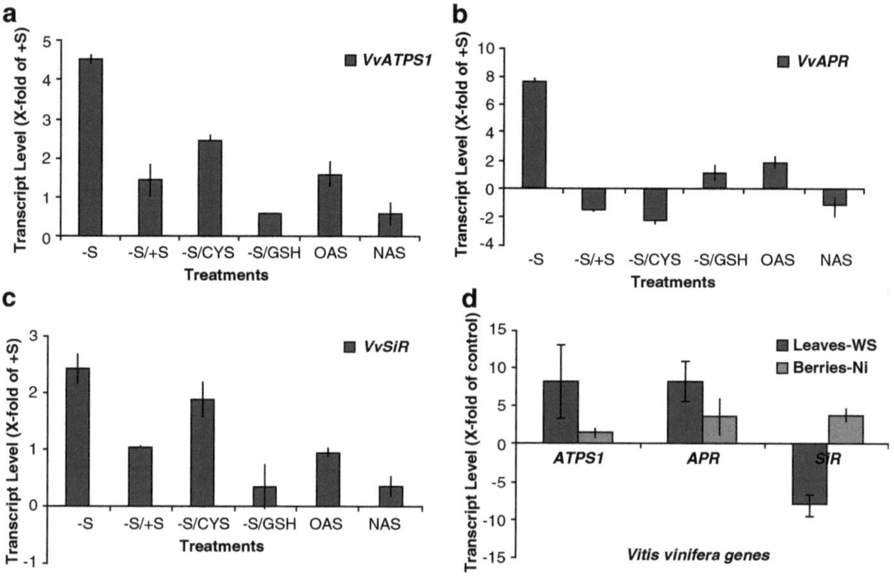

Fig. 3 Relative expression of ATP sulfurylase1 (*ATPS1*), APS reductase (*APR*) and Sulfide reductase (*SiR*) transcripts in grapevine cell culture (**a**, **b** and **c**) and leaves and berries (**d**). (**a**, **b** and **c**). Cells collected after 2 days in sulfur deficiency (−S) and sulfur sufficient (+S) medium, after which sulfur (S), cysteine (cys) and GSH were added to the –S medium and *O*-acetylserine (OAS) and *N*-acetylserine (NAS) to +S. Relative expression was compared to +S cells using RT-qPCR. (**d**) Water stress grapevine leaves and non-irrigated berries relative transcript level of *ATPS1*, *APR* and *SiR* was compared to control leaves and full irrigated berries by RT-qPCR

significantly decrease in cells after 1 day in S deficient medium. Cysteine added to cells in an S deficient medium also triggered a reversion of the up-regulation observed in the transcript level of *VvAPR* (Fig. 3b); apparently cysteine had a direct effect on the *VvAPR* transcription though we only detected a decrease in the amount of cysteine inside the cells 5 days after S limitation (Tavares et al. 2015). OAS is considered to act as a positive regulator of sulfate transporters, and commonly has similar effects under conditions of S deficiency (Takahashi et al. 2011). Consequently OAS added to *Vitis vinifera* cell culture showed the same up-regulation effect as S deficiency, although the magnitude was not so drastic (Fig. 3a, b and c).

In leaves from potted plants and berries collected from field plants, both grown under conditions of water stress the *VvATPS* and *VvAPR* transcript level increased. In contrast, *VvSiR* was down-regulated in leaves and up-regulated in berries (Fig. 3d). A change in ATPS activity in response to oxidative stress was reported by Kopriva et al. (2007). Similar results were obtained under conditions of abiotic stress (reviewed in Anjum et al. 2015). A high demand for GSH, an important S-compound in the response to oxidative stress, occurs under water stress. This may unfold an up-regulation effect in the mRNA of the first enzymes of sulfur assimilation. Likewise, the serine acetyltransferase (SAT) mRNA level was up-regulated under water stress conditions in *Vitis vinifera* leaves (Tavares et al. 2015).

As *VvSERAT2;1* was up-regulated in leaves of plants under water stress, this result prompted us to test the genes for the first enzymes of the S assimilation pathway (ATPS, APR and SiR) in leaves and berries of plants under water deficit. Although the patterns of transcripts expression were distinctly different between the two plant organs, the results suggest that some enzymes involved in sulfate metabolism are regulated by S-status and by environmental conditions, e.g. water deficit. Our study showed that major characteristics of ATPS, APR and SiR are very well conserved among plant species. However, it would be interesting to highlight differences between these enzymes in order to obtain a complete overview of S assimilation in different plant species.

References

Amâncio S, Tavares S, Fernandes JC, Sousa C (2009) Grapevine & sulfur: old partners, new achievements. In: Roubelakis-Angelakis KA (ed) Grapevine molecular physiology & biotechnology. Springer, Dordrecht, pp 31–52

Anjum NA, Gill R, Kaushik M, Hasanuzzaman M, Pereira E, Ahmad I, Gill SS (2015) ATP-sulfurylase, sulfur-compounds, and plant stress tolerance. Front Plant Sci 6:210

Bohrer AS, Yoshimoto N, Sekiguchi A, Rykulski N, Saito K, Takahashi H (2015) Alternative translational initiation of ATP sulfurylase underlying dual localization of sulfate assimilation pathways in plastids and cytosol in *Arabidopsis thaliana*. Front Plant Sci 5:750

Clarkson DT, Hawkesford MJ, Davidian J-C (1993) Membrane and long-distance transport of sulfate. In: De Kok LJ, Stulen I, Rennenberg H, Brunold C, Rausen WE (eds) Sulfur nutrition and sulfur assimilation in higher plants: fundamental environmental and agricultural aspects. SPB Academic Publishing, The Hague, pp 3–19

Clarkson DT, Diogo E, Amâncio S (1999) Uptake and assimilation of sulphate by sulphur deficient *Zea mays* cells: the role of O-acetyl-L-serine in the interaction between nitrogen and sulphur assimilatory pathways. Plant Physiol Biochem 37:283–290

Coito JL, Rocheta M, Carvalho L, Amâncio S (2012) Microarray-based uncovering reference genes for quantitative real time PCR in grapevine under abiotic stress. BMC Res Notes 5:220

Dai X, Zhao PX (2011) psRNATarget: a plant small RNA target analysis server. Nucleic Acids Res 39:W155–W159

Fletcher JS, Johnson FL, McFarlane JC (1990) Influence of greenhouse versus field testing and taxonomic differences on plant sensitivity to chemical treatment. Environ Toxicol Chem 9:169–776

Grimplet J, Deluc LG, Tillett RL, Wheatley MD, Schlauch KA, Cramer GR, Cushman JC (2007) Tissue-specific mRNA expression profiling in grape berry tissues. BMC Genomics 8:1

Guillaumie S, Fouquet R, Kappel C, Camps C, Terrier N, Moncomble D, Delrot S (2011) Transcriptional analysis of late ripening stages of grapevine berry. BMC Plant Biol 11:165

Hannah L, Roehrdanz PR, Ikegami M, Shepard AV, Shaw MR et al (2013) Climate change, wine, and conservation. Proc Natl Acad Sci U S A 110:6907–6912

Hatzfeld Y, Lee S, Lee M, Leustek T, Saito K (2000) Functional characterization of a gene encoding a fourth ATP sulfurylase isoform from *Arabidopsis thaliana*. Gene 248:51–58

Hell R, Kruse C (2007) Sulfur in biotic interactions of plants. In: Hawkesford MJ, De Kok LJ (eds) Sulfur in plants an ecological perspective. Springer, Dordrecht, pp 197–224

Kawashima CG, Yoshimoto N, Maruyama-Nakashita A, Tsuchiya YN, Saito K, Takahashi H, Dalmay T (2009) Sulphur starvation induces the expression of microRNA-395 and one of its target genes but in different cell types. Plant J 57:313–321

Kopriva S, Wiedemann G, Reski R (2007) Sulfate assimilation in basal land plants-what does genomic sequencing tell us? Plant Biol 9:556–564

Kopriva S, Mugford SG, Matthewman C, Koprivova A (2009) Plant sulfate assimilation genes: redundancy versus specialization. Plant Cell Rep 28:1769–1780

Leustek T, Saito K (1999) Sulfate transport and assimilation in plants. Plant Physiol 120:637–644

Leustek T, Martin MN, Bick JA, Davies JP (2000) Pathways and regulation of sulfur metabolism revealed through molecular and genetic studies. Annu Rev Plant Physiol Plant Mol Biol 51:141–165

Lopes CM, Santos TP, Monteiro A, Rodrigues ML, Costa JM, Chaves MM (2011) Combining cover cropping with deficit irrigation in a Mediterranean low vigor vineyard. Sci Hortic 129:603–612

Loudet O, Saliba-Colombani V, Camilleri C, Calenge F, Gaudon V, Koprivova A, Daniel-Vedele F (2007) Natural variation for sulfate content in *Arabidopsis thaliana* is highly controlled by APR2. Nat Genet 39:896–900

Lunn JE, Droux M, Martin J, Douce R (1990) Localization of ATP sulfurylase and *O*-acetylserine (thiol) lyase in spinach leaves. Plant Physiol 94:1345–1352

Marchler-Bauer A, Lu S, Anderson JB, Chitsaz F, Derbyshire MK, DeWeese-Scott C, Gwadz M (2011) CDD: a conserved domain database for the functional annotation of proteins. Nucleic Acids Res 39:D225–D229

Rocheta M, Becker JD, Coito JL, Carvalho L, Amâncio S (2014) Heat and water stress induce unique transcriptional signatures of heat-shock proteins and transcription factors in grapevine. Funct Integr Genomics 14:135–148

Rotte C, Leustek T (2000) Differential subcellular localization and expression of ATP sulfurylase and 5-adenylylsulfate reductase during ontogenesis of Arabidopsis leaves indicates that cytosolic and plastid forms of ATP sulfurylase may have specialized functions. Plant Physiol 124:715–724

Shibagaki N, Grossman A (2008) The state of sulfur metabolism in algae: from ecology to genomics. In: Hell R, Dahl C, Knaff D, Leustek T (eds) Advances in photosynthesis and respiration: sulfur metabolism in phototrophic organisms. Springer, Dordrecht, pp 231–267

Takahashi H, Kopriva S, Giordano M, Saito K, Hell R (2011) Sulfur assimilation in photosynthetic organisms: molecular functions and regulations of transporters and assimilatory enzymes. Annu Rev Plant Biol 62:157–184

Tavares S, Sousa C, Carvalho LC, Amâncio S (2008) De-repressed transporters are strongly repressed after sulfate addition to sulfur depleted *Vitis* cells. Int J Plant Sci 169:987–997

Tavares S, Vesentini D, Fernandes JC, Ferreira RB, Laureano O, Ricardo-Da-Silva JM, Amâncio S (2013) *Vitis vinifera* secondary metabolism as affected by sulfate depletion: diagnosis through phenylpropanoid pathway genes and metabolites. Plant Physiol Biochem 66:118–126

Tavares S, Wirtz M, Beier MP, Bogs J, Hell R, Amâncio S (2015) Characterization of the serine acetyltransferase gene family of *Vitis vinifera* uncovers differences in regulation of OAS synthesis in woody plants. Front Plant Sci 6:74

Vauclare P, Kopriva S, Fell D, Suter M, Sticher L, Von Ballmoos P, Brunold C (2002) Flux control of sulphate assimilation in *Arabidopsis thaliana*: adenosine 5′-phosphosulphate reductase is more susceptible than ATP sulphurylase to negative control by thiols. Plant J 31:729–740

Velasco R, Zharkikh A, Troggio M, Cartwright DA, Cestaro A et al (2007) A high quality draft consensus sequence of the genome of a heterozygous grapevine variety. PLoS One 2(12), e1326

Williams JS, Cooper RM (2004) The oldest fungicide and newest phytoalexin – a reappraisal of the fungitoxicity of elemental sulphur. Plant Pathol 53:263–279

Yi H, Ravilious GE, Galant A, Krishnan HB, Jez JM (2010) From sulfur to homoglutathione: thiol metabolism in soybean. Amino Acids 39:963–978

Impact of Sulfate Deprivation and H_2S Exposure on the Metabolites of the Activated Methyl Cycle in Chinese Cabbage

Mei-Hwei Tseng, Chao-Kai Yang, C. Elisabeth E. Stuiver, Chiu-Ching Chang, and Luit J. De Kok

Abstract The activated methyl cycle is a central metabolic pathway used to generate (and recycle) several important sulfur-containing metabolites including methionine, S-adenosylmethionine (SAM) and S-adenosylhomocysteine (SAH) and enable methylation. We have developed a precise and sensitive method for the simultaneous measurement of several sulfur metabolites based on liquid chromatography coupled with tandem mass spectrometry (LC-MS/MS) and ^{34}S-metabolic labeling of sulfur-containing metabolites including glutathione and the metabolites of the activated methyl cycle. Sulfate deprivation resulted in a decreased biomass production and content of glutathione, methionine, SAH of both shoot and root, and SAM of the root of Chinese cabbage. Foliarly absorbed H_2S may able to replace sulfate taken up by the root as sulfur source for growth and an atmospheric concentration of 0.2 $\mu l\ l^{-1}$ alleviated the decrease in the content of sulfur metabolites. The SAM content of the shoot was hardly affected upon sulfate-deprivation, resulting an increase in the SAM/SAH ratio, indicating a potential higher methylation capacity under this condition.

Seedlings of Brassicacea are characterized by their high growth rate (up to 0.4 g g^{-1} day^{-1}) and high sulfur demand; the sulfate uptake rate of some species may exceed 40 μmol g^{-1} fresh weight root day^{-1} (Shahbaz et al. 2010; Stuiver et al. 2014; Aghajanzadeh et al. 2016). The uptake of sulfate by the root is adjusted to the sulfur demand for growth, even at an external sulfate concentrations close to the K_m value

M.-H. Tseng (✉)
Department of Applied Physics and Chemistry, University of Taipei, Taipei 10048, Taiwan
e-mail: biomei@utaipei.edu.tw

C.-K. Yang
Department of Forensic Science, Central Police University, Taoyuan 33304, Taiwan

C. E.E. Stuiver • L.J. De Kok
Laboratory of Plant Physiology, Groningen Institute for Evolutionary Life Sciences, University of Groningen, P.O. Box 11103, 9700 CC Groningen, The Netherlands

Chiu-Ching Chang
Department of Applied Chemistry, National Chiao Tung University, Hsinchu 30050, Taiwan

L.J. De Kok et al. (eds.), *Sulfur Metabolism in Higher Plants - Fundamental, Environmental and Agricultural Aspects*, Proceedings of the International Plant Sulfur Workshop, DOI 10.1007/978-3-319-56526-2_19

of the high affinity sulfate transporters (approx. 5 μM; Koralewska et al. 2007, 2012). The interaction between atmospheric H_2S and pedospheric sulfate nutrition of plants has been extensively studied during the last three decades (De Kok 1990; De Kok et al. 2000, 2002, 2007). Foliarly absorbed H_2S was directly metabolized into cysteine and subsequently into other organic sulfur compounds, and exposure resulted in an increase in the content of water-soluble non-protein thiol content (*viz.* cysteine and glutathione) of the shoot (De Kok 1990; De Kok et al. 2000, 2002, 2007). H_2S exposure hardly affected the total sulfur and sulfate contents of *Brassica*, even not at relatively high atmospheric concentrations (De Kok et al. 2000), but resulted in a down-regulation of the uptake of sulfate by the root and reduction in the shoot (De Kok et al. 2000, 2002, 2007; Koralewska et al. 2008; Shahbaz et al. 2014; Aghajanzadeh et al. 2014, 2016). Upon sulfate-deprivation, the assimilated foliarly absorbed H_2S may replace sulfate as a sulfur source for growth of *Brassica* (De Kok et al. 2000, 2002, 2007; Buchner et al. 2004; Koralewska et al. 2008; Shahbaz et al. 2014; Aghajanzadeh et al. 2016). H_2S exposure hardly affected the up-regulated expression and activity of the high affinity sulfate transporters in sulfate-deprived *Brassica* (Koralewska et al. 2008; Shahbaz et al. 2014), and the decrease in shoot to root biomass partitioning upon sulfate deprivation remained largely unaffected upon H_2S exposure (Koralewska et al. 2008; Shahbaz et al. 2014; Aghajanzadeh et al. 2016). There is apparently a rather poor shoot to root signaling in Brassica of the regulation of both the sulfate transporters in the root and shoot to root biomass partitioning, indicating that both are determined by the sulfate concentration in the root environment rather than by the sulfur status of the plant itself.

Methionine is an essential metabolite in plants and all-living organisms (Ravanel et al. 1998). Apart from a role as a protein constituent, methionine is the precursor of *S*-adenosyl-L-methionine (SAM), the primary biological methyl-group donor (Roje 2006). The highly reactive methylated sulfur of SAM is used by a broad range of methyltransferases (Poel et al. 2013). A by-product of SAM-dependent transmethylation, *S*-adenosylhomocysteine (SAH) is released, which is recycled to methionine via homocysteine through the activated methyl cycle (Bürstenbinder and Sauter 2012). SAH strongly inhibited methyltransferase through competition with the substrate SAM (Barbes et al. 1990; Moffatt and Weretilnyk 2001). The ratio of cellular SAH and SAM is indicative for the methylation capacity of the cell (Fulnecek et al. 2011; Poel et al. 2013). In the current study the impact of sulfate deprivation and H_2S exposure on the content of the metabolites involved in the active methyl cycle was studied in Chinese cabbage.

Seeds of Chinese cabbage (*Brassica pekinensis* (Lour.) Rupr. cv. Kasumi F1; Nickerson Zwaan, Made, The Netherlands) were germinated in vermiculite in a climate controlled room. Day and night temperatures were 22 °C and 18 °C (±1 °C), respectively, relative humidity of 60–70% and a 14-h photoperiod at a photon fluence rate of 300 ± 20 μmol m^{-2} s^{-1} (400–700 nm) at plant height, supplied by Philips GreenPower LED (deep red/white 120) production modules. Ten-day-old seedlings were transferred to an aerated 25% Hoagland nutrient solution (pH 5.9) with 0.5 mM sulfate (+S, sulfate sufficient) or 0 mM sulfate (−S, sulfate deprived; all sulfate salts replaced by chloride salts) in 13 l stainless steel containers (ten sets of plants per container, three plants per set). Plants were exposed to 0.2 μl l^{-1} H_2S in 150 l cylindrical stainless steel cabinets (0.6 m diameter) with a poly(methylmethacrylate) top. The lids of the containers and the plant sets were sealed in order to prevent absorption of atmospheric H_2S by the nutrient solution. Day and night temperatures in the fumigation cabinets were 22 and 19 °C (± 2 °C), respectively, and relative humidity was 40–50%. The photoperiod was 14 h at a photon fluence rate of 300 ± 20 μmol m^{-2} s^{-1} (within the 400–700 nm range) at plant height, supplied by Philips GreenPower LED (deep red/white 120) production modules. The air exchange was 40 l min^{-1}, whereas a ventilator stirred the air inside the cabinets continuously. Pressurized H_2S diluted with N_2 (1 ml l^{-1}) was injected into the incoming air stream and the concentration in the cabinet was adjusted to the desired level using electronic mass flow controllers (ASM, Bilthoven, The Netherlands). The air exchange was 40 l min^{-1}, and a ventilator stirred the air inside the cabinets continuously. The H_2S concentration in the cabinets was monitored by an SO_2 analyzer (model 9850) equipped with a H_2S converter (model 8770; Monitor Labs, Measurement Controls Corporation, Englewood, CO, USA). Plants were harvested after H_2S exposure. Roots were separated from shoots, weighed, and were frozen immediately in liquid N_2 and stored at −80 °C until further use. For determination of the sulfur metabolites, plant tissue was freeze-dried at −60 °C for 48–72 h. Freeze-dried plant tissues were ground to powder with liquid nitrogen in a mortar with pestle. The sulfur metabolites were extracted as described in (Chang et al. 2013). For isotope dilution mass spectrometry analysis, the ^{34}S-labeled *Arabidopsis thaliana* tissue were extracted and added to the calibration standards, QC samples, and plant samples in a fixed ratio (Chang et al. 2013). Chromatographic separations of sulfur metabolites were performed on a Thermo Accela LC system using a Thermo Scientific Hypersil Gold aQ C18 (1.9 μm, 2.1 mm × 10 cm). Separations were performed under isocratic condition at a flow rate of 0.25 ml min^{-1}. The mobile phase was composed of 0.1% formic acid in water. ESI-MS/MS analysis was performed on a Thermo Scientific TSQ Quantum Ultra. Metabolites were detected in positive ionization mode using multiple reaction monitoring scanning mode. The spray voltage was set to 3.5 kV, the ion-transfer capillary temperature was set to 280 °C, the sheath gas pressure was set to 50 (arbitrary units), and the auxiliary gas pressure was set to 15 (arbitrary units). Collision energy set at 35% was used for each metabolite (Table 1).

An 11-day exposure of Chinese cabbage to sulfate-deprived condition resulted in a decreased plant biomass production (Table 2). The shoot growth was relatively more affected than the root growth resulting in a decrease in shoot/root ratio. Exposure of sulfate-sufficient plants to 0.2 μl l^{-1} H_2S hardly affected plant biomass production, but it alleviated the reduction in biomass production of sulfate-deprived plants. The latter demonstrated that similar to previous observations that at an atmospheric H_2S concentration $\geq$ 0.2 μl l^{-1} the foliarly absorbed sulfide fully could replace sulfate taken up by the sulfur source for growth (Koralewska et al. 2008; Shahbaz et al. 2014; Aghajanzadeh et al. 2014, 2016). H_2S exposure of sulfate-sufficient plants resulted in a slight increase in the total glutathione (expressed as GSH) content of the shoot and not of the root (Table 3). Sulfate

Table 1 Mass spectrometry parameters with transition pairs in MRM mode and normalized collision energy (%)

Compound name	Precursorion	Production	Collision energy
Methionine	150	56	15
Methionine S34	152	106	15
GSH	308	162	20
GSH S34	310	181	20
SAH	385	134	20
SAH S34	387	136	20
SAM	399	250	20
SAM S34	401	250	20
^{32}S GSSG	613	355	25
^{34}S GSSG	617	359	25

Table 2 Impact of H_2S and sulfate deprivation on biomass production of shoots and roots of Chinese cabbage

	+Sulfate		−Sulfate	
	0 μl l^{-1} H_2S	0.2 μl l^{-1} H_2S	0 μl l^{-1} H_2S	0.2 μl l^{-1} H_2S
Shoot (g FW)	3.98 ± 0.38b	5.39 ± 1.91b	1.12 ± 0.04a	4.69 ± 0.66b
Root (g FW)	0.78 ± 0.05b	1.05 ± 0.26b	0.33 ± 0.04a	1.10 ± 0.08c
Shoot/root ratio	5.13 ± 0.27b	5.29 ± 1.75ba	3.46 ± 0.44a	4.27 ± 0.27a

Ten-day-old seedlings were grown on a 25% Hoagland solution at 0 and 0.5 mM sulfate (+S) and exposed to 0.2 μl l^{-1} H_2S for 11 days. The initial fresh biomass of the shoot and root of Chinese cabbage was 0.171 ± 0.001 g and 0.057 ± 0.001 g, respectively. Data on plant yield (g FW) and shoot/root ratio represent the mean of three measurements with three plants in each (±SD) Means with different letters are significant different at $p < 0.01$ (unpaired Student's t-test)

deprivation resulted in a strong decrease in the glutathione content of both shoot and root but this decrease was largely alleviated upon H_2S exposure, though its content remained lower than that of the sulfate-sufficient plants. Sulfate deprivation also resulted in a substantial decrease in the methionine and SAH content of both shoot and root and SAM content of the root (Table 3). However, the SAM content of the shoot was hardly affected upon sulfate deprivation, resulting in a fourfold increase in the SAM/SAH ratio (Table 3). The SAM/SAH ratio in plant tissue has often been used as a reporter of the methylation capacity ("methylation index"; Groth et al. 2016). In this view the increase in SAM/SAH ratio in the shoot upon sulfate deprivation would indicate an increased methylation capacity. Exposure of sulfate-deprived plants to H_2S alleviated the decrease in SAM and SAH content of the shoots, which were comparable to that of shoots of sulfate-sufficient plants (Table 3). However, the methionine content was slightly lower than that of sulfate-sufficient plants. The contents of methionine, SAM and SAH sulfur of the roots were decreased upon sulfate deprivation, whereas the content of these sulfur metabolites in the root of sulfate-deprived H_2S-exposed plants was quite similar to that of sulfate-sufficient plants. The activated methyl cycle is a central metabolic

Table 3 Impact of H_2S and sulfate deprivation on the content of glutathione and metabolites of activated methyl cycle in shoots and roots of Chinese cabbage

	+Sulfate		−Sulfate	
	0 μl l^{-1} H_2S	0.2 μl l^{-1} H_2S	0 μl l^{-1} H_2S	0.2 μl l^{-1} H_2S
Shoot				
Glutathione	5.67 ± 0.22c	6.69 ± 0.42d	0.18 ± 0.01a	3.87 ± 0.16b
Methionine	0.224 ± 0.004c	0.260 ± 0.039c	0.038 ± 0.006a	0.179 ± 0.032b
SAM	0.123 ± 0.008ab	0.134 ± 0.011b	0.106 ± 0.011a	0.113 ± 0.007a
SAH	0.015 ± 0.001b	0.014 ± 0.001b	0.003 ± 0.001a	0.012 ± 0.001b
SAM/SAH ratio	8.2 ± 1.1a	9.6 ± 1.5a	35.3 ± 15.4b	9.4 ± 1.4a
Root				
Glutathione	4.62 ± 0.15c	5.17 ± 0.54c	0.55 ± 0.07a	2.79 ± 0.10b
Methionine	0.136 ± 0.014b	0.135 ± 0.024b	0.024 ± 0.004a	0.121 ± 0.004b
SAM	0.207 ± 0.002b	0.210 ± 0.024bc	0.076 ± 0.016a	0.231 ± 0.002c
SAH	0.019 ± 0.002bc	0.017 ± 0.002b	0.006 ± 0.003a	0.021 ± 0.002c
SAM/SAH ratio	10.9 ± 1.3a	12.3 ± 2.9a	12.3 ± 8.6a	11.0 ± 1.1a

For experimental details, see legends of Table 2. Data on metabolite content is expressed as μmol g^{-1} dry weight (freeze-dried plant material) and represent mean of three measurements with three plants in each (± SD). Means with different letters are significant different at $p < 0.01$ (unpaired Student's t-test)

pathway responsible for the methylation of cellular components and the recycling of sulfur-containing metabolites. The methylation of essential biological molecules, e.g. nucleic acids, hormones, lipids, proteins, is of crucial importance for many key biochemical processes (Chiang et al. 1996). Apparently the foliarly absorbed sulfide by sulfate-deprived Chinese cabbage upon exposure to an atmospheric level of 0.2 μl l^{-1} H_2S was sufficient for reduced sulfur requirement of the plant to support growth and to maintain the levels of the metabolites involved in the activated methyl cycle in both the shoot and the root.

Acknowledgements We would like to thank Dr. Ding-Tzai Li for all fruitful discussions and Mass Solution Technology Co., Ltd. for their excellent technical support.

References

Aghajanzadeh T, Hawkesford MJ, De Kok LJ (2014) The significance of glucosinolates for sulfur storage in Brassicaceae seedlings. Front Plant Sci 5:704

Aghajanzadeh T, Hawkesford MJ, De Kok LJ (2016) Atmospheric H_2S and SO_2 as sulfur sources for *Brassica juncea* and *Brassica rapa*: regulation of sulfur uptake and assimilation. Environ Exp Bot 124:1–10

Barbes C, Sanchez J, Yebra MJ, Robert-Gero M, Hardisson C (1990) Effects of sinefungin and *S*-adenosylhomocysteine on DNA and protein methyltransferases from *Streptomyces* and other bacteria. FEMS Microbiol Lett 57:239–243

Buchner P, Stuiver CEE, Westerman S, Wirtz M, Hell R, Hawkesford MJ, De Kok LJ (2004) Regulation of sulfate uptake and expression of sulfate transporter genes in *Brassica oleracea* as affected by atmospheric H_2S and pedospheric sulfate nutrition. Plant Physiol 136:3396–3408

Bürstenbinder K, Sauter M (2012) Early events in the ethylene biosynthetic pathway – regulation of the pools of methionine and Sadenosylmethionine. In: MT MM (ed) The plant hormone ethylene, Annual plant review, vol 44. Wiley, Oxford, pp 19–52

Chang YL, Hsieh CL, Huang YM, Chiou WL, Kuo YH, Tseng MH (2013) Modified method for determination of sulfur metabolites in plant tissues by stable isotope dilution-based liquid chromatography electrospray ionization tandem mass spectrometry. Anal Biochem 442:24–33

Chiang PK, Gordon RK, Tal J, Zeng GC, Doctor BP, Pardhasaradhi K, McCann PP (1996) *S*-Adenosylmethionine and methylation. FASEB J 10:471–480

De Kok LJ (1990) Sulfur metabolism in plants exposed to atmospheric sulfur. In: Rennenberg H, Brunold C, De Kok LJ, Stulen I (eds) Sulfur nutrition and sulfur assimilation in higher plants: fundamental environmental and agricultural aspects. Backhuys Publishers, Leiden, pp 111–130

De Kok LJ, Stuiver CEE, Westerman S, Stulen I (2000) Atmospheric H_2S as plant sulfur source: interaction with pedospheric sulfur nutrition – a case study with *Brassica oleracea* L. In: Brunold C, Rennenberg H, De Kok LJ, Stulen I, Schnug E, Davidian J-C (eds) Sulfur nutrition and sulfur assimilation in higher plants: molecular biochemical and physiological aspects. Verlag Paul Haupt, Bern, pp 41–55

De Kok LJ, Stuiver CEE, Westerman S, Stulen I (2002) Elevated levels of hydrogen sulfide in the plant environment: nutrient or toxin. In: Omasa K, Saji H, Youssefian S, Kondo N (eds) Air pollution and plant biotechnology. Prospects for phytomonitoring and phytoremediation. Springer, Tokyo, pp 3–11

De Kok LJ, Durenkamp M, Yang L, Stulen I (2007) Atmospheric sulfur. In: Hawkesford MJ, De Kok LJ (eds) Sulfur in plants: an ecological perspective. Springer, Dordrecht, pp 91–106

Fulnecek J, Matyasek R, Votruba I, Holy A, Krizova K, Kovarik A (2011) Inhibition of SAH-hydrolase activity during seed germination leads to deregulation of flowering genes and altered flower morphology in tobacco. Mol Gen Genomics 285:225–236

Groth M, Moissiard G, Wirtz M, Wang H, Garcia-Salinas C, Ramos-Parra PA, Smith DC, Zhai J, Hale CJ, Long JA, Hell R, Díaz de la Garza RI, Jacobsen SE (2016) MTHFD1 controls DNA methylation in *Arabidopsis*. Nat Commun 7:1–13

Koralewska A, Posthumus FS, CEE S, Buchner P, De Kok LJ (2007) The characteristic high sulfate content in Brassica oleracea is controlled by the expression and activity of sulfate transporters. Plant Biol 9:654–661

Koralewska A, Stuiver CEE, Posthumus FS, Kopriva S, Hawkesford MJ, De Kok LJ (2008) Regulation of sulfate uptake expression of the sulfate transporters Sultr1;1 and Sultr1;2, and APS reductase in Chinese cabbage (*Brassica pekinensis*) as affected by atmospheric H_2S nutrition and sulfate deprivation. Funct Plant Biol 35:318–327

Koralewska A, Posthumus FS, Stuiver CEE, Shahbaz M, Stulen I, De Kok LJ (2012) Regulation of the uptake of sulfate by Chinese cabbage (Brassica pekinensis) at various sulfate concentrations. In: De Kok LJ, Tausz M, Hawkesford MJ, Hoefgen R, McManus MT, Norton RM, Rennenberg H, Saito K, Schnug E, Tabe L (eds) Sulfur metabolism in plants – mechanisms and applications to food security and responses to climate change. Springer, Dordrecht/New York, pp 61–65

Moffatt BA, Weretilnyk EA (2001) Sustaining *S*-adenosyl-L-methionine-dependent methyltransferase activity in plant cells. Physiol Plant 113:435–442

Poel BV, Bulens I, Oppermann Y, Hertog MLA, Nicolai BM, Sauter M, Gerard AH (2013) *S*-Adenosyl-L-methionine usage during climacteric ripening of tomato in relation to ethylene and polyamine biosynthesis and transmethylation capacity. Physiol Plant 148:176–188

Ravanel S, Gakiere B, Job D, Douce R (1998) The specific features of methionine biosynthesis and metabolism in plants. Proc Natl Acad Sci U S A 95:7805–7812

Roje S (2006) *S*-Adenosyl-L-methionine: beyond the universal methyl group donor. Phytochemistry 67:1686–1698

Shahbaz M, Tseng M-H, Stuiver CEE, Koralewska A, Posthumus FS, Venema JH, Parmar S, Schat H, Hawkesford MJ, De Kok LJ (2010) Copper exposure interferes with the regulation of the uptake, distribution and metabolism of sulfate in Chinese cabbage. J Plant Physiol 167:438–446

Shahbaz M, Stuiver CEE, Posthumus FS, Parmar S, Hawkesford MJ, De Kok LJ (2014) Copper toxicity in Chinese cabbage is not influenced by plant sulphur status but affects sulphur metabolism-related gene expression and the suggested regulatory metabolites. Plant Biol 16:68–78

Stuiver CEE, Posthumus FS, Parmar S, Shahbaz M, Hawkesford MJ, De Kok LJ (2014) Zinc exposure has differential effects on uptake and metabolism of sulfur and nitrogen in Chinese cabbage. J Plant Nutr Soil Sci 177:748–757

Sulfate Transporters Involved in Cd-Induced Changes of Sulfate Uptake and Distribution in *Arabidopsis thaliana*

Chisato Yamaguchi and Akiko Maruyama-Nakashita

Abstract Cadmium (Cd) is a highly toxic element for living organisms, hence plants have evolved a variety of detoxification mechanisms. Phytochelatins and glutathione are low-molecular-weight sulfur compounds that function as chelators and play important roles in Cd detoxification. Previous studies have shown that the transcription of the genes involved in sulfate uptake and sulfur assimilation was increased in response to Cd stress. Recently, we reported that Cd-induced sulfate uptake is mainly attributed to the function of SULTR1;2, a high affinity sulfate transporter involved in sulfate uptake from the roots. Another distinct change in sulfate distribution induced by Cd treatment was preferential accumulation of sulfate to the shoots, which is due to the induction of root-to-shoot sulfate transport through xylem. In this study, we compared previous transcriptome data taken with Cd-treated plants to get suggestions about the SULTRs involved in Cd-induced sulfate distribution to shoots. In addition to the induction of *SULTR1;1* and *SULTR1;2* expressions, we found that the expression of *SULTR2;1*, a transporter involved in root-to-shoot transport of sulfate, and *SULTR4;1* and *SULTR4;2*, exporters of stored sulfate in vacuole, were increased in roots upon Cd treatment. These SULTRs were suggested as contributors to the increased distribution of sulfate to shoots under Cd exposure.

Cadmium (Cd) is a non-essential and toxic heavy metal for living organisms commonly contaminated in the ecosystems (Nawrot et al. 2006; Järup and Akesson 2009; Clemens et al. 2013; Choppala et al. 2014). In plants, Cd affects many physiological and metabolic processes, including photosynthesis, ion homeostasis, and mineral uptake, partly through the production of reactive oxygen species (Lin and Aarts 2012; Clemens et al. 2013; Choppala et al. 2014). Therefore, plants have evolved a variety of detoxification mechanisms to avoid these toxicities (Verbruggen et al. 2009; Yadav 2010; Lin and Aarts 2012; Chmielowska-Bąk et al. 2014; Choppala et al. 2014).

C. Yamaguchi • A. Maruyama-Nakashita (✉)
Laboratory of Plant Nutrition, Faculty of Agriculture, Kyushu University, Fukuoka 812-8581, Japan
e-mail: amaru@agr.kyushu-u.ac.jp

L.J. De Kok et al. (eds.), *Sulfur Metabolism in Higher Plants - Fundamental, Environmental and Agricultural Aspects*, Proceedings of the International Plant Sulfur Workshop, DOI 10.1007/978-3-319-56526-2_20

Low-molecular-weight sulfur compounds, glutathione (GSH) and phytochelatins (PCs) are known to contribute to heavy metal detoxification in plants (Cobbett 2000; Cobbett et al. 2002; Choppala et al. 2014; Seth et al. 2012). GSH or PCs chelate Cd, and these complexes are subsequently compartmentalized into vacuoles, thus lowering Cd levels in the cytosol (Cobbett 2000; Cobbett et al. 2002; Choppala et al. 2014). GSH is synthesized from cysteine, the end product of sulfate assimilation pathway (Leustek et al. 2000; Saito 2004), via two enzymatic reactions catalyzed by γ-glutamylcysteine (γ-Glu-Cys) synthase and glutathione synthase (Lu 2013). Then PCs are synthesized from GSH by the enzymatic conjugation of γ-Glu-Cys to GSH, which is catalyzed by phytochelatin synthase (Cobbett and Goldsbrough 2002). Phytochelatin synthase also catalyzes the sequential conjugation of the γ-Glu-Cys moiety to various PCs, which forms longer PC species with the general structure (γ-Glu-Cys)n-glycine (Gly) (n = 2 to 11) (Cobbett and Goldsbrough 2002).

The sulfur assimilation pathway is started from sulfate uptake. In Arabidopsis, two high-affinity sulfate transporters, SULTR1;1 and SULTR1;2, expressed in the epidermis and cortex of roots facilitate sulfate uptake from the rhizosphere (Takahashi et al. 2000; Vidmar et al. 2000; Shibagaki et al. 2002; Yoshimoto et al. 2002, 2007; Maruyama-Nakashita et al. 2003). Gene expression of both *SULTR1;1* and *SULTR1;2* is dramatically induced by Cd exposure (Herbette et al. 2006; Rouached et al. 2008; Besson-Bard et al. 2009; Jobe et al. 2012). Sulfate uptake activity was also increased by Cd treatment in wild-type Arabidopsis (WT), as reported in maize roots (Nocito et al. 2002, 2006). However, the Cd-induced increase of sulfate uptake activity was diminished in *SULTR1;2* knockout mutant, *sel1-10*, indicating that SULTR1;2 is the main contributor to Cd-induced sulfate uptake (Table 1; Fig. 1; Yamaguchi et al. 2016). Though not significant, Cd treatment tended to induce sulfate uptake activity and induce increased total sulfur and sulfate levels in *sel1-10* shoots, which suggests that the induction of *SULTR1;1* also contributes to the long-term accumulation of sulfate in Cd-treated plants.

Even though the sulfate acquisition was reduced in *sel1-10*, the growth of both WT and *sel1-10* plants were similarly inhibited by Cd treatment, probably because of the similar accumulation of PCs in shoots (Yamaguchi et al. 2016). Sulfate uptake activity, sulfate, and total sulfur content in *sel1-10* plants were less than 50% of those in the WT, both *sel1-10* and WT plants accumulated similar levels of thiols when treated with 20 μM $CdCl_2$, and upon treatment with 40 μM $CdCl_2$, *sel1-10* plants accumulated lower levels of thiols in roots, but still maintained similar thiol levels in shoots. In contrast, Cd-induced accumulation of cysteine and GSH observed in the WT was lower or absent in *sel1-10* plants. Such differences in induction of thiol compounds indicates that PC accumulation, especially in shoots, is the first priority for plants suffering from Cd stress.

In addition to the stimulation of sulfate uptake, Cd treatment increased sulfate accumulation in shoots (Yamaguchi et al. 2016). The Cd-induced sulfate accumulation was occurred only in shoots, whereas PC accumulation was enhanced in both root and shoot tissues, and was caused by the increased sulfate transport to shoots as demonstrated with the increased sulfate concentration in xylem sap under Cd

Table 1 Expression of several sulfate transporters (*SULTRs*) was influenced by Cd treatment in roots

Reference	(1)		(2)	(3)						(4)	(5)	(6)		(7)
Cd concentration (μM)	15	30	200	5			50			20	15	20	40	200
Treated priod	2 h	2 h	6 h	2 h	6 h	30 h	2 h	6 h	30 h	24 h	24 h	10 days		6 h
SULTR1;1	–	1.03	–	–	–	–	**0.96**	1.55	1.58	**0.89**	–	3.30	3.33	–
SULTR1;2	–	–	–	–	–	–	–	–	–	1.73	–	2.15	2.31	3.14
SULTR1;3	–	–	–	–	–	–	–	–	–	n.d.	–	2.13	2.56	–
SULTR2;1	**0.73**	1.48	–	–	1.08	**0.94**	–	1.40	3.39	n.d.	4.16	2.85	3.44	–
SULTR2;2	–	–	–	–	–	–	–	–	–	n.d.	–	–	–	–
SULTR3;1	–	–	–	–	–	–	–	–	–	n.d.	–	–	–	–
SULTR3;2	–	–	–	–	–	–	–	–	–	n.d.	–	–	–	–
SULTR3;3	–	–	–	–	–	–	–	–	–	n.d.	–	–	–	–
SULTR3;4	–	–	–	–	**0.65**	–	–	–	–	n.d.	–	1.45	1.43	–
SULTR3;5	–	–	–	–	***–0.60***	–	–	***–0.97***	***–0.80***	n.d.	–	–	–	–
SULTR4;1	–	–	–	–	–	–	–	**0.82**	2.05	n.d.	–	–	–	–
SULTR4;2	–	–	–	–	–	–	–	–	–	n.d.	–	1.27	2.08	–

Data previously taken with microarray and quantitative RT-PCR were summarized. Numbers in the "Reference" indicate the references in which the original data were reported; i.e. (1) Besson-Bard et al. (2009), (2) Li et al. (2010), (3) Herbette et al. (2006), (4) Rouached et al. (2008), (5) Zhao et al. (2009), (6) Yamaguchi et al. (2016), (7) Jobe et al. (2012). The values indicate log ratio (base 2) of fold changes of gene expression levels under Cd treatment compared with those without Cd ($p < 0.05$ by Bonferroni correction, $n = 3$, in Ref. (1); $p < 0.05$ by Student's *t*-test, $n = 3$, in Refs. (2 and 5); $p < 0.05$ by Student's *t*-test, $n = 2$, in Ref. (3); significant differences were determined from SD, $n = 4$ to 6, in Ref. (4); $p < 0.05$ by Tukey–Kramer test, $n = 4$, in Ref. (6); no statistical analysis, $n = 1$ in Ref. (7)). Data of the genes with log ratios less than 0.0, between 0.0 and 1.0, and more than 1.0 were shown in italic, plane, and bold, respectively. "n.d." means not determined. "–" means no significant difference was detected between Cd-treated and non-treated samples

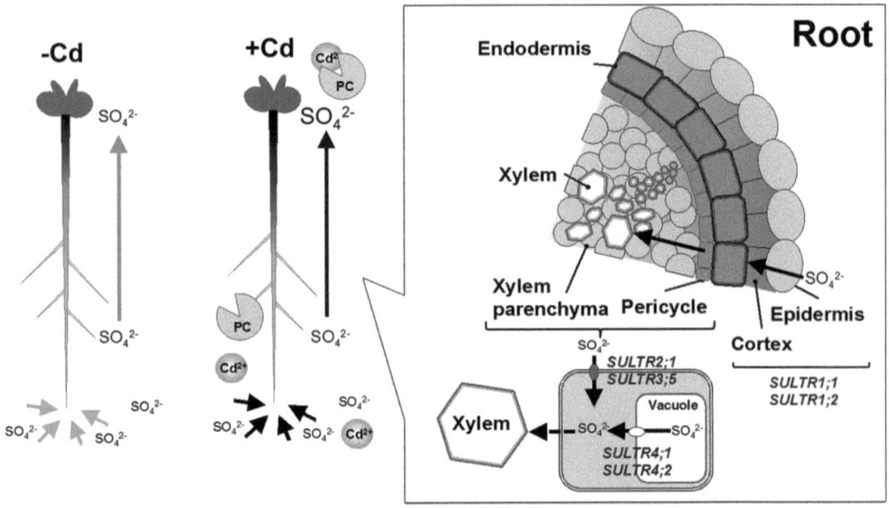

Fig. 1 Cd-induced alterations of sulfate uptake and distribution in *Arabidopsis thaliana*. Arrows indicate the direction of sulfate flow. Cd induces the increase of sulfate uptake mainly depends on SULTR1;2 activity. SULTR2;1, SULTR3;5 and SULTR4 is likely involved in root-to-shoot sulfate transport through xylem under Cd exposure. The mechanism of sulfate transport into xylem is unknown. Enhanced sulfate uptake and the higher distribution rate of sulfate to shoots could contribute to the active thiol synthesis observed in Cd-treated shoots, which results in promoting thiol-mediated Cd detoxification

exposure (Yamaguchi et al. 2016). However, the SULTRs responsible for the Cd-induced increase of root-to-shoot sulfate transport remain to be elucidated.

In Arabidopsis, there are several *SULTRs* involved in root-to-shoot transport of sulfate, i.e. *SULTR2;1*, *SULTR3;5*, *SULTR4;1*, and *SULTR4;2* (Fig. 1; Takahashi et al. 2000; Kataoka et al. 2004a, b; Kawashima et al. 2011; Maruyama-Nakashita et al. 2015). *SULTR2;1* and *SULTR3;5* synergistically mediate root-to-shoot transport of sulfate by retrieving apoplasmic sulfate released from xylem in roots (Takahashi et al. 2000; Kataoka et al. 2004a; Kawashima et al. 2011). In addition, increase of *SULTR2;1* expression in roots contributes to increasing root-to-shoot sulfate transport under sulfur deficiency (Kataoka et al. 2004a; Maruyama-Nakashita et al. 2015). SULTR4;1 and SULTR4;2 are the exporters of sulfate from vacuoles in the pericycle and xylem parenchyma cells of roots, and contribute to root-to-shoot sulfate transport by controlling the cytosolic concentration of sulfate in the cells around xylem vessels (Kataoka et al. 2004b).

In order to obtain any suggestions about the molecular machinery of Cd-induced increase of sulfate transport to shoots, the effects of Cd treatments on the expression of *SULTRs* in roots previously detected with microarray and quantitative RT-PCR were summarized in Table 1 (Herbette et al. 2006; Rouached et al. 2008; Besson-Bard et al. 2009; Zhao et al. 2009; Li et al. 2010; Jobe et al. 2012; Yamaguchi et al. 2016). Several studies of Cd-treated Arabidopsis have reported that the expression of *SULTR1;1* and/or *SULTR1;2* were induced by Cd treatment (Herbette et al. 2006;

Rouached et al. 2008; Besson-Bard et al. 2009; Jobe et al. 2012), in agreement with the stimulation of sulfate uptake upon Cd exposure. In addition, the increase of *SULTR2;1* transcript was often detected under Cd exposure. Transcript levels of *SULTR2;1* in roots were increased by Cd treatment in time- and concentration-dependent manner (Herbette et al. 2006; Besson-Bard et al. 2009; Zhao et al. 2009; Yamaguchi et al. 2016). The transcript levels of *SULTR4;1* and *SULTR4;2* were also increased in response to Cd treatment (Herbette et al. 2006; Yamaguchi et al. 2016). Though the expression of *SULTR3;5* was not increased by Cd treatment, SULTR3;5 also can take part because sulfate uptake activity of SULTR3;5 is controlled by the expression of SULTR2;1 (Kataoka et al. 2004a).

These results suggest that all SULTRs responsible for sulfate translocation to shoots are involved in Cd-induced increase of sulfate transport to shoots (Fig. 1). We need further evaluation to determine the molecular mechanisms how Cd treatment increases the sulfate transport to shoots and whether these SULTRs are involved in the enhanced sulfate transport, e.g. the analysis of sulfate translocation using the knockout mutants of these SULTRs. Then further exploration of the physiological meaning in enhanced sulfate transport to shoots by Cd treatment would be determined.

References

Besson-Bard A, Gravot A, Richaud P, Auroy P, Duc C, Gaymard F, Taconnat L, Renou JP, Pugin A, Wendehenne D (2009) Nitric oxide contributes to cadmium toxicity in Arabidopsis by promoting cadmium accumulation in roots and by up-regulating genes related to iron uptake. Plant Physiol 149:1302–1315

Chmielowska-Bąk J, Gzyl J, Rucińska-Sobkowiak R, Arasimowicz-Jelonek M, Deckert J (2014) The new insights into cadmium sensing. Front Plant Sci 5:245

Choppala G, Saifullah Bolan N, Bibi S, Iqbal M, Rengel Z, Kunhikrishnan A, Ashwath N, Ok YS (2014) Cellular mechanisms in higher plants governing tolerance to cadmium toxicity. Crit Rev Plant Sci 33:374–391

Clemens S, Aarts MG, Thomine S, Verbruggen N (2013) Plant science: the key to preventing slow cadmium poisoning. Trends Plant Sci 18:92–99

Cobbett CS (2000) Phytochelatins and their roles in heavy metal detoxification. Plant Physiol 123:825–832

Cobbett CS, Goldsbrough P (2002) Phytochelatins and metallothioneins: roles in heavy metal detoxification and homeostasis. Annu Rev Plant Biol 53:159–182

Cobbett CS, May MJ, Howden R, Rolls B (2002) The glutathione-deficient, cadmium-sensitive mutant, *cad2-1*, of *Arabidopsis thaliana* is deficient in gamma-glutamylcysteine synthetase. Plant J 16:73–78

Herbette S, Taconnat L, Hugouvieux V, Piette L, Magniette ML, Cuine S, Auroy P, Richaud P, Forestier C, Bourguignon J, Renou JP, Vavasseur A, Leonhardt N (2006) Genome-wide transcriptome profiling of the early cadmium response of Arabidopsis roots and shoots. Biochimie 88:1751–1765

Järup L, Akesson A (2009) Current status of cadmium as an environmental health problem. Toxicol Appl Pharmacol 238:201–208

Jobe TO, Sung DY, Akmakjian G, Pham A, Komives EA, Mendoza-Cózatl DG, Schroeder JI (2012) Feedback inhibition by thiols outranks glutathione depletion: a luciferase-based screen reveals glutathione-deficient γ-ECS and glutathione synthetase mutants impaired in cadmiuminduced sulfate assimilation. Plant J 70:783–795

Kataoka T, Hayashi N, Yamaya T, Takahashi H (2004a) Root-to-shoot transport of sulfate in Arabidopsis: evidence for the role of SULTR3;5 as a component of low-affinity sulfate transport system in the root vasculature. Plant Physiol 136:4198–4204

Kataoka T, Watanabe-Takahashi A, Hayashi N, Ohnishi M, Mimura T, Buchner P, Hawkesford MJ, Yamaya T, Takahashi H (2004b) Vacuolar sulfate transporters are essential determinants controlling internal distribution of sulfate in Arabidopsis. Plant Cell 16:2693–2704

Kawashima CG, Matthewman CA, Huang S, Lee BR, Yoshimoto N, Koprivova A, Rubio-Somoza I, Todesco M, Rathjen T, Saito K, Takahashi H, Dalmay T, Kopriva S (2011) Interplay of SLIM1 and miR395 in the regulation of sulfate assimilation in Arabidopsis. Plant J 66:863–876

Leustek T, Martin MN, Bick JA, Davies JP (2000) Pathways and regulation of sulfur metabolism revealed through molecular and genetic studies. Annu Rev Plant Physiol Plant Mol Biol 51:141–165

Li JY, Fu YL, Pike SM, Bao J, Tian W, Zhang Y, Chen CZ, Zhang Y, Li HM, Huang J, Li LG, Schroeder JI, Gassmann W, Gong JM (2010) The Arabidopsis nitrate transporter NRT1.8 functions in nitrate removal from the xylem sap and mediates cadmium tolerance. Plant Cell 22:1633–1646

Lin YF, Aarts MG (2012) The molecular mechanism of zinc and cadmium stress response in plants. Cell Mol Life Sci 69:3187–3206

Lu SC (2013) Glutathione synthesis. Biochim Biophys Acta 1830:3143–3153

Maruyama-Nakashita A, Inoue E, Watanabe-Takahashi A, Yamaya T, Takahashi H (2003) Transcriptome profiling of sulfur responsive genes in Arabidopsis reveals global effects of sulfur nutrition on multiple metabolic pathways. Plant Physiol 132:597–605

Maruyama-Nakashita A, Watanabe-Takahashi A, Inoue E, Yamaya T, Saito K, Takahashi H (2015) Sulfur-responsive elements in the 3′-nontranscribed intergenic region are essential for the induction of *SULFATE TRANSPORTER 2;1* gene expression in Arabidopsis roots under sulfur deficiency. Plant Cell 27:1279–1296

Nawrot T, Plusquin M, Hogervorst J, Roels HA, Celis H, Thijs L, Vangronsveld J, Van Hecke E, Staessen JA (2006) Environmental exposure to cadmium and risk of cancer: a prospective population-based study. Lancet Oncol 7:119–126

Nocito FF, Pirovano L, Cocucci M, Sacchi GA (2002) Cadmiuminduced sulfate uptake in maize roots. Plant Physiol 129:1872–1879

Nocito FF, Lancilli C, Crema B, Fourcroy P, Davidian JC, Sacchi GA (2006) Heavy metal stress and sulfate uptake in maize roots. Plant Physiol 141:1138–1148

Rouached H, Wirtz M, Alary R, Hell R, Arpat AB, Davidian JC, Fourcroy P, Berthomieu P (2008) Differential regulation of the expression of two high-affinity sulfate transporters, SULTR1.1 and SULTR1.2, in Arabidopsis. Plant Physiol 147:897–911

Saito K (2004) Sulfur assimilatory metabolism. The long and smelling road. Plant Physiol 136:2443–2450

Seth CS, Remans T, Keunen E, Jozefczak M, Gielen H, Opdenakker K, Weyens N, Vangronsveld J, Cuypers A (2012) Phytoextraction of toxic metals: a central role for glutathione. Plant Cell Environ 35:334–346

Shibagaki N, Rose A, McDermott JP, Fujiwara T, Hayashi H, Yoneyama T, Davies JP (2002) Selenate-resistant mutants of *Arabidopsis thaliana* identify Sultr1;2, a sulfate transporter required for efficient transport of sulfate into roots. Plant J 29:475–486

Takahashi H, Watanabe-Takahashi A, Smith FW, Blake-Kalff M, Hawkesford MJ, Saito K (2000) The roles of three functional sulfate transporters involved in uptake and translocation of sulfate in *Arabidopsis thaliana*. Plant J 23:171–182

Verbruggen N, Hermans C, Schat H (2009) Mechanisms to cope with arsenic or cadmium excess in plants. Curr Opin Plant Biol 12:364–372

Vidmar JJ, Tagmount A, Cathala N, Touraine B, Davidian JC (2000) Cloning and characterization of a root specific high-affinity sulfate transporter from *Arabidopsis thaliana*. FEBS Lett 475:65–69

Yadav SK (2010) Heavy metals toxicity in plants: an overview on the role of glutathione and phytochelatins in heavy metal stress tolerance of plants. South Afr J Bot 76:167–179

Yamaguchi C, Takimoto Y, Ohkama-Ohtsu N, Hokura A, Shinano T, Nakamura T, Suyama A, Maruyama-Nakashita A (2016) Effects of cadmium treatment on the uptake and translocation of sulfate in *Arabidopsis thaliana*. Plant Cell Physiol 57:2353–2366

Yoshimoto N, Takahashi H, Smith FW, Yamaya T, Saito K (2002) Two distinct high-affinity sulfate transporters with different inducibilities mediate uptake of sulfate in Arabidopsis roots. Plant J 29:465–473

Yoshimoto N, Inoue E, Watanabe-Takahashi A, Saito K, Takahashi H (2007) Posttranscriptional regulation of high-affinity sulfate transporters in Arabidopsis by sulfur nutrition. Plant Physiol 145:378–388

Zhao CR, Ikka T, Sawaki Y, Kobayashi Y, Suzuki Y, Hibino T, Sato S, Sakurai N, Shibata D, Koyama H (2009) Comparative transcriptomic characterization of aluminum, sodium chloride, cadmium and copper rhizotoxicities in *Arabidopsis thaliana*. BMC Plant Biol 9:32

A Glycine-Rich Protein Encoded by Sulfur-Deficiency Induced Gene Is Involved in the Regulation of Callose Level and Root Elongation

Anna Znój, Katarzyna Zientara-Rytter, Paweł Sęktas, Grzegorz Moniuszko, Agnieszka Sirko, and Anna Wawrzyńska

Abstract Glycine-rich proteins (GRPs) with characteristic repetitive glycine stretches are ubiquitous in organisms of all Kingdoms and have distinct functions. It is believed that Gly-rich domains serve mainly for interactions with other proteins. Previously, we identified the tobacco UP30 gene as strongly upregulated by sulfur deficiency. It encodes a protein highly similar to cdiGRP which affects tobacco defense response by elevating cell wall callose deposits thus blocking systemic movement of viruses. The closest *Arabidopsis thaliana* homologue of UP30 is GRP-3 (At2g05520). Here we report that GRP-3 is induced in Arabidopsis seedlings in both sulfur and nitrogen deficiency conditions. The transgenic Arabidopsis plants with changed GRP-3 expression (either overexpressing or with silenced GRP-3) tend to have longer roots than the wild type, especially in the conditions of sulfur deficiency. The effect could be alleviated by the addition of auxin to the media. Moreover, we observed the increased callose deposition in both Arabidopsis lines suggesting its negative effects on shoot-to-root movement of auxins in nutrient deficient conditions.

Glycine-rich proteins (GRPs) are characterized by the existence of several domains with little sequence conservation but highly enriched in glycine. Although the first genes encoding GRPs have been isolated from plants, proteins with characteristic repetitive glycine stretches exist in diverse organisms from cyanobacteria to animals (reviewed in Sachetto-Martins et al. 2000). The first reports described plant GRPs as cell wall associated proteins (Showalter 1993), however since then many other GRPs with different domain organizations, sub-cellular localizations as well as tissue specificity of expression were demonstrated. This clearly indicated that these proteins must be implicated in several independent physiological processes, though the specific functions of several so far characterized GRPs remain

A. Znój • K. Zientara-Rytter • P. Sęktas • G. Moniuszko • A. Sirko (✉) • A. Wawrzyńska
Institute of Biochemistry and Biophysics, Polish Academy of Sciences, Pawińskiego 5A St, 02-106 Warsaw, Poland
e-mail: asirko@ibb.waw.pl

L.J. De Kok et al. (eds.), *Sulfur Metabolism in Higher Plants - Fundamental, Environmental and Agricultural Aspects*, Proceedings of the International Plant Sulfur Workshop, DOI 10.1007/978-3-319-56526-2_21

speculative. The diversity of GRPs led even to the concept that they should rather not be considered as a family but as a wide group of proteins sharing a common motif (Sachetto-Martins et al. 2000). The presence of Gly-rich highly flexible domains may act as a molecular glue in the protein-protein interactions. It is therefore quite probable that GRPs are components of different multimolecular complexes, including homo- or hetero-dimerization between family members (Sachetto-Martins et al. 2000).

The *UP30* gene encoding protein belonging to GRP family has been identified as one of the strongly upregulated genes in the leaves of tobacco plants transferred for 2 days to the sulfur deficient medium (Wawrzynska et al. 2005). UP30 is a small protein containing Gly-rich, Tyr-rich and Pro-rich domains and a characteristic Cys-rich motif [-C-X3-CC-X6-C-X2-CC-] (Lewandowska et al. 2005). Subsequent experiments indicated that the protein fusion of UP30 and YFP is located in the cell wall (Fig.1a). The yeast two hybrid system allowed us to exclude the UP30-UP30 homodimerization (Fig. 1b); the interactions with other members of the GRP family were not checked.

UP30 belongs to Class II from the eight classes that the plant GRPs has been divided into based on the domain composition (Mangeon et al. 2010). Class II members have (beside optional signal peptide at the N-terminus) the Gly-rich and Cys-rich domains. Previously identified tobacco protein from the same class (cdiGRP) was proposed to be involved in plant defense. The gene encoding this protein is induced by cadmium and the elevated level of cdiGRP has been suggested to block the systemic spread of the turning vein-clearing tobamovirus (Ueki and Citovsky 2002) due to increasing the callose deposition in the plant cell wall (Ueki and Citovsky 2005). Interestingly, UP30 shares 81% amino acid sequence identity with cdiGRP allowing us to assume both proteins are functionally similar. The closest *Arabidopsis thaliana* homologue of UP30 is GRP-3. Both proteins have very similar domains/regions organization despite relatively low sequence conservation (only 29%; Fig. 2).

The *GRP-3* (At2g05520) gene was previously shown to be expressed in stems and leaves and only poorly in the roots, immature seed pods and flowers (de Oliveira et al. 1990). Such pattern of expression resembles the expression of UP30; the *UP30* mRNA was detected mostly in the leaves (Lewandowska et al. 2005). Likewise, the localization of GRP-3 in the apoplast (Gramegna et al. 2016) was an additional suggestion for the functional homology between UP30 and GRP-3. The GRP-3 protein interacts through its Cys-rich C-terminal motif with the cell wall receptor protein kinase WAK1. The binding of GRP-3 to WAK1 is critical for the structural integrity of the multimeric complex representing the activated signalosome of pathogen response (Park et al. 2001). On the other hand, GRP-3 acts as negative regulator of both elicitor-activated signaling cascade and response to wounding. It was proposed that GRP-3 may function in restoring the plant immune system to the baseline at the appropriate time after the pathogen attack (Gramegna et al. 2016).

To assay the function of GRP-3 during sulfur deficiency we have constructed the transgenic Arabidopsis plants constitutively overexpressing *GRP-3* (OX9 lines) and

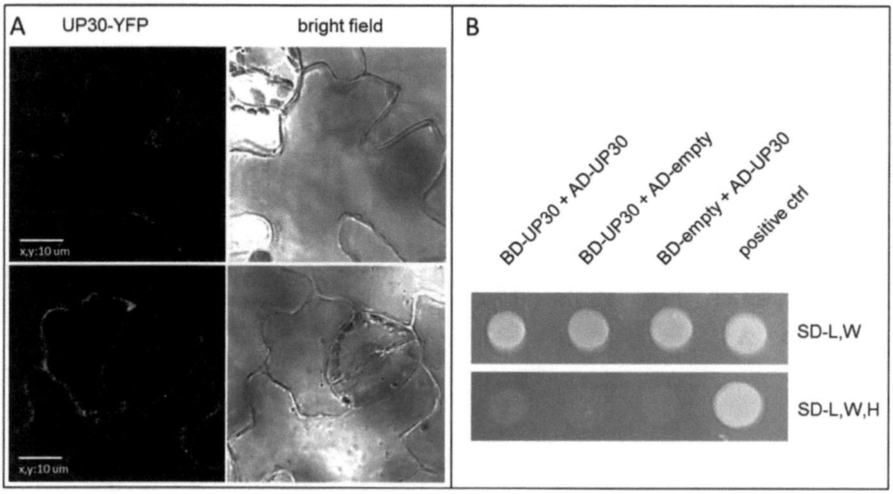

Fig. 1 UP30 protein localization in the cell wall of leaf epidermal cells when transiently expressed in *N. benthamiana* (**a**) and does not form dimers in the yeast two hybrid test (**b**)

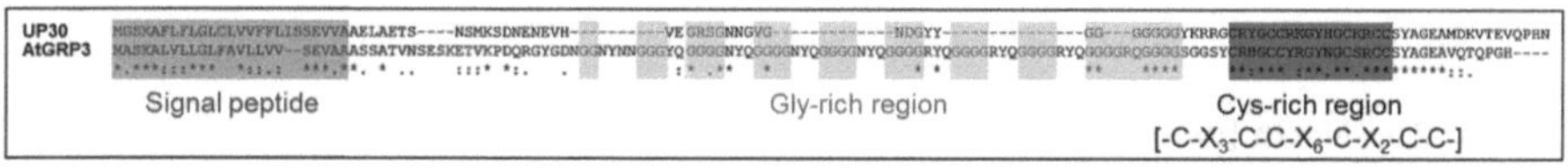

Fig. 2 Amino acid alignment of tobacco UP30 and GRP-3. Characteristic domains are marked

plants with silenced *GRP-3* due to constitutive expression in the antisense orientation (KD10 lines). The level of *GRP-3* mRNA in the transgenic plants was confirmed with semiquantitative RT-PCR. Additionally, the expression level was monitored in plants grown in S- or N-deficient conditions. The level of *GRP-3* transcript was induced in S- and N- conditions in the wild-type seedlings, it was high in the OX9 line and low (or undetected) in KD10 line (Fig. 3).

Interestingly, monitoring the growth of the seedlings in sulfur- deficient (S-), and sufficient (ctrl) media indicated that both overexpressors as well as knock-downs of *GRP-3* have longer roots that the wild-type, especially in the conditions of sulfur deficiency (Fig. 4).

Tobacco cdiGRP represents one of the factors that regulates callose accumulation in the plant vasculature (Ueki and Citovsky 2005). Callose is a 1,3-β-d-glucan, synthesized by callose synthase and degraded by 1,3-β-d-glucanase, which has been shown to accumulate around the neck region of the plasmodesmata. It is supposed to regulate the traffic through these channels and cell-to-cell communication. Therefore, in the next step we verified the level of callose in the *GRP-3* transformants. Surprisingly, higher level of callose was observed in the tissue of both OX9 and KD10 plants as compared to the wild type (Fig. 5a). It was especially significant for the stem and root vascular tissue, especially at the plasmodesmata region (Fig. 5b).

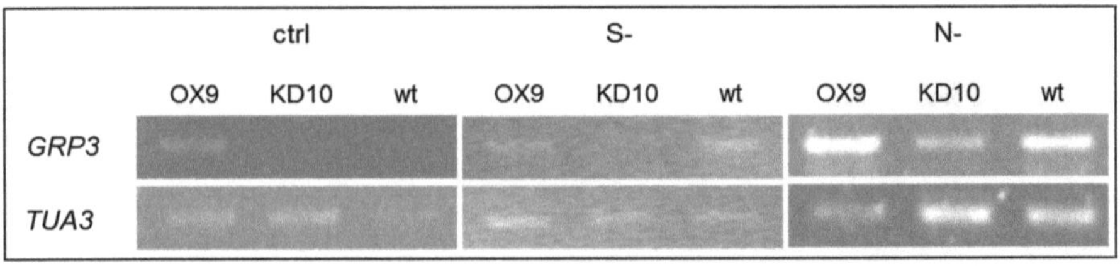

Fig. 3 The level of *GRP-3* expression in 2-week-old Arabidopsis seedlings grown in 0.5× Hoagland optimal (ctrl), or sulfur-deficient (S-) or nitrogen-deficient (N−) medium assayed with semi-quantitative RT-PCR. Tubuline (*TUA3*) was used as quantity control

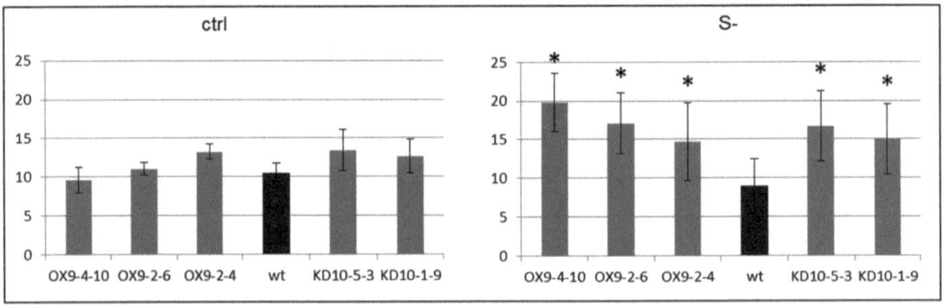

Fig. 4 The length [in mm] of the roots of the seedlings grown for 2 weeks in 0.5× Hoagland medium, either optimal (ctrl) or sulfur deficient (S-)

The plasmodesmata is a specific channel structure in plants that spans the cell wall connecting neighboring cells to allow cytoplasmic exchange and communication. The plasmodesmata contribution to auxins movement, their role as potential sites of receptor signaling and the impact of callose deposition in plasmodesmata on auxins function has been recently discovered (Jackson 2015). Auxins reduce the growth of roots by two types of mechanisms, either by reducing the extend of root growth (IAA, NAA) or the cell production rate (2,4-D) (Rahman et al. 2007). Therefore, we assumed that the OX9 and KD10 transformants have longer roots than the wild type in nutrient deficient conditions because the shoot-to-root movement of auxins in such plants is negatively affected due to extensive callose deposition in their plasmodesmata (Fig. 5). It has been previously shown that Arabidopsis plants grown in S- conditions have lower level of auxins, as indicated by the activity of the auxin response marker *DR5::GUS* in the roots (Dan et al. 2007). To check this we decided to compare the effect of NAA on the root elongation of the seedlings grown in S- conditions (Fig. 6). In this experiment the OX9 lines were used as plants overexpressing *GRP-3* and three T-DNA insertion SALK lines (SALK_084781, SALK_084877 and SALK_012941C) were used as KO *grp-3* mutants. These lines had no detectible transcript corresponding to *GRP-3* (not shown). The results shown in Fig. 6 indicated that addition of NAA to the S-deficient media negatively affected root length in all lines; however, the

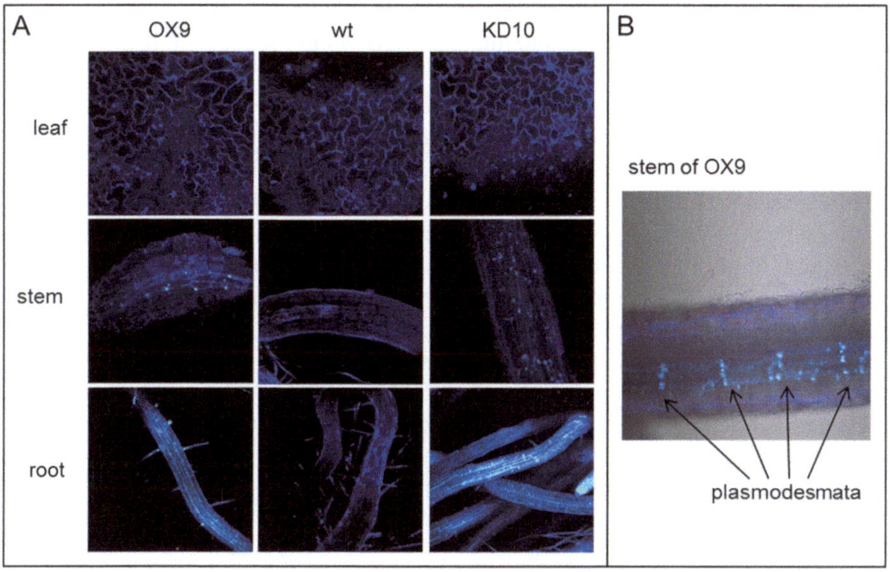

Fig. 5 Callose deposition in the 2-weeks-old Arabidopsis seedlings grown in 0.5× Hoagland optimal medium (wt, OX9–2-4 and KD10–1-9 line was used). The callose was stained with aniline blue for 1 h and next the tissues were observed under fluorescent confocal microscopy with 365 nm filter (1:40 magnification)

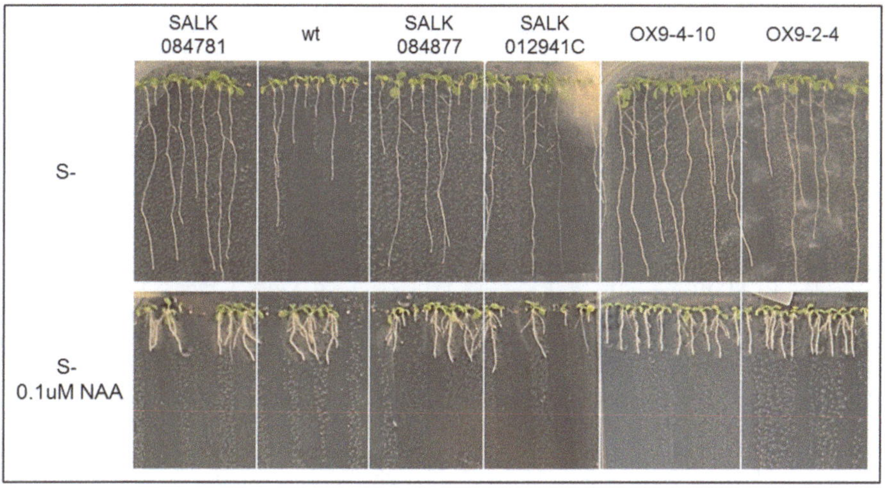

Fig. 6 Two-weeks-old Arabidopsis seedlings grown for 2 weeks in sulfur deficient 0.5× Hoagland medium (S-) with or without the addition of 0.1 μM NAA

difference in root lengths observed between the wild-type seedlings and the OX9 transformants and *grp3* insertional mutants was strongly reduced in the presence of auxin.

In conclusion, we have demonstrated that the GRP-3 protein encoded by the sulfur deficiency induced gene modulates the root length of the Arabidopsis seedlings in nutrient deficient conditions. The longer roots were observed in both, seedlings with increased level of *GRP-3* as well as with the reduced or abolished expression of *GRP-3*. An unbalanced level of GRP-3 results in increased callose deposition in plasmodesmata. Our findings are supported by the reports from other laboratories, where on one hand, the elevated expression of cdiGRP in tobacco resulted in callose accumulation (Ueki and Citovsky 2002, 2005), while on the other hand, the Arabidopsis *grp-3* mutants were reported to have higher level of callose than the wild-type in response to different elicitors (Gramegna et al. 2016). Results of the experiments presented in this work prompt us to hypothesize that GRP-3 is involved in the regulation of root growth in response to nutrient status of the plants through influencing auxin movement due to affecting callose deposits in plasmodesmata.

Acknowledgments We thank to Dimitris Bouranis, Agricultural University of Athens, Greece for the protocol of callose staining. This work was supported by the grant from the Polish Ministry of Science and Higher Education (W16/7.PR/2011).

References

Dan H, Yang G, Zheng ZL (2007) A negative regulatory role for auxin in sulphate deficiency response in *Arabidopsis thaliana*. Plant Mol Biol 63:221–235

De Oliveira DE, Seurinck J, Inze D, Van Montagu M, Botterman J (1990) Differential expression of five Arabidopsis genes encoding glycine-rich proteins. Plant Cell 2:427–436

Gramegna G, Modesti V, Savatin DV, Sicilia F, Cervone F, De Lorenzo G (2016) GRP-3 and KAPP encoding interactors of WAK1 negatively affect defense responses induced by oligogalacturonides and local response to wounding. J Exp Bot 67:1715–1729

Jackson D (2015) Plasmodesmata spread their influence. F1000Prime Rep 7:25

Lewandowska M, Wawrzynska A, Kaminska J, Liszewska F, Sirko A (2005) Identification of novel proteins of *Nicotiana tabacum* regulated by short term sulfur starvation. In: Saito K, De Kok LJ, Stulen I, Hawkesford MJ, Schnug E, Sirso A, Rennenberg H (eds) Sulfur transport and assimilation in plants in the postgenomic era. Backhuys Publishers, Leiden, pp 153–156

Mangeon A, Junqueira RM, Sachetto-Martins G (2010) Functional diversity of the plant glycine-rich proteins superfamily. Plant Signal Behav 5:99–104

Park AR, Cho SK, Yun UJ, Jin MY, Lee SH, Sachetto-Martins G, Park OK (2001) Interaction of the Arabidopsis receptor protein kinase Wak1 with a glycine-rich protein AtGRP-3. J Biol Chem 276:26688–26693

Rahman A, Bannigan A, Sulaman W, Pechter P, Blancaflor EB, Baskin TI (2007) Auxin actin and growth of the *Arabidopsis thaliana* primary root. Plant J 50:514–528

Sachetto-Martins G, Franco LO, de Oliveira DE (2000) Plant glycine-rich proteins: a family or just proteins with a common motif? Biochim Biophys Acta 1492:1–14

Showalter AM (1993) Structure and function of plant cell wall proteins. Plant Cell 5:9–23

Ueki S, Citovsky V (2002) The systemic movement of a tobamovirus is inhibited by a cadmium-ion-induced glycine-rich protein. Nat Cell Biol 4:478–486
Ueki S, Citovsky V (2005) Identification of an interactor of cadmium ion-induced glycine-rich protein involved in regulation of callose levels in plant vasculature. Proc Natl Acad Sci U S A 102:12089–12094
Wawrzynska A, Lewandowska M, Hawkesford MJ, Sirko A (2005) Using a suppression subtractive library-based approach to identify tobacco genes regulated in response to short-term sulphur deficit. J Exp Bot 56:1575–1590

Titles and Content of the Proceedings of the International Plant Sulfur Workshops (1990–2009)

Rennenberg H, Brunold C, De Kok LJ, Stulen I (1990) Sulfur Nutrition and Sulfur Assimilation in Higher Plants. SPB Academic Publishing, The Hague, ISBN 90-5103-038-X

De Kok LJ, Stulen I, Rennenberg H, Brunold C, Rauser WE (1993) Sulfur Nutrition and Assimilation in Higher Plants: Regulatory, Agricultural and Environmental Aspects. SPB Academic Publishing, The Hague, ISBN 90-5103-084-3

Cram WJ, De Kok LJ, Stulen I, Brunold C, Rennenberg H (1997) Sulphur Metabolism in Higher Plants: Molecular, Ecophysiological and Nutritional Aspects. Backhuys Publishers, Leiden, ISBN 90-73348-13-7

Brunold C, Rennenberg H, De Kok LJ, Stulen I, Davidian, J-C (2000) Sulfur Nutrition and Sulfur Assimilation in Higher Plants: Molecular, Biochemical and Physiological Aspects. Verlag Paul Haupt, Bern, ISBN 3-258-06239-0

Davidian J-C, Grill D, De Kok LJ, Stulen I, Hawkesford MJ, Schnug E, Rennenberg H (2003) Sulfur Transport and Assimilation in Plants: Regulation, Interaction, Signaling. Backhuys Publishers, Leiden, ISBN 90-5782-138-9

Saito K, De Kok LJ, Stulen I, Hawkesford MJ, Schnug E, Sirko A, Rennenberg H (2005) Sulfur Transport and Assimilation in Plants in the Post Genomic Era. Backhuys Publishers, Leiden, ISBN 90-5782-166-4

Sirko A, De Kok LJ, Haneklaus S, Hawkesford MJ, Rennenberg H, Saito K, Stulen I (2009) Sulfur Metabolism in Plants: Regulatory Aspects, Significance of Sulfur in the Food Chain, Agriculture and the Environment. Backhuys Publishers, Leiden/Margraf Publishers, Weikersheim, ISBN 978-3-8236-1547-7, ISBN 978-90-5782-215-5

L.J. De Kok et al. (eds.), *Sulfur Metabolism in Higher Plants - Fundamental, Environmental and Agricultural Aspects*, Proceedings of the International Plant Sulfur Workshop, DOI 10.1007/978-3-319-56526-2

Rennenberg H, Brunold C, De Kok LJ, Stulen I (1990) Sulfur Nutrition and Sulfur Assimilation in Higher Plants. SPB Academic Publishing, The Hague, ISBN 90-5103-038-X

Overview Papers

Brunold C – Reduction of sulfate to sulfide, pp 13–31
Cram WJ – Uptake and transport of sulfate, pp 3–11
De Kok LJ – Sulfur metabolism in plants exposed to atmospheric sulfur, pp 111–130
Ernst WHO – Ecological aspects of sulfur metabolism, pp 131–144
Giovanelli J – Regulatory aspects of cysteine and methionine synthesis, pp 33–48
Grill E, Winnacker E-L, Zenk MH – Phytochelatins, the heavy metal binding peptides of the plant kingdom, pp 89–95
Kleppinger-Sparace KF, Sparace SA, Mudd JB – Plant sulfolipids, pp 77–88
Kreis M – Synthesis, structure and evolutionary relationships of sulfur-rich seed proteins in higher plants, pp 67–76
Rennenberg H, Lamoureux GL – Physiological processes that modulate the concentration of glutathione in plant cells, pp 53–65
Schnug E – Glucosinolates – fundamental, environmental and agricultural aspects, pp 97–106
Thompson JF, Madison JT – The effect of sulfate and methionine on legume proteins, pp 145–158

Research Papers

Bell CI, Cram WJ, Clarkson DT – Turnover of sulfate in leaf vacuoles limits retranslocation under sulfur stress, pp 163–165
Bergmann L, Hell R – Determination and characterization of γ-glutamylcysteine synthetase from higher plants, pp 167–171
Bosma W, Kamminga G, De Kok LJ – H_2S-induced accumulation of sulfhydryl-compounds in leaves of plants under field and laboratory exposure, pp 173–175
Dietz K-J, Busch H – Intracellular compartmentation, utilization and transport of methionine in barley mesophyll cells, pp 177–180
Ferrari G, Cacco G, Bottacin A, Saccomani M – Energetics of sulfate uptake by plants as related to previous sulfate availability, pp 181–183
Goldsbrough PB, Mendum ML, Gupta SC – Effect of glutathione on phytochelatin synthesis and cadmium tolerance in tomato cells, pp 185–189
Grill D, Guttenberger H, Bermadinger E – Thiol content of plants depending on the altitude, pp 191–193
Gupta AS, Alscher RG, McCune DC – Ozone exposure, glutathione levels and photosynthesis in hybrid poplar, pp 195–197

De Kok LJ, Stulen, I, Rennenberg H, Brunold C, Rauser WE (1993) Sulfur Nutrition and Assimilation in Higher Plants: Regulatory, Agricultural and Environmental Aspects. SPB Academic Publishing, The Hague, ISBN 90-5103-084-3

Overview Papers

Research Papers

Cram WJ, De Kok LJ, Stulen I, Brunold C, Rennenberg, H (1997) Sulphur Metabolism in Higher Plants: Molecular, Ecophysiological and Nutritional Aspects. Backhuys Publishers, Leiden, ISBN 90-73348-13-7

Overview Papers

Research Papers

Kranner I, Grill D – The glutathione status during recovery of desiccated lichens: is desiccation tolerance correlated with a high reducing capacity for glutathione disulphide? pp 249–252

Kranner I, Grill D - Desiccation of lichens: changes in the glutathione status, pp 253–256

Kreuzweiser J, Herschback C, Rennenberg H – Sulphate uptake by mycorrhizal (*Laccaria laccata*) and non-mycorrhizal roots of beech (*Fagus sylvatica* L.) trees, pp 165–168

Lakkineni KC, Jagadish R, Abrol YP – Effect of thiol compounds on nitrate reductase activity in vivo on wheat (*Triticum aestivum* L.) leaves, pp 335–338

Lappartient AG, Leustek T, Touraine B – Are ATP sulphurylase mRNA and protein accumulated in roots of *Arabidopsis* following S stress? pp 207–210

Lappartient AG, Touraine B – Comparison between demand-driven regulation of ATP sulphurylase activity and responses to oxidative stress, pp 203–206

Logan H, Cathala N, Grignon C, Davidian J-C – ATP sulphurylase expression studies in yeast and in plants, pp 215–216

Mulholland MM, Corradini F, Otte ML – Dimethylsulphoniopropionate in *Spartina*-grasses, pp 279–280

Mullineaux PM, Wellburn AR, Baker NR, Creissen GP – Increased capacity for glutathione biosynthesis in the chloroplast paradoxically promotes oxidative stress in transgenic tobacco, pp 269–270

Nieder J, Stallmeyer B, Brinkmann H, Mendel RR – Molybdenum cofactor biosynthesis: identification of *Arabidopsis thaliana* cDNAs homologous to the *E. coli* sulphotransferase MoeB, pp 275–278

Noctor G, Arisis AC, Jouanin L, Foyer C – Regulation of glutathione synthesis in transformed poplars, pp 243–246

Poortinga AM, De Kok LJ – Uptake of atmospheric H_2S by *Spinacea oleracea* L. and consequences for thiol content and composition in shoots and roots, pp 285–288

Poortinga AM, Hoen G, De Kok LJ – Cysteine desulphhydrase of *Spinacia oleracea* L. may catalyse synthesis of cysteine, pp 233–234

Prosser I, Schneider A, Hawkesford MJ, Clarkson DT – Changes in nutrient composition, metabolite concentrations and enzyme activities in spinach in the early stages of S-deprivation, pp 339–342

Pultke U, Giesemann A, Weigel H-J – Translocation of sulphur in terrestrial ecosystems- a field approach using stable S-isotope analysis to trace fertiliser S, pp 319–322

Rose AB – Selenate-resistant mutants of *Arabidopsis*, pp 217–220

Saito K, Takahasi H, Takagi Y, Inoue K, Noji M – Molecular characterisation and regulation of cysteine synthase and serine acetyl-transferase from plants, pp 235–268

Schneider A, Bowsher CG, Hawkesford MJ – Purification of two tissue-specific isoforms of *O*-acetylserine (thiol) lyase from spinach (*Spinacia medania* L.) and influence by sulphur nutrition, pp 239–242

Brunold C, Rennenberg H, De Kok LJ, Stulen I, Davidian J-C (2000) Sulfur Nutrition and Sulfur Assimilation in Higher Plants: Molecular, Biochemical and Physiological Aspects. Verlag Paul Haupt, Bern, ISBN 3-258-06239-0

Overview Papers

Davidian J-C, Hatzfeld Y, Cathala N, Tagmount A, Vidmar JJ – Sulfate uptake and transport in plants, pp 19–40

De Kok LJ, Westerman S, Stuiver CEE, Stulen I – Atmospheric H_2S as plant sulfur source: interaction with pedospheric sulfur nutrition – a case study with *Brassica oleracea* L., pp 41–58

Droux M, Gakiére B, Denis L, Ravanel S, Tabe L, Lappartient AG, Job D – Methionine biosynthesis in plants: biochemical and regulatory aspects, pp 73–92

Foyer CH, Rennenberg H – Regulation of glutathione synthesis and its role in abiotic and biotic stress defence, pp 127–154

Höfgen R, Hesse H – Molecular engineering of sulfate assimilation, pp 109–126

Kredich NM – Foreword: Plant sulfur, pp xv–xx

Leustek T, Bick JA – The evolution of sulfur assimilation in plants, pp 1–18

Okanenko A – Sulfoquinovosyl diacylglycerol in higher plants: biosynthesis and physiological function, pp 203–220

Paszewski A, Natorff R, Piotrowska M, Brzywczy J, Sieńko M, Grynberg M, Pizzinini E, Turner G – Regulation of sulfur amino acid biosynthesis in *Aspergillus nidulans*: physiological and genetical aspects, pp 93–108

Rauser WE – The role of thiols in plants under stress, pp 169–186

Saito K, Takahashi H, Noji M, Inoue K, Hatzfeld Y – Molecular regulation of sulfur assimilation and cysteine synthesis, pp 59–72

Städler E – Secondary sulfur metabolites influencing herbivorous insects, pp 187–202

Walbot V, Mueller LA, Silady RA, Goodman CD – Do glutathione *S*-transferases act as enzymes or carrier proteins for their natural substrates? pp 155–168

Research Papers

Amáncio S, De Carvalho L – Is the sulfhydryl status a marker for autotrophic competence of *in vitro* woody plants, pp 359–360

Anisimova M, Haneklaus S, Schnug E – Significance of sulfur for soil organic matter, pp 239–244

Awazuhara M, Hirai MY, Hayashi H, Chino M, Naito S, Fujiwara T – *O*-acetyl-L-serine content in rosette leaves of *Arabidopsis thaliana* affected by S and N nutrition, pp 331–334

Blake-Kalff M, Riley N, Hawkesford MJ – Utilization of sulfur by oilseed rape in response to sulfur deficiency, pp 245–248

Bork C, Hell R – Expression patterns of the sulfite reductase gene from *Arabidopsis thaliana*, pp 299–302

Boursis F, Roje S, Nuccio ML, Sisher DB, Tarczynski MC, Li C, Herschbach C, Rennenberg H, Pimenta MJ, Sen T-L, Gage DA, Hanson A – *S*-methylmethionine has a major role in phloem sulfur transport, and is synthesized by a novel methyltransferase, pp 238–284

Bruns I, Sutter K, Neumann D, Krauss G-J – Glutathione accumulation – a specific response of mosses to heavy metal stress, pp 389–394

Büchert T, Fritz G, Kroneck PMH – Biochemical and molecular properties of adenosine 5′-phophosulfate reductase from sulfate-reducing bacteria, pp 221–222

Davies J, Yildiz F, Wykoff D, Grossman A – Sensing, signaling and surviving sulfur stress in *Chlamydomonas*, pp 249–250

De Carvalho L, Amâncio S – Glutathione content and glutathione reductase activity in leaves of *in vitro* grapevine plantlets acclimatized under different light conditions, pp 365–368

Droux M, Lappartient AG, Ducerf S, Douce R, Job D – Structural and kinetic properties of the cysteine synthase multi-enzyme complex, pp 311–312

Edwards R – Towards understanding the function of glutathione transferases in plant metabolism, pp 369–374

Flückiger H, Jecklin S, von Ballmoos P, Suter M, Brunold C – Effect of hydrogen peroxide induced osmotic stress in maize roots, pp 401–404

Fricker M, May M, White N, Meyer A – Modelling glutathione pathways from *in vivo* imaging, pp 353–354

Fujiwara T, Mori T, Sasaki T, Ishiwatari Y, Fukuda A, Hanaoka H, Mano H, Hayashi H, Chino M – Thioredoxin h in rice phloem sap – a model for cell-to-cell and long distance transport of proteins and RNS in higher plants, pp 285–288

Gakiére B, Denis L, Droux M, Ravanel S, Job D – Methionine synthesis in higher plants: sense strategy applied to cystathionine γ-synthase and cystathionine β-lyase in *Arabidopsis thaliana*, pp 313–316

Giesemann A, Boyle C – Sulfur translocation in a plant-pathogen system: evidence through stable S-isotope analysis, pp 413–418

Giesemann A, Schulz H, Jung K, Gehre M, Schüürmann G, Weigel H-J – Natural isotope variations of sulfur as an indication of atmospheric depositions on pine stands, pp 235–238

Gondet L, Ullmann P, Bach TJ – Regulation of glutathione metabolism in plants, pp 361–364

Gotor C, Romero LC, Barroso C, Dominguez JR, Gutiérrez-Alcalá G, Vega JM – *Atcys-3a* gene coding for the cytosolic *O*-acetylserine(thiol)lyase is regulated by stress conditions in *Arabidopsis thaliana*, pp 379–380

Hatzfeld Y, Saito K – Regulation of sulfate transporter genes in *Arabidopsis thaliana*, pp 269–272

Hawkesford MJ, Prosser IM – The plant sulfate transporter family, pp 263–264

Hesse H, Harms K, von Ballmoos P, Brunold C, Willmitzer L, Höfgen R – Serine acetyl-transferase: a bottleneck for cysteine and glutathione synthesis? pp 321–324

Hopkins L, Hawkesford MJ – S-supply, free sulfur and free amino acid pools in potato (*Solanum tuberosum* L. cv Désirée), pp 259–262

Imsande J – Iron-sulfur clusters and the coordination of plant growth, pp 229–230

Inouse K, Noji M, Katagiri T, Shinozaki K, Saito K – Subcellular localization and feedback inhibition of serine acetyltransferase, pp 327–330

Jost R, Hell R – Regulation and expression of sulfate transporter genes in *Arabidopsis thaliana*, pp 273–276

Kopriva S, Muheim R, Koprivova A, Trachsel N, Catalano C, Suter M, Brunold C – Light regulation of APS reductase in *Arabidopsis thaliana*, pp 309–310

Kopriva S, Suter M, Weber M, Schürmann P, Brunold C – Characterization of APS reductase enzyme purified from *Arabidopsis thaliana*, pp 223–224

Koysan A, Taran N, Musienko M – Sulfoxide accumulation in garlic bulbs by lipoic acid exposure, pp 429–430

Kranner I, Zorn M, Kurokawa T, Kinoshita T, Yamamoto Y, Yoshimura I – Responses of glutathione to desiccation in cultures mycobionts and photobionts isolated from the lichen *Cladonia vulcani*, pp 375–378

Lambert R, Lombardie M, Peeters A, Toussaint B – Utilization in agriculture of anhydrite produced by the desulfurization of industrial smoke, pp 255–258

Lappartient AG, Ducerf S, Drouz M – Regulation of serine acetyltransferase activity by cysteine in pea leaves, pp 325–326

Manabe T, Hasumi A, Sugiyama M, Yamazaki M, Saito K – Alliinase (*S*-alk(en)yl-L-cysteine sulfoxide lyase) from *Allium tuberosum* (Chinese chive), pp 419–420

Manninen S, Huttunen S, Ilvesniemi H, Altimir – Changes in sulfur and nitrogen concentrations of Scots pine needles during the growing season of 1997, pp 335–336

Meyer AJ, Fricker M – Quantitative imaging of glutathione in live cells by laser scanning microscopy, pp 351–352

Mulholland MM, Otte ML – The effects of sulfate concentrations on dimethylsulfonio-propionate (DMSP) levels in *Spartina anglica*, pp 421–424

Müller M, Tausz M, Grill D – Subcellular distribution of glutathione in living root protoplasts of *Allium cepa* L. studied by histochemical staining and image analysis, pp 355–358

Nieder J, Kinghorn JR, Mendel R-R – Isolation of a novel cystathionine-γ-synthase-like protein in *Aspergillus nidulans*, pp 291–294

Noji M, Saito M, Aono M, Saji H, Saito K – Modulation of cysteine biosynthesis and resistance to SO_2 in transgenic tobacco overexpressing cysteine synthase in cytosol and chloroplasts, pp 289–290

Papenbrock J, Schmidt A – Initial characterization of a rhodanese-like protein from *Arabidopsis*, pp 303–304

Pfanz H, Grau K, Wittman C, Zelgert W – Do land snails prefer specific sulfur compounds? pp 425–428

Quaggiotti S, Abrahamsohn C, Malagoli M – Different adaptation to sulfur and nitrogen shortages by two maize hybrids, pp 337–340
Ravina CG, Vega JM, Gotor C – Regulation of cysteine biosynthesis by sulfur nutrition in *Chlamydomonas reinhardtii*, pp 295–296
Rennenberg H, Will B – Phytochelatin production and cadmium accumulation in transgenic poplar (*Populus tremula* × *P. alba*), pp 393–398
Schnug E, Haneklaus S – Significance of interactions between sulfur and nitrogen supply for growth and quality of crop plants, pp 345–350
Schnug E, Haneklaus S, Schimmel H, Linsinger T, Mabon N, Marlier M, Wathelet J-P – Standard reference materials for total sulfur and glucosinolate, pp 251–254
Schulz H, Härtling S, Schüürmann G – Effects of SO_2-reduction on sulfur metabolism and vitality of Scots pine trees at three field sites in eastern Germany, pp 305–308
Schürmann P, Schwendtmayer C, Dai S, Ramaswamy S, Eklund H – Structure and function of ferredoxin:thioredoxin reductase, pp 231–234
Spielhofer P, Kopriva S, Täschler S, Oberer H, Suter M, Brunold C – Stimulation of key enzymes in sulfate assimilation by safener CGA 173072 may induce higher protection against heavy metal stress in *Zea mays* L., pp 387–388
Stulen I, Posthumus F, Amâncio S, Masselink-Beltman I, Müller M, De Kok LJ – Mechanism of H_2S phytotoxicity, pp 381–384
Suter M, Weber M, Kopriva S, Kuhlemeier C, Schürmann P, Brunold C – Adenosine 5′-phosphosulfate sulfotransferase and adenosine 5′-phosphosulfate reductase are identical enzymes, pp 225–228
Tabe L, Drouz M – Sulfur metabolism in developing seeds of transgenic narrow leaf lupin expressing a sulfur-rich protein, pp 317–320
Takahashi H, Watanabe A, Sasakura N, Asanuma W, Nakamura M, Saito K – Molecular characterization of the sulfate transporter gene family in *Arabidopsis thaliana*, pp 265–268
Tausz M, Wonisch A, Grill D, Bytnerowicz A, Arbaugh MJ, Peters J, Soledad Jiménez MS, Morales D – The significance of the glutathione redox system in responses of conifers to natural and anthropogenic stress, pp 385–386
Van der Zalm E, Schneider A, Rennenberg H – Characteristics and regulation of sulfate uptake and xylem loading by poplar roots (*Populus tremula* × *P. alba*), pp 277–282
von Ballmoos P, Kocsy G, Suter M, Brunold C – Buthionine sulfoximine (BSO) reduces chilling tolerance of a chilling tolerant maize genotype, pp 399–400
Westerman S, Baczynska M, de Kok LJ, Stulen I – The impact of H_2S on growth and nitrate reductase activity in *Brassica oleracea* L., pp 341–344
Wirtz M, Berkowitz O, Hell R – Analysis of plant cysteine synthesis *in vivo* using *E. coli* as a host, pp 297–298
Xiang C, Bertrand D – Glutathione synthesis in *Arabidopsis*: multilevel controls coordinate responses to stress, pp 409–412
Zorn M, Pfeifhofer MW, Grill D, Kranner I – Glutathione and hydroxymethylglutathione have similar functions in protection from oxidative stress, pp 405–408

Davidian J-C, Grill D, De Kok LJ, Stulen I, Hawkesford MJ, Schnug E, Rennenberg H (2003) Sulfur Transport and Assimilation in Plants: Regulation, Interaction, Signaling Backhuys Publishers, Leiden, ISBN 90-5782-138-9

Overview Papers

Research Papers

Saito K, De Kok LJ, Stulen I, Hawkesford MJ, Schnug E, Sirko A, Rennenberg H (2005) Sulfur Transport and Assimilation in Plants in the Post Genomic Era Backhuys Publishers ISBN 90-5782-166-4

Overview Papers

Smith FW, Diatloff E – Sulfate transport processes in plants, pp 3–12

Takahashi H – Functions and regulation of plant sulfate transporters, pp 13–22

Anderson JW – Regulation of sulfur distribution and redistribution in grain plants, pp 23–32

Schmidt A – Open questions in sulfur metabolism – an update: neglected areas of sulfur metabolism in plants, pp 61–72

Nagami Y, Onouchi H, Haraguchi Y, Sakurai R, Nakamoto M, Nagao N, Naito S – Autoregulated mRNA stability of the gene for cystathionine γ-synthase, the key-step enzyme of methionine biosynthesis in *Arabidopsis*, pp 73–80

Nikiforova VJ, Bielecka M, Gakière B, Krueger S, Rinder J, Kempa S, Hesse H, Hoefgen R – Effect of sulfur availability on *Arabidopsis thaliana* primary metabolism, pp 139–152

Rennenberg H, Peuke AD – Improved phytoremediation of contaminated soils by changes in sulfur metabolism, pp 201–208

Hell, R, Kruse C, Jost R, Lipschis M – Molecular analysis of sulfur-based defense reactions in plant-pathogen interactions, pp 209–216

Haneklaus S, Walker KC, Schnug E – A chronicle of sulfur research in agriculture, pp 249–256

Research Papers

Durenkamp M, De Kok LJ – Sulfur metabolism in onion (*Allium cepa* L): accumulation of sulfur compounds upon exposure to H_2S, pp 127–132

Fourcroy P, Rouached H, Berthomieu P, El Kassis E, Cathala N, Catherinot V, Labesse G, Davidian J-C – Involvement of the STAS domain in the regulation of the *Arabidopsis thaliana* sulfate transporter Sultr1;2, pp 39–42

Grzam A, Meyer AJ, Hell R – Subcellular compartmentation and degradation of glutathione-conjugates during detoxification of xenobiotics, pp 241–244

Higashi Y, Hirai MY, Fujiwara T, Naito S, Noji M, Saito K – Proteomics of Arabidopsis seeds grown under sulfur-deficient condition, pp 177–178

Hirai M Y, Fujikawa Y, Yano M, Goodenowe DB, Yamazaki Y, Kanaya S, Nakamura Y, Kitayama M, Suzuki H, Sakurai N, Shibata D, Saito K – Clustering of sulfur-assimilation gene-family members for comprehensive prediction of their functions, pp 183–186

Howarth JR, Barraclough PB, Hawkesford MJ – A highly specific sulfate-deficiency induced gene (*sdi1*) from wheat, pp 161–164

Ide Y, Kato T, Sato S, Tabata S, Naito S, Fujiwara T – Screening of *Arabidopsis* enhancer-trap lines for sulfur-responsive genes in seeds, pp 173–176

Imai S, Tsuge N, Tomotake M, Masamura N, Sawada H, Nagata T, Kumagai H – The production of a lachrymatory factor in onion is mediated by a novel enzyme, pp 123–126

Jalilehvand F – Sulfur speciation in intact plant leaves by XANES spectroscopy, pp 53–58

Kaminishi A, Doken H, Nomura K, Kita N – Divergence of alliinase among *Allium* species, pp 117–122

Kataoka T, Takahashi H – Sulfate transporters involved in redistribution and translocation of sulfate in root vasculature of *Arabidopsis thaliana*, pp 45–48

Kita N, Kawata T, Jose N, Kaminishi A, Nomura K, Kitaura T – Inter- and intraspecific divergence of the myrosinase gene among *Brassica* species, pp 113–116

Krauss G-J, Krauss G, Menge S, Wesenberg D – Cadmium stress response at the thiol peptide level in mosses and fungi, pp 233–236

Kruse C, Kopp B, Hartmann M, Jost R, Hell R – Sulfate-based performance of *Arabidopsis thaliana* in response to pathogen infection, pp 217–220

Lewandowska M, Wawrzyńska A, Kamińska J, Liszewska F, Sirko A – Identification of novel genes of *Nicotiana tabacum* regulated by short-term sulfur starvation, pp 153–156

López-Martin MC, Romero LC, Gotor C – Differential involvement of cysteine biosynthesis genes in cadmium tolerance in *Arabidopsis thaliana*, pp 225–228

Lu C, Howarth JR, Barraclough PB, Edwards KJ, Hawkesford MJ – Transcriptome analysis of wheat responses to sulfur nutrition, pp 157–160

Maruyama-Nakashita A, Takahashi H – Transcriptional regulation of Sultr1;1 and Sultr1;2 in *Arabidopsis* root, pp 43–44

Meyer AJ, Cairns NG, Pasternak M, Cobbett CS – *In vivo* imaging of glutathione in Arabidopsis embryos, pp 49–52

Morioka R, Kanaya S, Yano M, Minato K, Nakamura Y, Hirai MY, Saito K – Statistical analysis of the adaptation process of *Arabidopsis thaliana* to the sulfur deficiency, pp 195–197

Nakamura S, Maruyama K, Watanabe A, Hattori H, Chino M – Response of glutathione in the sieve tube of *Brassica napus* L to cadmium treatment, pp 229–232

Nakamura Y, Shinbo Y, Asahi H, Altaf-Ul-Amin M, Kurokawa K, Oikawa A, Kimura A, Ogura T, Morishita Y, Sakurai N, Suzuki H, Nishi T, Hirai MY, Yano M, Saito K, Ohta D, Shibata D, Kitayama M, Kanaya S – Metabolix: metabolome profiling system based on mass spectrometry, pp 179–182

Nakano Y, Sekiya J – γ-Glutamyltransferase and glutathione catabolism in *Raphanus sativus* L, pp 107–110

Nakayama Mo, Kurisu G, Hase T – Structure and characterization of the active site of maize sulfite reductase, pp 81–82

Noji M, Higashi Y, Watanabe M, Saito K – T-DNA insertion mutant of ATP sulfurylase, the first enzyme of sulfur assimilatory metabolism, in *Arabidopsis thaliana,* pp 73–80

North- K, Koprivova A, Kopriva S – Investigating the effects of abiotic stress on the sulfate assimilation pathway in *Arabidopsis*, pp 221–224

Ohkama-Ohtsu N, Kasajima I, Ide Y, Naito S, Fujiwara T – Identification and characterization of genes involved in regulation of sulfur deficiency induced expression of genes in *Arabidopsis thaliana*, pp 169–172

Riemenschneider A, Bonacina E, Schmidt A, Papenbrock J – Isolation and characterization of a second d-cysteine desulfhydrase-like protein from *Arabidopsis*, pp 103–106

Riemenschneider A, Riedel K, Hoefgen R, Papenbrock J, Hesse H – *O*-Acetylserine (thiol)lyase regulates cysteine homeostasis in potato plants (*Solanum tuberosum* L), pp 95–98

Saitoh T, Toyota H, Nakayama M, Ikegami T, Kurisu G, Hase T – NMR study of the interaction between plant sulfite reductase and ferredoxin, pp 83–86

Sato S, Naito Y, Noji M, Saito K, Tomita M – Mathematical model of sulfur metabolism in higher plants using the e-cell system, pp 187–190

Sekiya J, Okumura R, Otoshi H – The enzyme that catalyzes hydroxymethylglutathione synthesis in rice (*Oryza sativa* L) leaves, pp 111–112

Shibagaki N, Grossman A – Approaches using yeast cells to probe the function of STAS domain in Sultr1;2, pp 33–36

Shinbo Y, Nakamura Y, Altaf-Ul-Amin M, Asahi H, Kurokawa K, Suzuki H, Sakurai N, Oikawa A, Kitayama M, Yano M, Hirai MY, Saito K, Ohta D, Shibata D, Kanaya S – KNApSAcK: Natural product database system for searching relationships between metabolites and species, pp 191–194

Sugimoto K, Sato N, Watanabe A, Tsuzuki M – Effects of sulfur deprivation on sulfolipid metabolism in *Chlamydomonas reinhardtii*, pp 133–135

Sugiyama A, Sekiya J – Homoglutathione synthesis and herbicide tolerance of transgenic tobacco plants expressing soybean homoglutathione synthetase, pp 245–248

Watanabe M, Noji M, Kusano M, Kato T, Tabata S, Saito K – Functional analysis of serine acetyl-transferase genes (*serat*) involved in sulfur assimilation in *Arabidopsis thaliana*, pp 99–102

Wesenberg D, Rother M, Bleuel C, Menge S, Krauss G-J – Single cell sampling and analysis of cadmium-exposed *Physcomitrella patens* by thiol peptide profiling and quantitative PCR, pp 237–240

Wirtz M, Lindermayr C, Durner J, Hell R – Post-translational modification of cytosolic OAS-TL from *Arabidopsis thaliana*, pp 91–94

Yano M, Hirai MY, Kusano M, Kitayama M, Kanaya S, Saito K – Studies for elucidation of jasmonate signaling in the response of *Arabidopsis* to sulfur deficiency, pp 165–168

Yoshimoto N, Saito K, Takahashi H – High-affinity sulfate transporters in *Arabidopsis*: physiological roles and regulation of transport activities, pp 37–38

Sirko A, De Kok LJ, Haneklaus S, Hawkesford MJ, Rennenberg H, Saito K, Stulen I (2009) Sulfur Metabolism in Plants: Regulatory Aspects, Significance of Sulfur in the Food Chain, Agriculture and the Environment Backhuys Publishers, Leiden/Margraf Publishers, Weikersheim ISBN 978-3-8236-1547-7; ISBN 978-90-5782-215-5

Overview Papers

Research Papers

Bromke MA, Höfgen R, Hesse Holger – Amino acids in *Thalassiosira pseudonana*, pp 289–292

Buchner P, Hawkesford MJ – The wheat sulfate transporter family in relation to other cereals, pp 163–168

Cabannes E, Buchner P, Broadley MR, White PJ, Hawkesford MJ – Sulfate/selenate transporters in selenium-accumulating plants, pp 175–178

Dixon DP, Edwards R – Identification of novel sulfur-containing metabolites bound to *Arabidopsis* glutathione transferase, pp 221–224

Frerigmann H, Berger B, Gigolashvili T, Flügge U-I – Interaction of Myb and β -HLH transcription factors in the regulation of indolic glucosinolate biosynthesis, pp 231–234

Haas F, Queiroz R, Bauer A, Schanné M, Wirtz M, Hell R – High resolution gene expression analysis of the sulfur-dependent transcriptome, pp 147–152

Haneklaus S, Bloem E, Aljmli F, Buechs W, Schnug E – Influence of sulfur fertilization on the insect inventory in oilseed rape during the vegetation period, pp 141–146

Haneklaus S, Bloem E, Schnug E – Sulfur induced resistance (SIR): Biological and environmentally sound concept for disease control, pp 129–134

Heeg C, Wirtz M, Hell R – Impact of abiotic stress on plants inactivated in cysteine synthesis in distinct compartments, pp 207–210

Hesse H, Kosmehl T, Höfgen R – Regulatory impacts of an AP2 transcription factor on plant metabolism, pp 157–160

Hirai MY, Sawada Y, Araki R, Saito K – Omics-based identification of the genes involved in glucosinolate biosynthesis, pp 239–245

Hockin N, Malin G, Kopriva S – Sulfur metabolism in marine phytoplankton: biochemical process to climate, pp 285–288

Honsel A, Kojima M, Sakakibara H, Dürr J, Herschbach C, Rennenberg H – The influence of sulfur depletion on the expression of sulfur metabolism related genes and on the phytohormone profile of poplars (*Populus tremula* × *P. alba*), pp 69–72

Howarth JR, Barraclough PB, Hawkesford MJ – Sulfur and nitrogen remobilisation during wheat grain-filling and related shifts in amino acid metabolism, pp 109–112

Hubberten H-M, Hesse H, Höfgen R – Lateral root growth in sulfur-enriched patches, pp 105–108

Ide Y, Kasajima I, Kimura T, Ohkama-Ohtsu N, Fujiwara T – Isolation and characterization of *Arabidopsis* mutants with altered responses to sulfur deficiency, pp 153–156

Iwona S, Rüdiger H, Markus W – Possible connection of sulfur and C2-subunit metabolism by N-terminal acetylation of proteins, pp 211–216

Klikocka – Hanna Sulfur supply in Polish agriculture, pp 45–48

Kopriva S – Sulfur assimilation in lower plants and algae: surprising lessons from sequenced genomes, pp 293–298

Koprivova A, Durenkamp M, Kopriva S – Control of root growth by glutathione, pp 101–104

Koralewska A, Stuiver CEE, Posthumus FS, Hawkesford MJ, De Kok LJ – The up-regulated sulfate uptake capacity, but not the expression of sulfate transporters is strictly controlled by the shoot sink capacity for sulfur in curly kale, pp 61–68

Kørner C, Knillmann S, Ehlting B, Wennrich R, Brüggemann N, Wesenberg D, Krauss G-J, Rennenberg H – Significance of copper for the uptake and detoxification of atrazine by poplar tree species, pp 265–272

Lewandowska M, Sirko A – Identification of additional genes regulated by sulfur shortage in tobacco, pp 73–78

Lewandowska, M, Bajda A, Świeżewska E, Agnieszka S – Influence of short term sulfur starvation on photosynthesis-related compounds and processes in tobacco, pp 79–84

Marty L, Wirtz M, Meyer AJ, Hell R – Cellular redox homeostasis, glutathione reduction and pathogen defence in *Arabidopsis thaliana*, pp 135–140

Matthewman C, Mugford SG. Kopriva Stanislav – Bio-computational analysis of differential expression of individual ATP sulfurylase genes in *Arabidopsis thaliana*, pp 185–188

Mugford SG, Reichelt M, Yoshimoto N, Nakazato Y, Noji M, Takahashi H, Saito K, Gershenzon J, Kopriva S – Characterization of the APS kinase gene family in *Arabidopsis thaliana*, pp 189–194

Papenbrock J, Bartels A, Hartmann F, Hartmann J, Triulzi T – Reduced sulfur in the plant cell – enzymatic formation and functional role, pp 217–220

Paszewski A, Natorff R, Sieńko M, Brzywczy J, Lewandowska I – Sulfur metabolism in *Aspergillus nidulans*: Pathways and regulation, pp 279–284

Podleśna A, Kocoń A – The effects of sulfur and nitrogen on yield and quality of winter oilseed rape, pp 123–128

Schiavon M, Agostini G, Pittarello M, Dalla Vecchia F, Pastore P, Malagoli M – Interactions between chromate and sulfate affect growth, photosynthesis and ultrastructure in *Brassica juncea* (L) Czern, pp 259–264

Siyiannis VF, Protonotarios VE, Chorianopoulou SN, Zechmann B, Müller M, Hawkesford MJ, Bouranis DL – Effect of sulfate deprivation on pectins and maize nodal roots, pp 97–100

Stuiver CEE, Koralewska A, Posthumus FS, De Kok LJ – Impact of sulfur deprivation on root formation, and activity and expression of sulfate transporters in Chinese cabbage, pp 89–92

Sugimoto K, Matsui K, Takabayashi J, Sato N, Tsuzuki M – Change of the phospholipid metabolism in chloroplasts under sulfur-starved conditions in a green alga, *Chlamydomonas reinhardtii*, pp 85–88

Tavares S, Sousa C, Amâncio S – Sulfate transporters in *Vitis*: Roles and expression, pp 169–174

Thomas LA, Leung S, McCallum JA, McManus MT – Control of sulfur assimilation in onion (*Allium cepa* L), pp 55–60

Vandermeiren K, De Bock M, Horemans N, Gielen B – Impact of tropospheric ozone on glucosinolate and vitamin C content of oilseed rape and broccoli, pp 245–252

Index

L.J. De Kok et al. (eds.), *Sulfur Metabolism in Higher Plants - Fundamental, Environmental and Agricultural Aspects*, Proceedings of the International Plant Sulfur Workshop, DOI 10.1007/978-3-319-56526-2

MIX
Papier aus verantwortungsvollen Quellen
Paper from responsible sources
FSC® C105338

If you have any concerns about our products,
you can contact us on
ProductSafety@springernature.com

In case Publisher is established outside the EU,
the EU authorized representative is:
Springer Nature Customer Service Center GmbH
Europaplatz 3, 69115 Heidelberg, Germany

Printed by Libri Plureos GmbH
in Hamburg, Germany